# Reviews in Modern Astronomy

## 18

From Cosmological Structures to the Milky Way

## Further Titles in Astronomy

Phillipps, S.
**The Structure and Evolution of Galaxies**
296 pages, 2005, Hardcover
ISBN 0-470-85506-1

Stahler, S. W., Palla, F.
**The Formation of Stars**
865 pages, 2004, Softcover
ISBN 3-527-40559-3

Schielicke, R. E. (ed.)
**Reviews in Modern Astronomy**
**Vol. 16: The Cosmic Circuit of Matter**
330 pages, 2003, Hardcover
ISBN 3-527-40451-1

Schielicke, R. E. (ed.).
**Reviews in Modern Astronomy**
**Vol. 17: The Sun and Planetary Systems - Paradigms for the Universe**
324 pages, 2004, Hardcover
ISBN 3-527-40476-7

Foukal, P. V.
**Solar Astrophysics**
480 pages, 2004, Hardcover
ISBN 3-527-40374-4

Liddle, A.
**An Introduction to Modern Cosmology**
188 pages, 2003, Hardcover
ISBN 0-470-84834-0

Cotera, A., Markoff, S., Geballe, T. R., Falcke, H. (eds.)
**Galactic Center Workshop 2002**
**The central 300 parsecs of the Milky Way**
Astronomische Nachrichten/
Astronomical Notes
Supplementary Issue 1/2003
650 pages, 2004, Hardcover
ISBN 3-527-40466-X

Rüdiger, G., Hollerbach, R.
**The Magnetic Universe**
**Geophysical and Astrophysical Dynamo Theory**
343 pages, 2004, Hardcover
ISBN 3-527-40409-0

**Astronomische Nachrichten /**
**Astronomical Notes**
ISSN 0004-6337

**Astronomische Nachrichten /**
**Astronomical Notes 324 (2004)**
Suppl. Issue 2: Short Contributions presentetd at the Annual Scientific Meeting of the Astronomische Gesellschaft in Berlin, 2002
ISSN 0004-6337

**Astronomische Nachrichten /**
**Astronomical Notes 324 (2004**
Suppl. Issue 3: Short Contributions presentetd at the Annual Scientific Meeting of the Astronomische Gesellschaft in Freiburg, 2003
ISSN 0004-6337

Edited by Siegfried Röser

# Reviews in Modern Astronomy 18

## From Cosmological Structures to the Milky Way

WILEY-VCH Verlag GmbH & Co. KGaA

**Edited on behalf of the Astronomische Gesellschaft by**

Dr. Siegfried Röser
Astronomisches Rechen-Institut
Mönchhofstraße 12 - 14
69120 HEIDELBERG
Germany

**Cover**
Cluster of Galaxies 1ESO657-55
(VLTUT1 + FORS1)




**Library of Congress Card No.: applied for**
**British Library Cataloging-in-Publication Data:**
A catalogue record for this book is available from the British Library.

**Bibliographic information published by Die Deutsche Bibliothek**
Die Deutsche Bibliothek lists this publication in the Deutsche Nationalbibliografie; detailed bibliographic data is available in the Internet at http://dnb.ddb.de.



**Printing** Strauss GmbH, Mörlenbach
**Binding** Litges & Dopf Buchbinderei GmbH, Heppenheim

Printed in the Federal Republic of Germany
Printed on acid-free paper

**ISBN-13:** 978-3-527-40608-1
**ISBN-10:** 3-527-40608-5

# Preface

The annual series *Reviews in Modern Astronomy* of the ASTRONOMISCHE GESELLSCHAFT was established in 1988 in order to bring the scientific events of the meetings of the Society to the attention of the worldwide astronomical community. *Reviews in Modern Astronomy* is devoted exclusively to the Karl Schwarzschild Lectures, the Ludwig Biermann Award Lectures, the invited reviews, and to the Highlight Contributions from leading scientists reporting on recent progress and scientific achievements at their respective research institutes.

The Karl Schwarzschild Lectures constitute a special series of invited reviews delivered by outstanding scientists who have been awarded the Karl Schwarzschild Medal of the Astronomische Gesellschaft, whereas excellent young astronomers are honoured by the Ludwig Biermann Prize.

Volume 18 continues the series with twelve invited reviews and Highlight Contributions which were presented during the International Scientific Conference of the Society on "From Cosmological Structures to the Milky Way", held at Prague, Czech Republic, September 20 to 25, 2004.

The Karl Schwarzschild medal 2004 was awarded to Professor Riccardo Giacconi, Washington, D.C., USA. His lecture with the title "The Dawn of X-Ray Astronomy" opened the meeting.

The talk presented by the Ludwig Biermann Prize winner 2004, Dr Falk Herwig, Los Alamos, USA, dealt with the topic "The Second Stars".

Other contributions to the meeting published in this volume discuss, among other subjects, X-ray astronomy, cosmology, galaxy evolution, star formation and the Galactic Centre.

The presentation by Alvaro Giménez on the Hot Topic "The Future of ESA's Space Programme" is not printed in Reviews in Modern Astronomy; the speaker emphasized that the ESA programme is so much in progress at present, that an interested reader will be better informed and continuously updated through ESA's web-pages.

The editor would like to thank the lecturers for stimulating presentations. Thanks also to the local organizing committee from the Astronomical Institute of the Charles University of Prague, Czech Republic, chaired by Martin Šolc.

Heidelberg, April 2005 *Siegfried Röser*

The ASTRONOMISCHE GESELLSCHAFT awards the **Karl Schwarzschild Medal**. Awarding of the medal is accompanied by the Karl Schwarzschild lecture held at the scientific annual meeting and the publication.

Recipients of the Karl Schwarzschild Medal are

1959 Martin Schwarzschild:
Die Theorien des inneren Aufbaus der Sterne.
Mitteilungen der AG 12, 15

1963 Charles Fehrenbach:
Die Bestimmung der Radialgeschwindigkeiten
mit dem Objektivprisma.
Mitteilungen der AG 17, 59

1968 Maarten Schmidt:
Quasi-stellar sources.
Mitteilungen der AG 25, 13

1969 Bengt Strömgren:
Quantitative Spektralklassifikation und ihre Anwendung
auf Probleme der Entwicklung der Sterne und der Milchstraße.
Mitteilungen der AG 27, 15

1971 Antony Hewish:
Tree years with pulsars.
Mitteilungen der AG 31, 15

1972 Jan H. Oort:
On the problem of the origin of spiral structure.
Mitteilungen der AG 32, 15

1974 Cornelis de Jager:
Dynamik von Sternatmosphären.
Mitteilungen der AG 36, 15

1975 Lyman Spitzer, jr.:
Interstellar matter research with the Copernicus satellite.
Mitteilungen der AG 38, 27

1977 Wilhelm Becker:
Die galaktische Struktur aus optischen Beobachtungen.
Mitteilungen der AG 43, 21

1978 George B. Field:
Intergalactic matter and the evolution of galaxies.
Mitteilungen der AG 47, 7

1980 Ludwig Biermann:
Dreißig Jahre Kometenforschung.
Mitteilungen der AG 51, 37

1981 Bohdan Paczynski:
Thick accretion disks around black holes.
Mitteilungen der AG 57, 27

1982 Jean Delhaye:
Die Bewegungen der Sterne
und ihre Bedeutung in der galaktischen Astronomie.
Mitteilungen der AG 57, 123

1983 Donald Lynden-Bell:
Mysterious mass in local group galaxies.
Mitteilungen der AG 60, 23

1984 Daniel M. Popper:
Some problems in the determination
of fundamental stellar parameters from binary stars.
Mitteilungen der AG 62, 19

1985 Edwin E. Salpeter:
Galactic fountains, planetary nebulae, and warm H I.
Mitteilungen der AG 63, 11

1986 Subrahmanyan Chandrasekhar:
The aesthetic base of the general theory of relativity.
Mitteilungen der AG 67, 19

1987 Lodewijk Woltjer:
The future of European astronomy.
Mitteilungen der AG 70, 21

1989 Sir Martin J. Rees:
Is there a massive black hole in every galaxy.
Reviews in Modern Astronomy 2, 1

1990 Eugene N. Parker:
Convection, spontaneous discontinuities,
and stellar winds and X-ray emission.
Reviews in Modern Astronomy 4, 1

1992 Sir Fred Hoyle:
The synthesis of the light elements.
Reviews in Modern Astronomy 6, 1

1993 Raymond Wilson:
Karl Schwarzschild and telescope optics.
Reviews in Modern Astronomy 7, 1

1994 Joachim Trümper:
X-rays from Neutron stars.
Reviews in Modern Astronomy 8, 1

1995 Henk van de Hulst:
Scaling laws in multiple light scattering under very small angles.
Reviews in Modern Astronomy 9, 1

1996 Kip Thorne:
Gravitational Radiation – A New Window Onto the Universe.
Reviews in Modern Astronomy 10, 1

1997 Joseph H. Taylor:
Binary Pulsars and Relativistic Gravity.
not published

1998 Peter A. Strittmatter:
Steps to the LBT – and Beyond.
Reviews in Modern Astronomy 12, 1

1999 Jeremiah P. Ostriker:
Historical Reflections
on the Role of Numerical Modeling in Astrophysics.
Reviews in Modern Astronomy 13, 1

2000 Sir Roger Penrose:
The Schwarzschild Singularity:
One Clue to Resolving the Quantum Measurement Paradox.
Reviews in Modern Astronomy 14, 1

2001 Keiichi Kodaira:
Macro- and Microscopic Views of Nearby Galaxies.
Reviews in Modern Astronomy 15, 1

2002 Charles H. Townes:
The Behavior of Stars Observed by Infrared Interferometry.
Reviews in Modern Astronomy 16, 1

2003 Erika Boehm-Vitense:
What Hyades F Stars tell us about Heating Mechanisms
in the outer Stellar Atmospheres.
Reviews in Modern Astronomy 17, 1

2004 Riccardo Giacconi:
The Dawn of X-Ray Astronomy
Reviews in Modern Astronomy 18, 1

The **Ludwig Biermann Award** was established in 1988 by the ASTRONOMISCHE GESELLSCHAFT to be awarded in recognition of an outstanding young astronomer. The award consists of financing a scientific stay at an institution of the recipient's choice. Recipients of the Ludwig Biermann Award are

1989 Dr. Norbert Langer (Göttingen),
1990 Dr. Reinhard W. Hanuschik (Bochum),
1992 Dr. Joachim Puls (München),
1993 Dr. Andreas Burkert (Garching),
1994 Dr. Christoph W. Keller (Tucson, Arizona, USA),
1995 Dr. Karl Mannheim (Göttingen),
1996 Dr. Eva K. Grebel (Würzburg) and
Dr. Matthias L. Bartelmann (Garching),
1997 Dr. Ralf Napiwotzki (Bamberg),
1998 Dr. Ralph Neuhäuser (Garching),
1999 Dr. Markus Kissler-Patig (Garching),
2000 Dr. Heino Falcke (Bonn),
2001 Dr. Stefanie Komossa (Garching),
2002 Dr. Ralf S. Klessen (Potsdam),
2003 Dr. Luis R. Bellot Rubio (Freiburg im Breisgau),
2004 Dr. Falk Herwig (Los Alamos, USA).

# Contents

*Karl Schwarzschild Lecture*

# The Dawn of X-Ray Astronomy[1]

Riccardo Giacconi

Associated Universities, Inc., 1400 16th Street NW, Suite 730,
Washington, DC 20036, USA

## 1 Introduction

The development of rockets and satellites capable of carrying instruments outside the absorbing layers of the Earth's atmosphere has made possible thc observation of celestial objects in the x-ray range of wavelength.

X-rays of energy greater than several hundreds of electron volts can penetrate the interstellar gas over distances comparable to the size of our own galaxy, with greater or lesser absorption depending on the direction of the line of sight. At energies of a few kilovolts, x-rays can penetrate the entire column of galactic gas and in fact can reach us from distances comparable to the radius of the universe.

The possibility of studying celestial objects in x-rays has had a profound significance for all astronomy. Over the x-ray to gamma-ray range of energies, x-rays are, by number of photons, the most abundant flux of radiation that can reveal to us the existence of high energy events in the cosmos. By high energy events I mean events in which the total energy expended is extremely high (supernova explosions, emissions by active galactic nuclei, etc.) or in which the energy acquired per nucleon or the temperature of the matter involved is extremely high (infall onto collapsed objects, high temperature plasmas, interaction of relativistic electrons with magnetic or photon fields).

From its beginning in 1962 until today, the instrumentation for x-ray astronomical observations has improved in sensitivity by more than 9 orders of magnitude, comparable to the entire improvement from the capability of the naked eye to those of the current generation of 8- or 10-meter telescopes. All categories of celestial objects, from planets to normal stars, from ordinary galaxies to quasars, from small groups of galaxies to the furthest known clusters, have been observed. As a result of these studies it has become apparent that high energy phenomena play a fundamental role in the formation and in the chemical and dynamical evolution of structures on all scales. X-ray observations have proved of crucial importance in discovering important aspects of these phenomena. It was from x-ray observations that we obtained the first evidence for gravitational energy release due to infall of matter onto a collapsed object such as a neutron star or black hole. It was the x-ray emission from the high temperature plasmas in clusters of galaxies that revealed this high temperature

[1] Reprinted with permission of the Nobel Foundation. 

*Reviews in Modern Astronomy 18.* Edited by S. Röser
 ISBN 3-527-40608-5.

component of the Universe, which more than doubled the amount of "visible" matter (baryons) present in clusters.

Table 1: Estimates of fluxes from sources outside the solar system. From Giacconi, Clark and Rossi, 1960.

| Source | Maximum Wavelength | Mechanism for emission | Estimated Flux |
|---|---|---|---|
| Sun | $< 20$ Å | Coronal emission | $\sim 10^6\ cm^{-2}\ s^{-1}$ |
| Sun at 8 light years | $< 20$ Å | Coronal emission | $2.5 \times 10^{-4}\ cm^{-2}\ s^{-1}$ |
| Sirius if $L_X \sim L_{OPT}$ | $< 20$ Å | ? No convective zone | $0.25\ cm^{-2}\ s^{-1}$ |
| Flare stars | $< 20$ Å | Sunlike flare? | ? |
| Peculiar $A$ stars | $< 20$ Å | $B \sim 10^4$ gauss Large B Particle acceleration | ? |
| Crab nebula | $< 25$ Å | Synchrotron. $E_E \geq 10^{13}$ eV in $B = 10^{-4}$ gauss. Lifetimes? | ? |
| Moon | $< 23$ Å | Fluorescence | $0.4\ cm^{-2}\ s^{-1}$ |
| Moon | $\sim 20$ Å | Impact from solar wind Electrons $\phi_E = 0 - 10^{13}\ cm^{-2}\ s^{-1}$ | $0 - 1.6 \times 10^3\ cm^{-2}\ s^{-1}$ |
| SCO X-1 | 2-8 Å | ? | $28 \pm 1.2\ cm^{-2}\ s^{-1}$ |

The prospects for future studies of the universe in x-rays are equally bright. The advent of new and even more powerful experimental techniques, such as non-dispersive high resolution spectroscopy and x-ray telescopes capable of focusing increasingly higher energies over wider fields, ensures a wide opportunity for new astronomical discoveries.

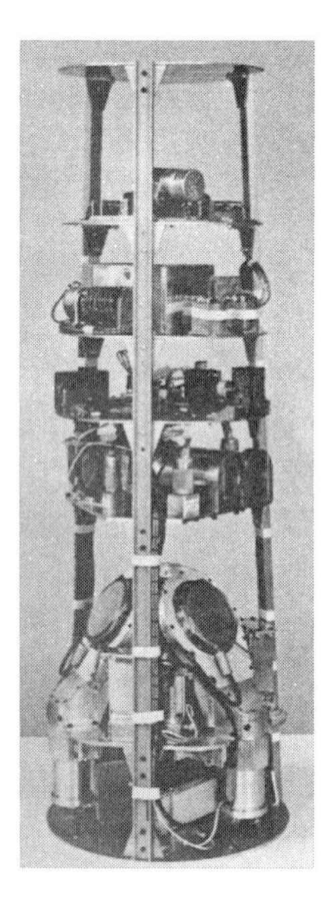

Figure 1: The payload of the June 12, 1962, AS&E rocket. From Giacconi and Gursky, 1974, p.9.

## 2 The Beginning of X-Ray Astronomy

There had been solar x-ray observations for about 10 years by the Naval Research Laboratory (NRL) group led by Herbert Friedman and several failed attempts to find x-ray emissions from stellar objects (Hirsh 1979) when a group at AS&E (a small private research corporation in Cambridge, Massachusetts) started work in 1959 to investigate the theoretical and experimental possibilities for carrying out x-ray astronomy. Giacconi, Clark, and Rossi (Giacconi et al. 1960) published a document: "A Brief Review of Experimental and Theoretical Progress in X-Ray Astronomy," in which we attempted to estimate expected x-ray fluxes from several celestial sources.

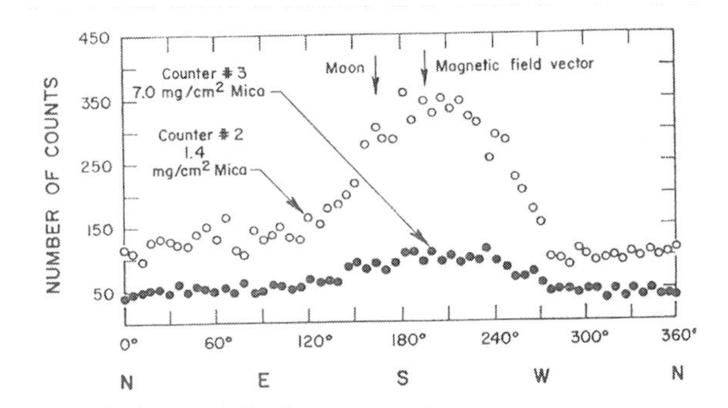


Figure 2: The first observation of Sco X–1 and of the X–ray background in the June, 12, 1962, flight. (From Giacconi, Gursky, Paolini, and Rossi, 1962.

The results are summarized in Table 1. The Sun produced $10^6$ x-ray photons $cm^{-2}s^{-1}$ at Earth which could easily be detected with the then-available counters with sensitivities of about 10 - $10^2$ photons $cm^{-2}s^{-1}$. But if all the stars emitted x-rays at the same rate as the Sun, we would expect fluxes at earth as small as $10^{-4}$ photons $cm^{-2}s^{-1}$. Other possible sources, such as supernova remnants, flare stars, peculiar A stars, etc., were considered, and great uncertainty had to be assigned to the estimates of their x-ray fluxes. It seemed that the brightest source in the night sky could be the Moon, due to fluorescent emission of lunar material under illumination of solar x-rays.

We designed an experiment capable of detecting 0.1-1 photon/$cm^{-2}s^{-1}$, 50 to 100 times more sensitive than any flown before. This increase in sensitivity was due to larger area, an anticoincidence shield to reduce particle background, and a wide solid angle to increase the probability of observing a source during the flight. The payload shown in Fig. 1 was successful in detecting the first stellar x-ray source in the flight of June 12, 1962 (Giacconi et al. 1962). An individual source (Sco X-1) dominated the night sky and was detected at $28 \pm 1.2$ counts $cm^{-2}s^{-1}$, just below the threshold of previous experiments (Fig. 2). No exceptionally bright or conspicuous visible light or radio object was present at that position (which, however, was very poorly known). An early confirmation of our result came from the rocket flight of April 1963 by the NRL group led by Friedman, which also discovered x–ray emission from the Crab Nebula (Bowyer et al. 1964a).

The truly extraordinary aspect of the discovery was not that an x-ray star had been found but its extraordinary properties. The x-ray radiation intensity from the Sun is only $10^{-6}$ of its visible light intensity. In Sco X-1, the x-ray luminosity is $10^3$ times the visible light intensity and it was later determined that the intrinsic luminosity is $10^3$ the entire luminosity of the Sun! This was a truly amazing and new type of celestial object. Furthermore, the physical process by which the x-rays were emitted on Sco X-1 had to be different from any process for x-ray generation we knew in the laboratory since it has not been possible on Earth to generate x-rays with 99.9% efficiency.

Many rocket flights carried out by several groups in the 60's were able to find new stellar sources and the first extragalactical sources. The NRL group and the Lockheed group (led by Phil Fisher) continued to carry out mostly broad surveys, with the notable exception of the Crab occulation experiment by NRL in 1964 (Bowyer et al. 1964b).

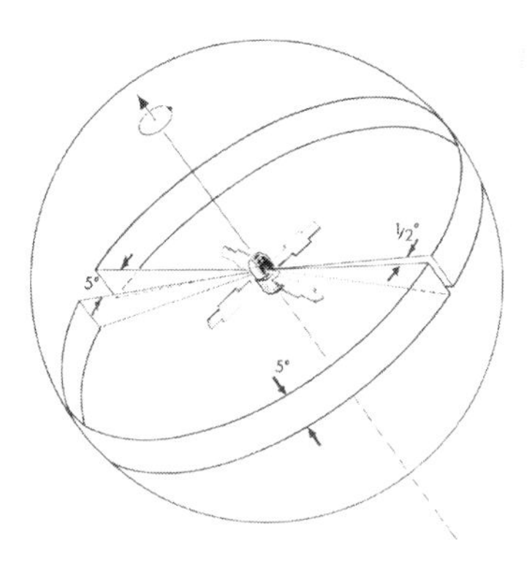


Figure 3: The fields of view of the detectors on the *UHURU* satellite. From Gursky, 1970. Reprinted in Giacconi and Gursky, 1974, p.5.

The AS&E group concentrated on the detailed study of individual x-ray sources. Most significant was the series of rocket flights which culminated in the identification of the optical counterpart of Sco X-1. First the group determined that Sco X-1 could not have the thermal spectrum that would be expected from neutron stars (Giacconi et al. 1965), which implied that an optical counterpart should have magnitude 13. In a first rocket flight to measure the angular size of the source it was found to be less than 7 arc sec (Oda et al. 1965). Thus the source had to be a visible star and not a diffused nebulosity. This led to the sophisticated measurement of the location of Sco X-1 by an AS&E-MIT group led by Herbert Gursky (Gursky et al. 1966), with sufficient precision to enable its identification with a 13$^{th}$ magnitude star (Sandage et al. 1996) which had spectral characteristics similar to an old nova. This renewed interest in a binary star model for Sco X–1 (Burbidge 1967) and Shklovsky proposed a binary containing a neutron star (Shklovsky 1967). However, the absence of x-ray emission from other novas, the lack of indications from either the optical spectra or the x-ray data of a binary system, and the general belief that the supernova explosion required to form the neutron star would disrupt the binary system did not lead to the general acceptance of the idea. The discovery by Hewish of pulsars in 1967 turned the attention of the theorists to pulsar models for the x-ray emitters. But such models also were not quite persuasive given the lack of observed x-ray pulsations. The solution to the riddle of Sco X–1 and similar sources was not achieved until the launch of the *UHURU* satellite, the first of a generation of x-ray observatories.

The proposal to launch a "scanning satellite," which eventually became *UHURU*, was contained in a document written by Herb Gursky and myself and submitted to NASA on September 25, 1963. In this document we described a complete program of x-ray research culminating in the launch of a 1.2 meter diameter x-ray telescope in 1968. This youthful dream was not fully realized until the launch of the *Chandra* X-Ray Observatory in 1999, which, not by chance, had a 1.2 meter diameter mirror. But while the difficult technology development that made x-ray telescopes possible was being carried out, the most fundamental advances in x-ray astronomy were made with relatively crude detectors mounted on orbiting satellites.

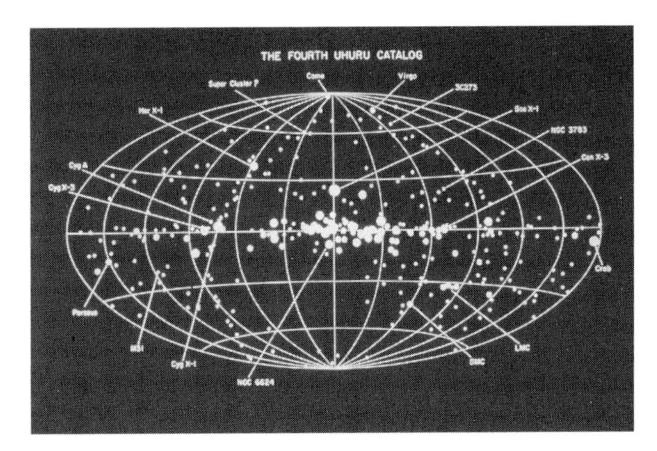


Figure 4: The x–ray sources observed by *UHURU* plotted in galactic coordinates. The size of the dot is proportional to intensity on a logarithmic time scale. From Gursky and Giacconi, 1974, p. 156.

## 3 Discoveries with *UHURU*

The total amount of time which was available for observation of the x-ray sky during the 60's was about one hour: five minutes above 100 km for each of about a dozen launches. The next step which led us from the phenomenological discoveries to those of great astrophysical relevance occurred on December 12, 1970, when *UHURU*, the first of the Small Astronomy Satellite series, was launched from the Italian S. Marco platform in Kenya. *UHURU* was a small satellite (Fig. 3) which we had labored over at AS&E for seven years between conception, development, testing, and integration.

It was the first observatory entirely dedicated to x-ray astronomy and it extended the time of observation from minutes to years or by 5 orders of magnitude (Giacconi et al. 1971). The field of view of the detector on board the satellite slowly rotated, examining a 5° band of the sky that shifted 1° a day. In three months all the sky could be studied systematically and many new sources could be localized with a precision of about 1 arc minute, often permitting the identification of the x-ray sources with a visual or radio counterpart. This in turn led to an evaluation of the distance, the intrinsic luminosity, and the physical characteristics of the celestial object from which the x-rays originated. Among the 300 new sources which were discovered, we were able to identify binary x-ray sources, supernovas, galaxies, active galaxies, quasars, and clusters of galaxies (Fig. 4). But even more important from a certain point of view was the ability, which was provided by the control system, to slow down the satellite spin and spend a very long time on an individual source to study its temporal variations. It was this special ability which permitted the solution of the fundamental unresolved problem of x-ray astronomy until then, namely, the nature of the energy source capable of producing the large intrinsic luminosity of the stellar x-ray sources.

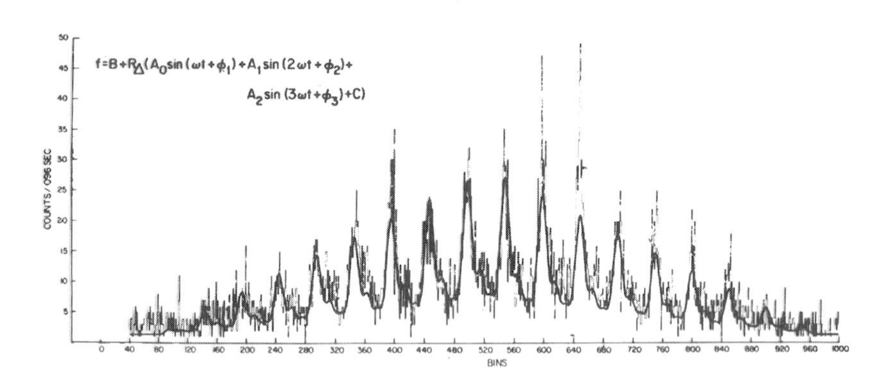


Figure 5: X-ray pulsations of Cen X–3. From Schreier, 1972.

## 3.1 The Binary X-Ray Sources

In summary, inspection of the data revealed that some x-ray sources (Her X-1 and Cen X-3) (Fig. 5) were regularly pulsating with periods of seconds (Schreier et al. 1972), while others (Cyg X-1) were pulsating with an erratic behavior with characteristic times of less than a tenth of a second as first noted by Minoru Oda, who was a guest at AS&E at the time. Ethan Schreier and I noticed that the average intensity of Cen X-3 was modulated over the span of days and that the period of pulsation itself was changing as a function of the phase of the average intensity, which exhibited occulations (Fig. 6). The explanation for this behavior soon became clear: we were observing a stellar x-ray source orbiting a normal star (Fig. 7). The variation of the pulsation period was then due to the Doppler effect. In 1967 Hewish had discovered pulsars in the radio domain. Was this x-ray source a pulsar in orbit about a normal star? This seemed difficult to accept at the time. A pulsar is a neutron star whose formation is believed to be due to the collapse of a star at the end of its life. The concomitant explosion was believed to disrupt any binary system in which it took place. Joseph Taylor had not yet discovered the binary pulsar.

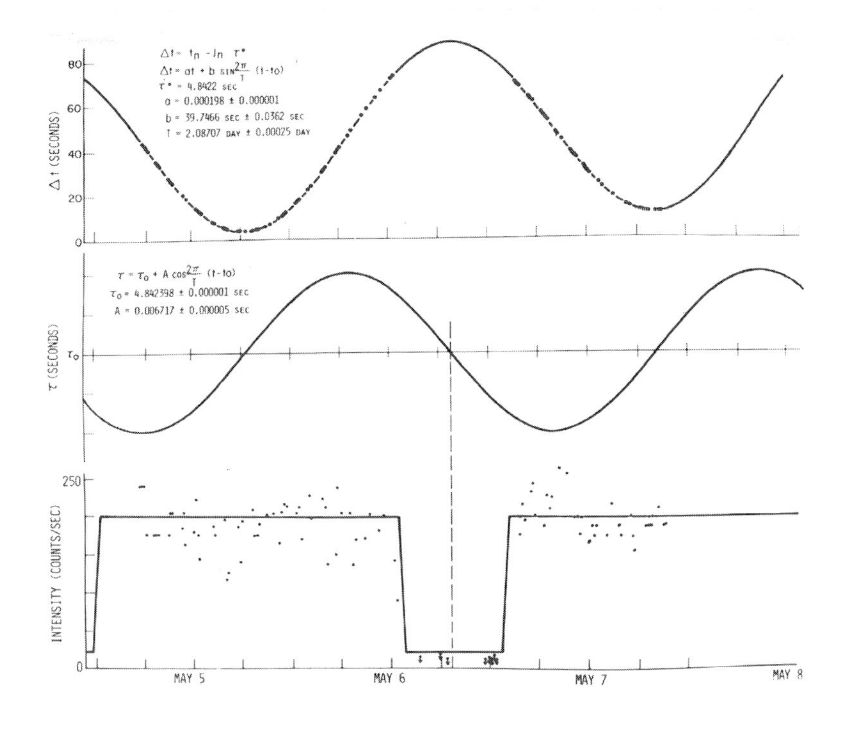


Figure 6: Period variations and occulations of Cen X–3. From Schreier, 1972.

But a new, unexpected, and important finding came to light: the period of pulsation was decreasing rather than increasing with time (Fig. 8). This was true not only in Cen X-3, but also, as Harvey Tananbaum found, the same behavior in Her X-1 (Tananbaum et al. 1972a). Now this was truly embarrassing! In a pulsar the loss of electromagnetic energy occurs at the expense of the kinetic energy of rotation. But in the x-ray sources the neutron star was acquiring rather than losing energy! The explanation was found in the interaction of the gas in the normal star with the collapsed star. Gas from the outer layer of the atmosphere of the normal star can fall into the strong gravitational field of the collapsed star and acquire energies of order of 0.1 $mc^2$ per nucleon. The accelerated nucleons in turn heat a shock of very high

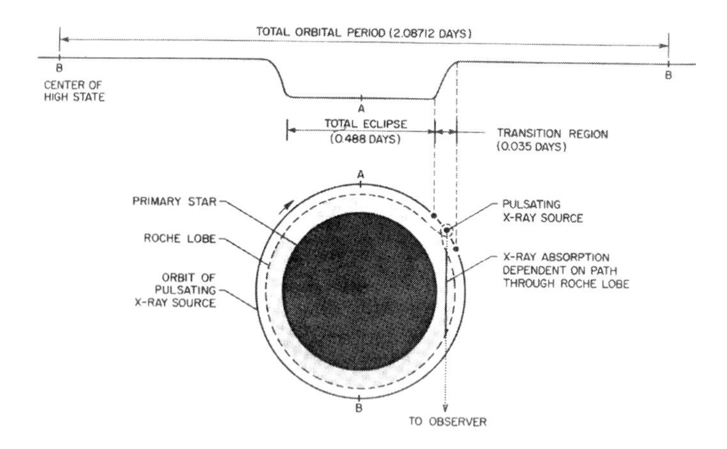


Figure 7: A model of the Cen X–3 binary system. Illustration of R. Giacconi.

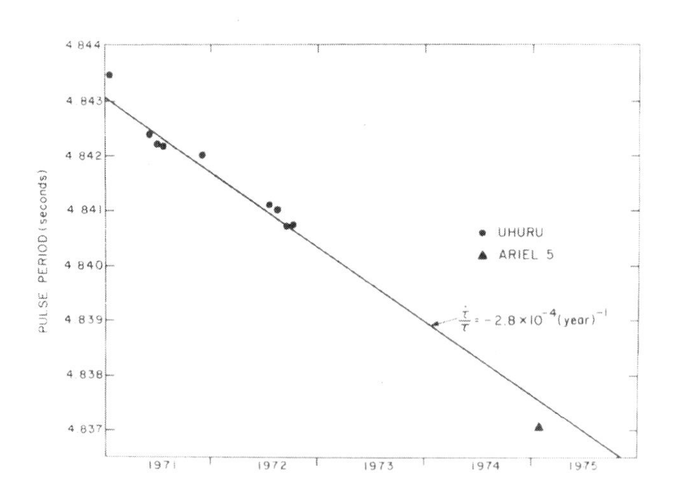


Figure 8: Annual change of Cen X–3 pulsation period. Illustration of R. Giacconi.

temperature above the surface of the neutron star, which emits the observed x-rays (Fig. 9). It is this material infall that gives energy to the collapsed object. This is the model for Sco X–1 and most of the galactic x–ray sources. In the case of a neutron star with a strong magnetic field ($10^{12}$ Gauss), the ionized plasma is confined to the poles of the rotating neutron star, generating the observed periodicities (Fig. 10). For a black hole, there exists no surface with particular structures and therefore the pulsation occurs chaotically (Fig. 11).

The observation by *UHURU* of rapid variability in Cyg X–1 reported by Oda (Oda et al. 1971) was soon followed by the rocket flights of the GSFC and MIT groups. These observations clarified that the observed pulsations were not periodic

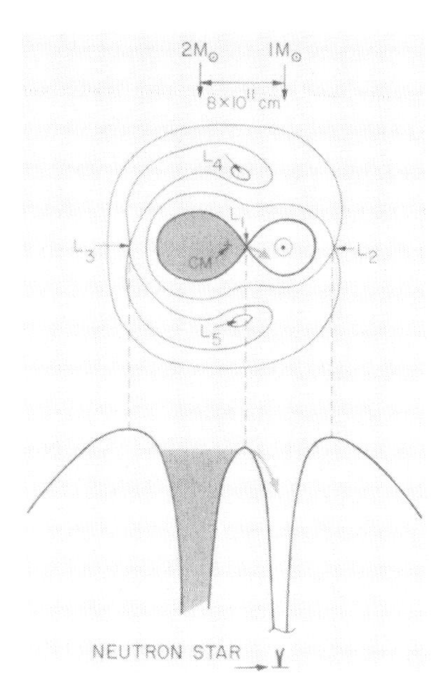


Figure 9: Representation of the equipotentials in the gravitational field of a typical binary x–ray source. Top view and cross section. Illustration of R. Giacconi.

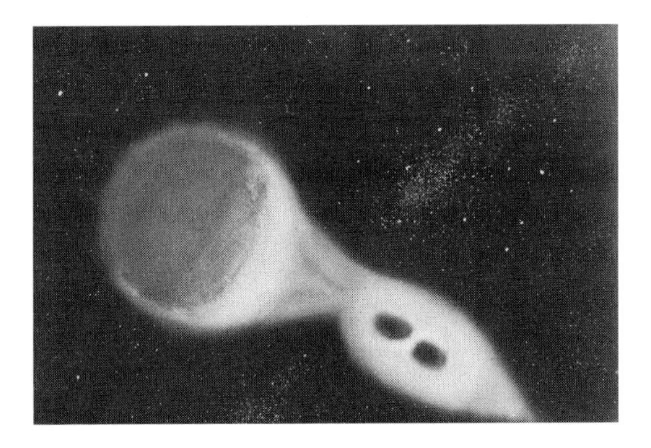

Figure 10: An artist's conception of Her X–1 with accretions occurring at the poles of the magnetic field of the neutron star. Illustration of R. Plourde.

but chaotic (Holt 1971; Rappaport et al. 1971a). By 1974, the GSFC group had achieved a temporal resolution of 1 millisecond and showed large chaotic fluctuations occurring even on this time scale (Rothschild et al. 1974). Such behavior could be expected to occur if the compact object in the binary system (the x-ray source) was a black hole rather than a neutron star. Stimulated by these findings, the search for optical or radio counterparts had become intense in 1971–1972. *UHURU* had obtained a considerably improved position for Cyg X–1 which was made available

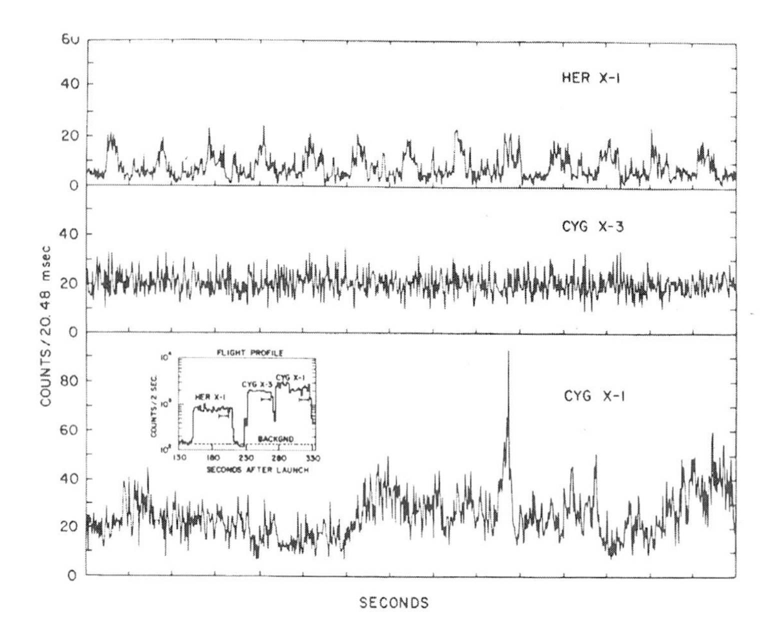


Figure 11: Comparison of the time variability of Her X–1, Cyg X–3, and Cyg X–1. Courtesy of R. Rothschild.

to Hjellming and Wade to aid in the search for a radio counterpart (Tananbaum et al. 1971). More refined positions were obtained by the Japanese group led by Oda (Miyamoto et al. 1971) and by the MIT group (Rappaport et al. 1971b) by use of modulation collimators. Hjellming and Wade (1971) and Braes and Miley (1971) reported the discovery of a radio counterpart. The precise radio location led to the optical identification by Webster and Murdin (Webster and Murdin 1972) and by Bolton (1972) of Cyg X–1 with the 5.6 day binary system HDE 226862. The identification of the radio source with Cyg X–1 was confirmed by the observation of a correlated x-ray radio transition in Cyg X–1 (Tananbaum et al. 1972b). Spectroscopic measurements of the velocity of HDE 226862 also permitted Webster and Murdin to establish that Cyg X–1 was indeed in a binary system. The estimated mass for the compact object was greater than 6 solar masses. Rhoades and Ruffini had shown in 1972 that black holes would have masses greater than 3.4 times the mass of the Sun (Rhoades and Ruffini 1974).

Thus we could reach conclusions regarding Cyg X–1: the Cyg X–1 x–ray emitter is a compact object of less than 30 km radius due to the rapidity of the pulsations and the fact that the pulsations are so large that they must involve the whole object (Giacconi 1974). The object has mass greater than that allowed by our current theories for neutron stars. Therefore the object is the first candidate for a black hole (Fig. 12). Currently there are at least six candidates for galactic x–ray sources containing a black hole (Tanaka 1992).

Figure 12: Artist's conception of Cyg X–1. Illustration of L. Cohen.

Table 2: Consequences of the discovery of binary x-ray systems.

| |
|---|
| • Existence of binary stellar systems containing a neutron star or a black hole |
| • Existence of black holes of stellar mass |
| • Measure of the mass, radius, moment of inertia, and equation of state for neutron stars (Density $10^{15}$ gr/cm$^3$) |
| • A new source of energy due to gravitational infall (100 times more efficient per nucleon than fusion) |
| • A model (generally accepted) for the nucleus of active galaxies and quasars |

The consequences of the discovery of binary source x-rays have had far reaching consequences (Table 2). We had proven the existence of binary systems containing a neutron star and of systems containing a black hole. Black holes of solar mass size existed. The binary x-ray sources have become a sort of physical laboratory where we can study the mass, moment of inertia, and equation of state for neutron stars (density $10^{15}$gr/cm$^3$). We had found a new source of energy for celestial objects: the infall of accreting material in a strong gravitation field. For a neutron star the energy liberated per nucleon is of order of 50 times greater than generated in fusion. The above model (of accretion of gas on a collapsed object) has become the standard explanation for the internal engines of quasars and all active nuclei. Recent data seem to confirm the model of accretion on a massive central black hole of $> 10^7$ solar masses as the common denominator among all the active galaxies.

## 3.2 The Discovery of High-Temperature Intergalactic Gas

The establishment of variability in the x-ray universe, the discovery of the existence of neutron stars and black holes in binary systems, and the discovery of accretion as a dominant energy source were only the first major accomplishments of *UHURU*. Among others was a second very important discovery of *UHURU* and x-ray astronomy, both because of its intrinsic interest and for its consequences in the field of cosmology, the detection of emission from clusters of galaxies. This emission is not simply due to the sum of the emission from individual galaxies, but originates in a thin gas which pervades the space between galaxies. This gas was heated in the past during the gravitational contraction of the cluster to a temperature of millions of degrees and contains as much mass as that in the galaxies themselves (Gursky et al. 1972). In one stroke the mass of baryons contained in the clusters was more than doubled. This first finding with *UHURU*, which could detect only the three richest and closest galaxy clusters and with a poor angular resolution of ½ a degree, were followed and enormously expanded by the introduction of a new and powerful x-ray observatory, *Einstein*, which first utilized a completely new technology in extrasolar x-ray astronomy: grazing incidence telescopes.

# 4 X–ray Telescopes

Here I must make a short technical diversion to explain the revolution brought about in x-ray astronomy by the telescope technology. When contemplating the estimates made in 1959, I was persuaded that to ultimately succeed in x-ray astronomy, we had to develop new systems quite different from those then in use. Friedman had developed for solar studies a Geiger counter with a thin window which allowed the x-ray to penetrate the interior of the gas volume of the counter. The counter could not decide either the direction of the incoming x-ray or its energy. In order to improve the directional sensitivity, x-ray astronomers used collimators, that is, mechanical baffles which defined a field of view typically of $1^\circ$. To improve sensitivity we developed anticoincidence shields against spurious particles and enlarged the area. This is what was done in the discovery rocket of 1962. In 1970 *UHURU* had a very similar detection system. Its improvement in sensitivity was due to the much larger area (800 $cm^2$ instead of 10) and the much longer time of observation. This led to an increase in sensitivity of about $10^4$. It should be noted, however, that in the presence of a background noise, the sensitivity improvement was only proportional to the square root of the area. Thus, further improvement would have required satellites of football-stadium size. Furthermore, all attempts to gain angular resolution by clever systems of baffles (such as the modulation collimators) led to intrinsically insensitive experiments.

The solution which occurred to me as early as 1959 was to use a telescope just as it is done in visible light astronomy (Giacconi and Rossi 1960). This has the great advantage that the flux from a large area of collection is focused onto a small detector, therefore improving both the flux and the signal to noise ratio. In addition, high angular resolution can be obtained within a field which is imaged at once, without scanning or dithering motions, therefore yielding an enormous improvement in exposure time for each source in the field.

The only problem is that an x-ray telescope had to be invented and the technology necessary for its fabrication had to be developed. It ultimately took about

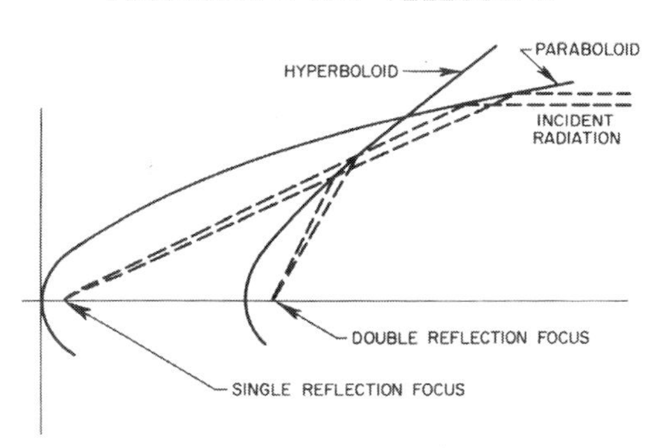


Figure 13: Principle of x–ray grazing incidence telescope. Illustration of R. Giacconi.

20 years between the conception of the x-ray telescope in 1959 and its first use for stellar x-ray astronomy in 1979. An x-ray telescope is quite different from a visible light telescope since the wavelength of x-ray photons is comparable to atomic dimensions. According to Lorentz' dispersion theory, it is clear that the index of refraction of x-rays is less than one, which makes optical systems based on refraction essentially impractical, as was realized by Roentgen himself in his classical experiment of 1895. However, x-rays can be efficiently externally reflected by mirrors, provided only that the reflection takes place at very small angles with respect to the mirror's surface. Hans Wolter had already discussed in the 40's and 50's the possibility of using images formed by reflection for microscopy. He showed that using a double reflection from a system of coaxial mirrors consisting of paraboloid and hyperboloid, one could achieve systems with a reasonably large field (1°) corrected for spherical aberrations and coma. Theoretically, therefore, the system was feasible, although the difficulties of construction given the tiny dimension of the systems for microscopy were impossible to overcome (Fig. 13). I persuaded myself, however, that in the corresponding optical designs to be used for telescopes, which required much larger scales (meters rather than microns), such difficulties would not be severe.

In our 1960 paper we described a system that could achieve sensitivities of $5 \times 10^{-14}$ erg cm$^{-2}$s$^{-1}$ and angular resolutions of 2 arc minutes. The improvement in sensitivity was $10^6$ - $10^7$ times greater than for any detector then current and the improvement in angular resolution about a factor of $10^3$. Unfortunately it took a long time to develop this technology (Fig. 14) (Giacconi et al. 1969). The first primitive pictures of the Sun with an x-ray telescope were obtained in 1965. Giuseppe Vaiana took over leadership of our solar physics program in 1967. In 1973 a high resolution x-ray telescope studied the Sun over a period of many months with a field large enough to image the disk and nearby corona and with angular resolution finer than 5 arc seconds (Fig. 15) (Vaiana and Rosner 1978).

It was not until 1979 that a fully instrumented x-ray telescope suitable for the detection and study of the much weaker stellar fluxes could be launched. The new

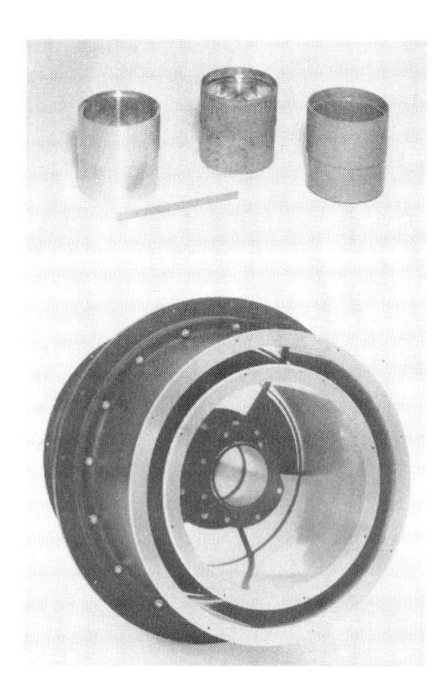

Figure 14: Several early telescope realizations. From Giacconi et al., 1969.

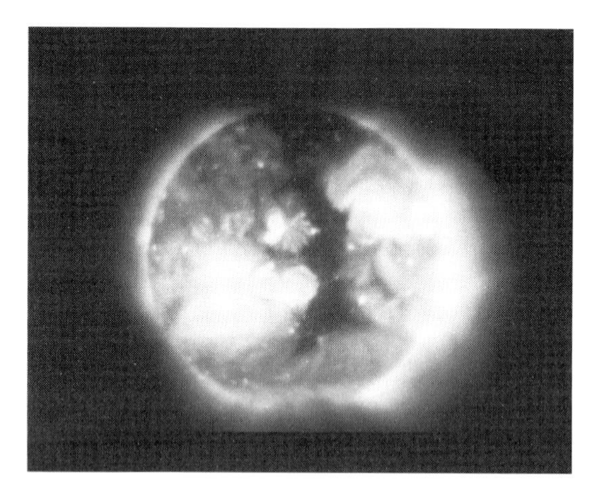

Figure 15: Picture of Sun in x–rays from Skylab. Courtesy of L. Golub, Harvard-Smithsonian Center for Astrophysics.

satellite, which became known as *Einstein,* was a real astronomical observatory (Fig. 16) (Giacconi et al. 1974). In the focal plane of the telescope one could use image detectors with angular resolutions of a few arc seconds, comparable to those used in visible light. The sensitivity with respect to point sources was increased by $10^3$ with respect to *UHURU* and $10^6$ with respect to Sco X-1. Spectroscopy could be carried out with a spectral resolving power of 500. This substantial technical improvement made possible the detection of all types of astro-

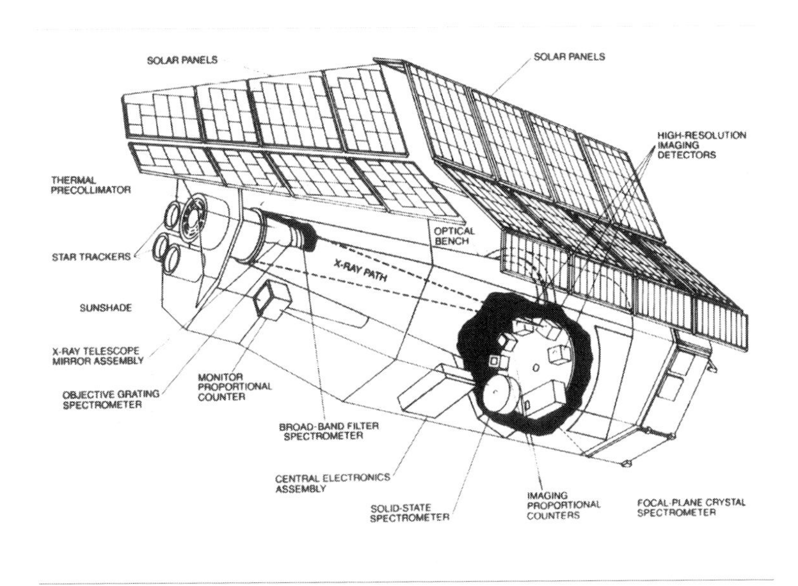


Figure 16: The Observatory *Einstein* schematic representation. From Giacconi et al., 1979.

physical phenomena (Table 3). Auroras due to the Jovian Belts, main sequence stars of all types, novae and supernovae were detected. Binary x-ray sources could be studied anywhere in our own galaxy as well as in external galaxies (Fig. 17). Normal galaxies as well as galaxies with active galactic nuclei, such as Seyferts and B Lac, could be detected at very great distances. The most distant quasars ever detected in visible light or radio could be conveniently studied. The sources of the mysterious, isotropic extragalactic background could begin to be resolved.

Table 3: Classes of celestial objects observed with the *Einstein* Observatory.

- Aurora on Jupiter
- X-ray emission from all types of main-sequence stars
- Novas and supernovas
- Pulsars
- Binary x-ray sources and supernovas in extragalactic sources
- Normal galaxies
- Nuclei of active galaxies
- Quasars
- Groups and clusters of galaxies
- Sources of the extragalactic x-ray background

But to come back to the study of the intergalactic plasma, it is in the study of x-ray emissions from clusters of galaxies that the *Einstein* observations have had some of the most profound impact. The ability to image the hot plasma has given

Figure 17: X–ray binaries in the galaxy M31. From the *Einstein* Observatory.

us the means to study in detail the distribution of the gravitational potential which contains both gas and galaxies. This study could only be carried out with some difficulty by studying the individual galaxies which, even though rather numerous in rich clusters, did not yield sufficient statistical accuracy. The x-ray study has revealed a complex morphology with some clusters exhibiting symmetry and a central maximum of density, which demonstrates an advanced stage of dynamical evolution; but many others show complex structures with two or more maxima. This morphology shows that the merging of the clusters' substructures is not yet completed. The relative youth of many of these clusters had not been sufficiently appreciated previously. The discovery of x-ray emission in clusters is therefore used to study one of the most interesting open questions of modern cosmology, namely, the formation and development of structures in the early epoch of the life of the Universe. Piero Rosati (Rosati et al. 2002) has pushed this work to very distant clusters ($z \sim 1.2$) by use of the *ROSAT* Satellite, a splendid successor to *Einstein*, built at the Max Planck Institute for Extraterrestrial Research by Joachim Trümper and Günther Hasinger and their group.

# 5 Current Research with *Chandra*

Harvey Tananbaum and I reproposed the concept of a 1.2 meter diameter telescope to NASA in 1976. Tananbaum, who had been project scientist on *UHURU* and scientific program manager on *Einstein*, gave leadership to the *Chandra* program and brought it to successful conclusion after I left Harvard in 1981. *Chandra* has met or exceeded all of our expectations. Comparison of the pictures of the Crab

Nebula pulsar by *Einstein* and by *Chandra* (Fig. 18) shows the great improvement in the sensitivity and angular resolution achieved (Tananbaum and Weisskopf 2001).

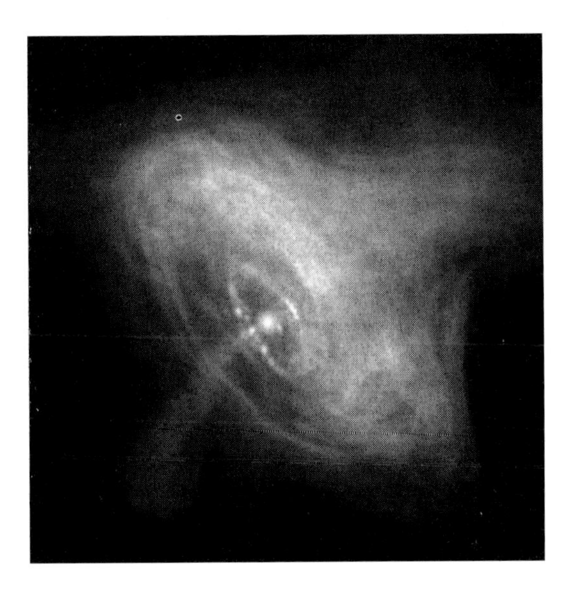

Figure 18: The x–ray image of the Crab Nebula obtained from *Chandra*. (Courtesy of NASA).

I was able to use *Chandra* for one million seconds to solve the problem of the sources of the x-ray background, a problem that had remained unsolved since its discovery in 1962 (Fig. 19) (Giacconi et al. 2002). Owing to the great sensitivity and angular resolution of *Chandra*, we were able to resolve the apparently diffused emission into millions of individual sources. They are active galactic nuclei, quasars, and normal galaxies. The gain in sensitivity that this represents is illustrated in Fig. 20. In Fig. 21, a reminder that x-ray astronomy now involves many groups in the world with distinct and essential contributions.

# 6 The Future of X-Ray Astronomy

I would like to conclude by attempting to answer the simple question: Why is x-ray astronomy important? The reason is that this radiation reveals the existence of astrophysical processes where matter has been heated to temperatures of millions of degrees or in which particles have been accelerated to relativistic energies. The x-ray photons are particularly suited to study these processes because they are numerous, because they penetrate cosmological distances, and because they can be focused by special telescopes. This last property significantly distinguishes x-ray from $\gamma$-ray astronomy. However, in a more fundamental way, high energy astronomy has great importance in the study of the Universe, because high energy phenomena play a crucial role in the dynamics of the Universe.

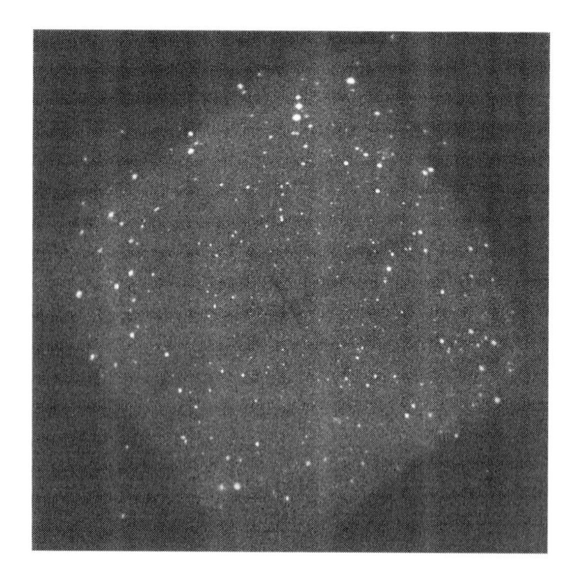

Figure 19: The one million second exposure in the deep x-ray field from *Chandra* (CDFS-Chandra Deep Field South).

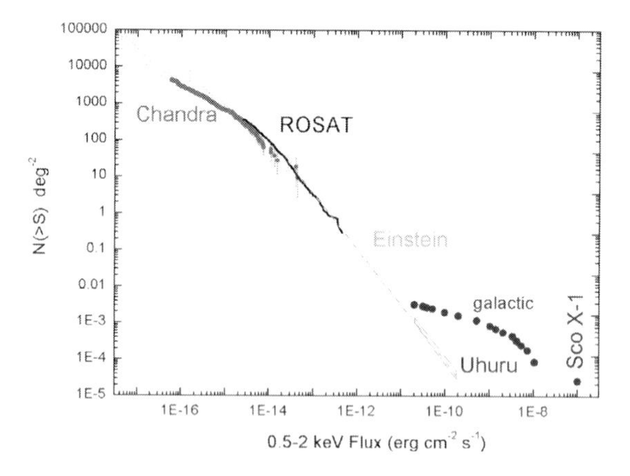


Figure 20: The sensitivity change from 1962 to 2000 (Sco X–1 to *Chandra*). (Courtesy of Günther Hasinger).

Gone is the classical conception of the Universe as a serene and majestic ensemble whose slow evolution is regulated by the consumption of the nuclear fuel. The Universe we know today is pervaded by the echoes of enormous explosions and rent by abrupt changes of luminosity on large energy scales. From the initial explosion to formation of galaxies and clusters, from the birth to the death of stars, high energy phenomena are the norm and not the exception in the evolution of the Universe.

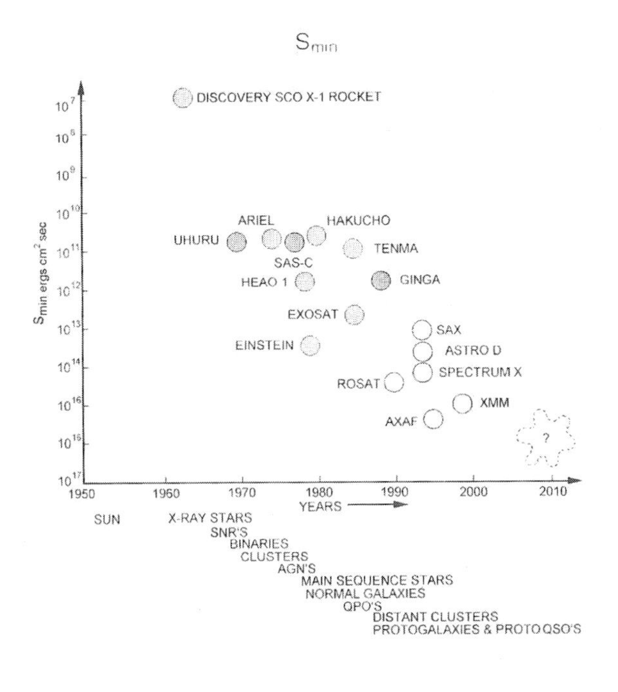


Figure 21: The sensitivity achieved by various x–ray missions (1970 to 2000).(Illustration of R. Giacconi).

## References

Bolton, C. 1972, Nature 235, 271.

Bowyer, S. et al. 1964a, AJ, 69, 135.

Bowyer, S. et al. 1964b, Science 146, 912.

Braes, L. and Miley, G. 1971, Nature 232, 246.

Burbidge, G. R. 1967, "Theoretical Ideas concerning x-ray sources," Proceedings IAU Symposium No. 31 (New York Academic Press, New York 1967), pp. 465-466.

Giacconi, R. 1974, "Binary X-Ray Sources", Proceedings of IAU Symposium No. 64 (D. Reidel Publishing Co., 1974) pp. 147-180.

Giacconi, R. and Rossi, B. 1960, J. of Geophys. Res. 65, 773.

Giacconi, R. Clark, G.W., Rossi, B.B. 1960, "A brief review of Experimental and Theoretical Progress in X-ray Astronomy," Technical Note of American Science and Engineering, ASE-TN-49, Jan. 15.

Giacconi, R., Gursky, H., Waters, J.et al. 1965, Nature 207, 573.

Giacconi, R. et al. 1962, Phys. Rev. Letters, 9, 439.

Giacconi, R., et al. 1969, Space Science Reviews 9, 3.

Giacconi, R. et al. 1971, ApJ 165, L27.

Giacconi, R. et al. 1979, ApJ 230, 540.

Giacconi, R. et al. 2002, ApJS 139, Issue 2, 369.

Gursky, H. et al. 1966, ApJ 146, 310.

Gursky, H. et al. 1972, ApJ 173, 99.

Hirsh, R.F. 1979, "Science, Technology and Public Policy: the case of X-ray astronomy, 1959-1972", Ph.D. Thesis at the University of Wisconsin, Madison

Hjellming, R.M. and Wade, C.M. 1971, ApJ 168, L21.

Holt, S. 1971, ApJ 166, L65.

Miyamoto, S. et al. 1971, ApJ 168, L11.

Oda, M. et al. 1965, Nature 205, 554.

Oda, M. et al, 1971, ApJ 166, L1.

Rappaport, S. et al. 1971a, ApJ 168, L43.

Rappaport, S. et al. 1971b, ApJ 168, L17.

Rhoades, C. and Ruffini, R. 1974, Phys. Rev. Lett. 32, 324.

Rosati, P. et al. 2002, Annual Review of Astronomy and Astrophysics 40, 539.

Rothschild, R. et al. 1974, ApJ 189, L13.

Sandage, A. et al. 1966, ApJ 146, 316.

Shklovsky, I. S. 1967, ApJ 148, L1-L4.

Schreier, E. et al. 1972, ApJ 172, L79.

Tanaka, Y. 1992, "Black Hole X-ray Binaries", Proceedings of ISAS Symposium on Astrophysics, (Inst. of Space and Astronomical Science, Tokyo) p. 19.

Tananbaum, H. et al, 1971 , ApJ 165, L37.

Tananbaum, H. et al. 1972a, ApJ 174, L143.

Tananbaum, H. et al. 1972b, ApJ 177, L5.

Tananbaum, H. and Weisskopf, M. 2001, in "New Century of X-Ray Astronomy", ASP Proceedings Vol. 251, (Astron. Soc. Pacific, San Francisco) p.4.

Vaiana, G.S. and Rosner, R. 1978, Annual Review of Astronomy and Astrophysics, 16, 393.

Webster, B. and Murdin, P. 1972, Nature, 235, 37.

*Ludwig Biermann Award Lecture*

# The Second Stars

Falk Herwig

Theoretical Astrophysics Group
Los Alamos National Laboratory, Los Alamos, NM 87544, USA
fherwig@lanl.gov

## Abstract

*The ejecta of the first probably very massive stars polluted the Big Bang primordial element mix with the first heavier elements. The resulting ultra metal-poor abundance distribution provided the initial conditions for the second stars of a wide range of initial masses reaching down to intermediate and low masses. The importance of these second stars for understanding the origin of the elements in the early universe are manifold. While the massive first stars have long vanished the second stars are still around and currently observed. They are the carriers of the information about the first stars, but they are also capable of nuclear production themselves. For example, in order to use ultra or extremely metal-poor stars as a probe for the r process in the early universe a reliable model of the s process in the second stars is needed. Eventually, the second stars may provide us with important clues on questions ranging from structure formation to how the stars actually make the elements, not only in the early but also in the present universe. In particular the C-rich extremely metal-poor stars, most of which show the s-process signature, are thought to be associated with chemical yields from the evolved giant phase of intermediate mass stars. Models of such AGB stars at extremely low metallicity now exist, and comparison with observation show important discrepancies, for example with regard to the synthesis of nitrogen. This may hint at burning and mixing aspects of extremely metal-poor evolved stars that are not yet included in the standard picture of evolution, as for example the hydrogen-ingestion flash. The second stars of intermediate mass may have also played an important role in the formation of heavy elements that form through slow neutron capture reaction chains (s process). Comparison of models with observations reveal which aspects of the physics input and assumptions need to be improved. The s process is a particularly useful diagnostic tool for probing the physical processes that are responsible for the creation of elements in stars, like for example rotation. As new observational techniques and strategies continue to penetrate the field, for example the multi-object spectroscopy, or the future spectroscopic surveys, the extremely metal-poor stars will play an increasingly important role to address some of the most fundamental and challenging, current questions of astronomy.*

*Reviews in Modern Astronomy 18.* Edited by S. Röser
 ISBN 3-527-40608-5

# 1 Introduction

One of the most intriguing questions of astrophysics and astronomy is the origin of the elements in the early universe, and how this relates to the first formation of structure. A few million years after the Big Bang the epoch of structure formation emerged and the first stars were born from the initial, pristine baryonic matter. Without any elements heavier than helium to provide cooling, the first stars that formed from the baryonic matter trapped in the emerging mini-dark matter halos were probably very massive, greater than 30 $M_{\odot}$(e.g. Abel et al. 2002). These massive stars burned through their available fuel in about one to two million years, exploded as supernovae, and dispersed the first elements heavier than helium into the nascent universe, or collapsed into black holes. These first events of stellar evolution influenced their early universe neighborhood, and determined under which conditions and with which initial abundance low-mass stars with masses like the Sun eventually formed. These take about 100 to several 1000 times longer to form than more massive stars. Thus, by the time the first low-mass stars formed there was probably already a small amount of heavier elements present. Such low-mass stars are really the second stars, stars which formed from a non-primordial abundance distribution. These second stars are important because while the massive first stars have long since vanished, some second stars are still around. Their importance lies in their capacity as carriers of the information about the formation and evolution of those long gone first generations of stars. The second stars are also important because of the nuclear production they are capable of by themselves (for example the first s process in the second stars). Observations of such second stars are now emerging in increasing quantity and with high-resolution spectroscopic abundance determination (Beers & Christlieb 2005).

There are a number of strategies to find ultra and extremely metal-poor stars. The most effective appears to be spectroscopic surveys. Most of the EMP and UMP stars have been discovered in the HK survey (Beers et al. 1992) and the Hamburg/ESO survey (HES Christlieb et al. 2001). Here, I designate somewhat arbitrarily ultra metal-poor (UMP) stars as those with $[\mathrm{Fe/H}] \leq -3.5$ ($Z \leq 6 \cdot 10^{-6}$) and extremely metal-poor (EMP) stars as those with $[\mathrm{Fe/H}] \leq -1.8$ ($Z \leq 3 \cdot 10^{-4}$). EMP stars include the most metal-poor globular clusters. UMP stars are rare and hard to find in the spectroscopic surveys. Only approximately a dozen have been reported so far, with the present record holder HE 0107-5240 at $[\mathrm{Fe/H}] = -5.4$ (Christlieb et al. 2002, Christlieb et al. 2004). At and below this metallicity spectral features are so weak that even the 10m class telescopes are barely suitable for the job.

The formation mode of the second stars depends on a complex history of events, each of which may leave signatures in the observed abundances of EMP and in particular the UMP stars. These events include the ionizing radiation of the very first primordial stars and the expansion of the associated H II regions, their effect on possibly primordial star formation in neighboring mini-dark matter halos, as well as possibly the metal enrichment from a pair-instability supernova, depending on the mass of the first primordial stars. These processes and the cosmological assumption that fix the underlying dark-matter structure will eventually determine the abundance distribution of the EMP and UMP stars now discovered and analyzed. The most re-

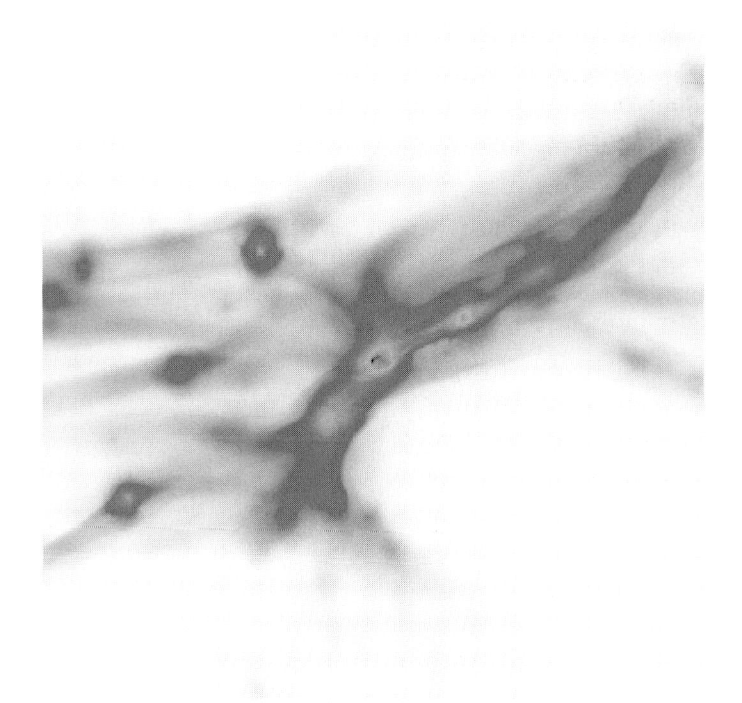

Figure 1: Distribution of baryonic matter clustering around a cosmological dark matter halo in a hydrodynamics and N-body simulation at redshift $z \approx 17$ (O'Shea et al. 2005). The projection volume of 1.5 proper kiloparsecs on a side is centered on the halo where the first population III star in this region will form. This halo is surrounded by other halos in which the second generations of primordial and/or metal-enriched stars will form. Star formation in these neighboring halos is effected by the ionizing radiation from the first star formation in the center halo, and may include primordial or ultra metal-poor intermediate mass stars.

cent studies confirm that the very first primordial stars are massive (O'Shea et al. 2005). However in these calculations the effect of the ionization from star formation in nearby mini-dark matter halos is taken into account, and it is found that the following generation of primordial stars may include intermediate mass stars.

The most metal-poor stars are potentially an extremely powerful tool to study star formation and evolution in the early universe (Beers & Christlieb 2005). These stars with metallicities reaching in excess of 5 orders of magnitude in metallicity below the solar metallicity often show abundance distributions that are very different from the solar abundance distribution. The solar abundance distribution is the result of a long galactic chemical evolution, reflecting the cumulative result of mixing and nuclear processing of many generations of stars of a wide range of masses. The most metal-poor stars are very old and have formed very early, at the onset of galactic chemical evolution. It is therefore reasonable to assume that the abundance pattern in these second stars are the result of only few different sources and corresponding stellar nuclear production sites. The goal is to identify these nuclear production sites and the mechanisms and processes through which the corresponding stellar ejecta have been brought together in the star formation cloud of the second stars. An important part of the problem is to account for self-pollution or external pollution events that may have altered the stellar abundances between the time of second star formation and observation.

As the number of EMP and UMP stars with detailed abundance continuously grows one can discern certain abundance patterns. A very unusual sub-group are the carbon-rich EMP and UMP stars. Observations indicate that many of these stars are polluted by Asymptotic Giant Branch (AGB) binary companions. This implies that reliable models of AGB stars including their heavy element production through the s process at extremely low metallicity are required to understand these stars, and to disentangle the AGB contribution from other sources that have contributed to the observed abundance patterns. In the following section the carbon-rich EMP and UMP stars will be discussed in more detail. In § 3 current models of EMP intermediate mass stars and their uncertainties are presented. Such models have important applications for the emerging field of near-field cosmology, which uses galactic and extra-galactic stellar abundances to reconstruct the processes that lead to the formation of our and other galaxies (§ 4). Then, ongoing work – both observational and theoretical – to determine the nucleosynthesis of nitrogen at extremely low metallicity is described (§ 5) and § 6 covers the hydrogen-ingestion flash in born-again giants, because this type of events may be important for EMP AGB stars as well, and may indeed be a major source of nitrogen in these stars. § 7 deals with the $s$ process, and how it can be used as a diagnostic tool to improve the physics understanding of stellar mixing. Finally, and before some forward lookgin concluding remarks (§ 9) the importance of nuclear physics input for the question of C-star formation will be discussed in § 8.

# 2 The C-rich most metal-poor stars

About $20\ldots30\%$ of all EMP and UMP stars show conspicuous enrichment of the CNO elements, most notably C. For example, CS 29497-030 (Sivarani et al. 2004) has $[\mathrm{Fe/H}] = -2.9$, $[\mathrm{C/Fe}] = 2.4$, $[\mathrm{N/Fe}] = 1.9$, $[\mathrm{O/Fe}] = 1.7$, and $[\mathrm{Na/Fe}] = 0.5$. Others, like LP 625-44 ($[\mathrm{Fe/H}] = -2.7$ Aoki et al. 2002) or CS 22942-019 ($[\mathrm{Fe/H}] = -2.7$ Preston & Sneden 2001), have a similar overabundance of C, but smaller N ($[\mathrm{N/Fe}] = 1.0$ for LP 625-44) and larger Na ($[\mathrm{Na/Fe}] = 1.8$ for LP 625-44). Many of the heavier elements, like Ti, Cr, Mn or Zn in CS 29497-030 are either not overabundant or rather somewhat underabundant compared to the solar distribution.

Large overabundances of N and Na are not predicted by standard models of massive stars, neither for Pop III (Heger & Woosley 2002) nor at larger metallicity (Woosley & Weaver 1995). Primary production of N requires to expose the He-burning product C to H-burning again. Such a condition is encountered in massive AGB stars during hot-bottom burning (Herwig 2005). Na is made in C-burning in massive stars but its final abundance is sensitive to the neutron excess, and thus scales less than linear with metallicity. If primary $^{14}$N is exposed to He-burning it will lead to the production of $^{22}$Ne. Either an additional proton collected in hot-bottom burning, or a neutron from the $^{22}\mathrm{Ne}(\alpha, \mathrm{n})^{25}\mathrm{Mg}$ reaction in the He-shell flash of AGB stars can then lead to primary $^{23}$Na production. This process leads to significant Na overabundances in standard AGB stellar models of very low or zero metallicity (Herwig 2004b, Siess et al. 2004).

The overabundances of C, N and Na are not the only indication that the abundances in C-rich EMP stars are at least partly caused by AGB stellar evolution. All of the objects mentioned above, and in fact most C-rich EMP stars show in some cases significant overabundances of the s-process elements. In addition they reveal their binarity through radial velocity variations, corresponding to binary periods of a few days in the case of HE 0024-2523 (Lucatello et al. 2003) to more typically one to a dozen years. In fact, in their statistical analysis Lucatello et al. (2004) find that observations of radial velocities of CEMP-s stars (C-rich EMP stars with s-process signature) obtained to date is consistent with a 100% binarity rate of CEMP-s stars.

HE 0107-5420 is currently the intrinsically most metal-poor star with $[\mathrm{Fe}/\mathrm{H}] = -5.3$. However, the total metal abundance in terms of elements heavier than He is quite large. In particular the CNO elements are significantly overabundant compared to the solar abundance ratios, with $[\mathrm{C}/\mathrm{Fe}] = +4.0$, $[\mathrm{N}/\mathrm{Fe}] = +2.4$ (Christlieb et al. 2004), and $[\mathrm{O}/\mathrm{Fe}] = +2.3$ (Bessell et al. 2004). The large overabundance alone indicate that only a few nuclear sources were involved in creating this abundance pattern. Together with the low Fe abundance the large CNO abundances require a primary production of the CNO elements. A primary nucleosynthesis production chain in a star is based only on the primordial elements H and He. A secondary production in contrast requires heavier elements already to be present in the initial element mix at the time of star formation. For example, standard models of massive stars predict a secondary production of N from CNO cycling of C and O that is initially present from earlier generations of stars.

HE 0107-5420 is a good example to discuss the possible nuclear production sites that may be important at the lowest metallicities. Certainly the production of the first massive Pop III stars plays an important role. Umeda & Nomoto (2003) have discussed the observed abundance pattern within the framework of mixing and fallback of the ejecta during the explosion of supernovae that eventlually form a black hole. They show that the mixing and fallback parameters can be chosen in a way that accounts for much of the observed pattern. For example, their model correctly reflects the large observed Fe/Ni ratio, and the large C/Fe and C/Mg ratios. The model does predict a significant N overabundance but it quantitatively falls short of matching the observed value. Their model predicts the charactereistic odd-even effect that reflects the small (or missing) neutron-excess of the primordial nuclear fuel. One of the observed species that hints that HE 0107-5420 may not have such a pronounced odd-even pattern is Na. The observerved overabundance is $[\mathrm{Na}/\mathrm{Fe}] = 0.8$, more than 1.5 dex above the model prediction from the mixing and fallback supernova. Unfortunately, for the heavy trans-iron elements only upper limits could be derived so far, which prohibits definite access to the s- and r-process abundances in HE 0107-5420. At this point Na and to a lesser extent N are the major indicators that some additional sources are part of the nuclear production site inventory that caused the abundance pattern in this most metal-poor star.

Without better guidance from ab-initio coupled structure and star formation simulations – such as shown in Fig. 1 – several scenarios have to be considered. Material from the first supernova could be ejected from the corresponding halo and injected into a neighbouring halo. This metal enrichment could lead to the formation of an intermediate mass star, for example a very massive AGB star, maybe even a super-AGB star that eventually ends in a ONeMg core-collapse supernova. As shown in § 3 a contribution from such stars can account for both Na as well as N, but more detailed models of this particular scenario need to be done for quantitative comparison. A low-mass star like HE 0107-5420 could then form from the combined nuclear production of these two sources, a roughly $25\,M_{\odot}$ black hole forming supernova and a massive AGB or super-AGB star.

Alternatively the low-mass star forms from the ejecta of just one supernova, and this initial abundance pattern is modified by self-polution. HE 0107-5420 is in fact a giant on the first ascent. Observations, in particular of globular cluster red giants show that mixing and nuclear processing of C into N below the deep convective envelope lead to secondary N production (Denissenkov & VandenBerg 2003, Weiss et al. 2004). This could explain the quantitative discrepancy between the observed N abundance and the model by Umeda & Nomoto (2003). However, the rather large carbon isotopic ratio $^{12}C/^{13}C \sim 60$ implies that self-polution can not account for the observed Na abundance, in particular assuming the absence of $^{22}Ne$ which is a secondary nucleosynthesis product in massive stars.

This leads to the external polution scenario that is frequently considered in this context. As mentioned above many CEMP stars are in fact in binary systems, and the abundance signatures seen in these C-rich very metal-poor stars could be – in part – the result of mass transfer from the AGB progenitor of a white dwarf companion. For HE 0107-5420 Suda et al. (2004) discuss this possibility in detail. For the evolution of primordial AGB stars they consider the possibility of the H-ingestion flash, a process in which the He-shell flash convection zone reaches outward into the H-rich envelope, resulting in a peculiar nucleosynthesis and mixing regime. Such events are known at solar metallicity to cause a significant fraction of young white dwarfs to evolve for a short period of time back to the AGB (born-again evolution, § 6), and the most prominent representative of this class of objects is Sakurai's object (Duerbeck et al. 2000).

# 3 Stellar evolution models for the second stars

As discussed in the previous section the evolution of low- and intermediate mass stars at zero or extremely or ultra metal-poor abundance is essential for studying the most metal-poor stars and their cosmological origin and environment. This section deals first with the evolution of AGB stars that may have formed from Big Bang material. A possibly important evolutionary phase - the H-ingestion flash triggered by the He-shell flash - has been observed in models of Pop III and even in models of non-zero but ultra-low metal content (§ 3.2). Finally the evolution and yields of extremely metal-poor TP-AGB stars is described. The evolution of Pop III massive stars is discussed, for example, by Heger & Woosley (2002).

## 3.1 Pop III AGB evolution

Simulations show that many of the overall properties of Pop III intermediate mass stellar models (Chieffi et al. 2001, Siess et al. 2002) probably apply to ultra metal-poor stars as well. At $Z = 0$ the initial thermal pulses may show peculiar convective mixing events at the core-envelope boundary during or after the actual He-shell flash. But eventually the models enter a phase of rather normal thermal pulse AGB evolution, with regular thermal pulse cycles, third dredge-up, hot-bottom burning and mass loss. Many uncertainties of the Pop III stellar evolution is related to the possibility of flash-burning that is not yet well understood (§ 3.2).

$Z = 0$ stellar evolution is different from evolution at higher metal content because CNO catalytic material for H-burning is initially absent. Burning hydrogen via the pp-chain requires a higher temperature than H-burning via the CNO cycle. Eventually the triple-$\alpha$ process will provide some C, and a mass fraction as low as $10^{-10}$ is sufficient to switch H-core burning to CNO cycling. During He-core burning some C is produced outside the convective core and a tail of carbon reaches from the convective core out toward the H-shell. After the end of He-core burning, and depending on the initial mass, this carbon and some nitrogen is mixed into the envelope. According to Chieffi et al. (2001) this second dredge-up raises the envelope C-abundance above $10^{-7}\,\mathrm{M}_\odot$ for $M_{ini} > 6\,\mathrm{M}_\odot$. For such a C-abundance the H-shell is fully supported by CNO cycling. As a result the thermal pulse AGB evolution is like that of the ultra metal-poor cases.

Models with initial masses below $6\,\mathrm{M}_\odot$ show a peculiarity that is unknown in more metal-rich models. After an initial series of weak thermal pulses the H-shell forms a convection zone when it re-ignites after the He-shell flash (Chieffi et al. 2001, Siess et al. 2002, Herwig 2003). The lower boundary of this convection zone is highly unstable, because it coincides with the opacity discontinuity that marks the core-envelope interface. Even small amounts of mixing will drive flash-like burning and deep mixing, as protons enter a $^{12}$C-rich zone. It seems that the details of the evolution of the H-shell convection zone is not important as this one-time event leads to deep dredge-up and enrichment of the envelope with a sufficient amount to support regular CNO cylce burning in the H-shell thereafter. Accordingly, these lower-mass cases will evolve like extremely metal-poor stars too, and the $Z = 0$ models discussed here can be used as proxies for what is in reality born as an ultra metal-poor star.

The effect of rotationally induced mixing processes may alter the evolution of low- and intermediate-mass stars and very low metallicity even before the the AGB and lead to a qualitatively different thermal pulse evolution. Meynet & Maeder (2002) describe models for metallicity $Z = 10^{-5}$ in which $^{12}$C and $^{16}$O is mixed out of the He-burning core and up to the location of the H-shell. This leads to an increase of CNO material, in particular $^{14}$N in the envelope that could result in a rather normal thermal pulse evolution without any peculiar hydrogen convection zone. Whether the rotationally induced mixing processes before the AGB or the subsequent thermal pulse AGB evolution with the associated dredge-up events dominate the final CNO yields is not yet clear.

The difference between a real Pop III and an extremely metal-poor thermal pulse AGB star is the initial absence of Fe and other elements heavier than Al in the Pop III star. This does not preclude that a Pop III AGB stars can not generate $s$-process-elements by the slow neutron capture process. Goriely & Siess (2001) have used the $s$-process-framework established for solar-like metallicities (§ 7) and ran network calculations with an initial abundance appropriate for the interior of a $Z = 0$ TP-AGB star. They used the models of Siess et al. (2002) as stellar evolution input. The major uncertainty was the mixing for the formation of the neutron donor species $^{13}$C, and the unknown feedback of such mixing to the thermodynamic structure evolution. Herwig (2004a) and Goriely & Siess (2004) have shown that this feedback is likely very important. Nevertheless, the $s$-process-models for $Z = 0$ are instructive, because they show that if a sufficient neutron source is available, the $s$ process can be based on seed nuclei with lower mass number than Fe, for example C. C is produced in a primary mode in TP-AGB stars. For such a $Z = 0$ $s$ process the Fe/Ni ratio should be markedly sub-solar, approximately close to the quasi steady-state value of $\sim 3$, that is given by the ratio of the neutron cross sections of Ni and Fe (the isotopic minimum for each element). HE 0107-5240 with $[\mathrm{Fe/H}] = -5.3$ (Christlieb et al. 2004) shows a larger than solar Fe/Ni ratio, which is evidence for the contribution of supernovae to the abundace distribution of this star.

## 3.2 The hydrogen-ingestion flash in ultra metal-poor or metal-free AGB stars

The peculiar H-convective episode described in the previous section was observed in AGB models with of $5\,\mathrm{M}_\odot$ and $Z = 0$ by Herwig (2003). During the following thermal pulse the He-shell flash convection zone reaches out into the H-rich envelope. This leads to the H-ingestion flash (HIF). Protons are mixed on the convective time scale into the He-shell flash region which is with depth increasingly hot. The hydrogen flash-burning leads to a separate convection zone as shown in Fig. 2. Convective H-burning in this layer is characterized by a large abundance of $^{12}$C and protons. The protons have been mixed into this region from the envelope. Accordingly the $^{14}$N abundance in this layer is very large, up to $10\%$ by mass in a layer of $\approx 10^{-3}\,\mathrm{M}_\odot$. The models show that after both the He-shell flash and the HIF have subsided a deep dredge-up episode will mix this $^{14}$N-rich layer into the envelope. This may account for a substantial $^{14}$N production and part of the observational pattern observed in EMP stars that are believed to be poluted by EMP AGB stars (§ 5). A HIF was also found in the $2\,\mathrm{M}_\odot$, $Z = 0$ model, but Herwig (2003) did not find the HIF in models of $Z \geq 10^{-5}$.

AGB stars are highly non-linear systems. The occurrence of the HIF depends on many details of the models, including the physics of mixing, opacities, the numerical resolution and nuclear reaction rates. Small changes in any of these ingredients, that may go unnoticed during a more robust stellar evolution phase (like the main-sequence), can be the deciding factor here whether a HIF occurs or not. Currently, theoretical models do not agree if and in which initial mass and metallicity regime

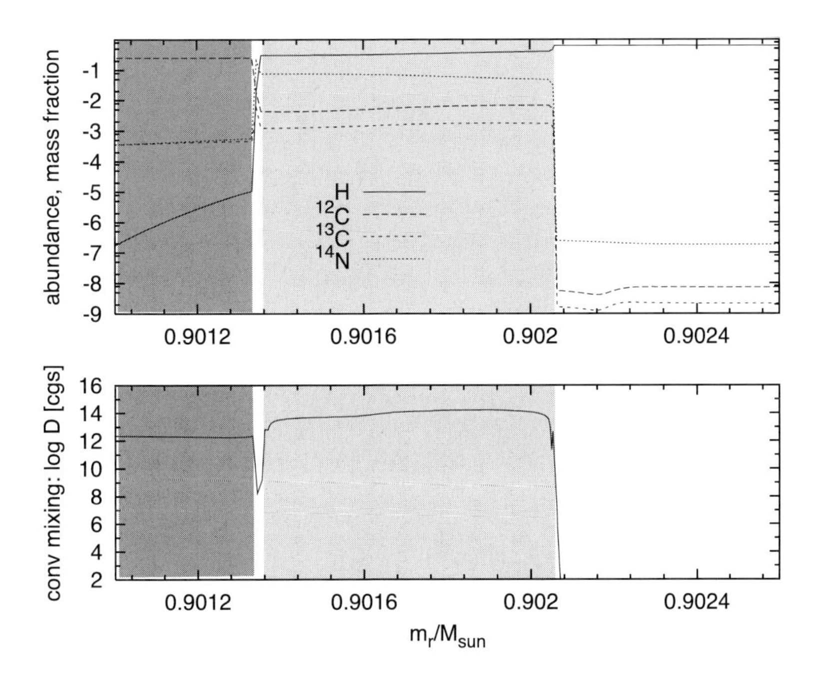


Figure 2: Burning and mixing in the H-ingestion flash during the $10^{\rm th}$ thermal pulse of a $5\,{\rm M}_{\odot}$, $Z = 0$ sequence (Herwig 2003). Shown are the top layers of the He-shell flash convection zone (dark grey), and the H-ingestion flash-driven convection zone (light grey). Top panel: abundance profiles; bottom panel: convective mixing coefficient.

the HIF occurs (Fujimoto et al. 2000, Chieffi et al. 2001, Siess et al. 2002, Herwig 2003). However, it is important to better understand this situation because the HIF may play an important role in the $s$ process in extremely metal-poor stars (Iwamoto et al. 2004). Hydrogen ingested into the He-shell flash or pulse-driven convection zone (PDCZ) may lead to $^{13}$C production which could release neutrons. Suda et al. (2004) propose that the abundance pattern of the currently most metal-poor star – HE 0107-5420 – is in part due to the nucleosynthesis in a thermal pulse AGB star with HIF. The details of the nucleosynthesis in a HIF depend sensitively on the physics of simultaneous rapid nuclear burning and convective mixing. The hydrogen-ingestion flash can also be a source of lithium, distinctly different from conditions during hot-bottom burning (Herwig & Langer 2001).

The evolution of born-again stars, like Sakurai's object, provide valuable additional constraints with regard to the HIF in AGB stars (§ 6). The very late thermal pulse is associated with a HIF (Iben et al. 1983, Herwig et al. 1999). By reproducing the evolution of stars like Sakurai's object one can gain some confidence in models of the HIF in metal-poor AGB stars. Herwig (2001) showed that models in which the convective hydrogen ingestion occurs with lower velocities than predicted by the

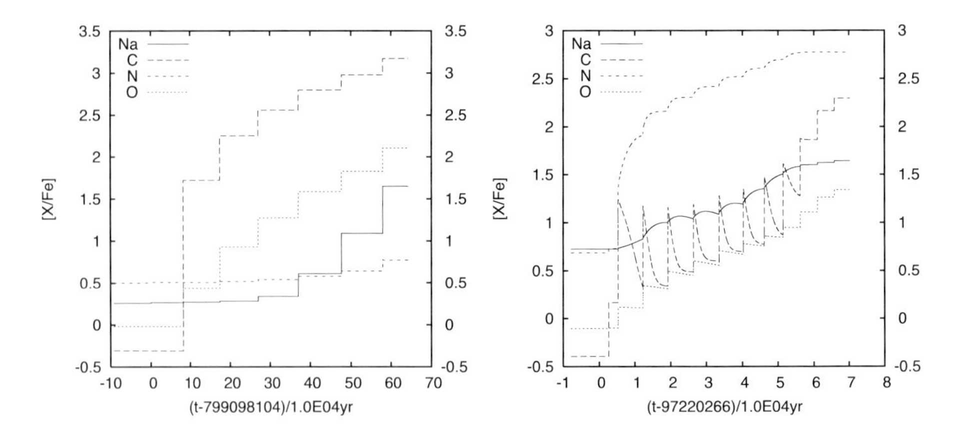


Figure 3: Surface evolution of C, N, O and Na for 2 and 5 $M_\odot$ AGB evolution models at $[Fe/H] = -2.3$ (Herwig 2004b). The ejecta of the $2\,M_\odot$ case are released after $0.8$Gyr, whereas the $5\,M_\odot$ case releases the ejecta after $0.1$Gyr. The abundance evolution is determined by the interplay of dredge-up and hot-bottom burning. Each step in the curves corresponds to a dredge-up event after a thermal pulse. Hot-bottom burning can be obseved in the 5 $M_\odot$ case as the decrease of C and increase of N (and to a lesser extent Na) inbetween dredge-up events. Note the dichotomy in the C/N ratio in the low-mass vs. high-mass AGB star.

mixing-length theory (Boehm-Vitense 1958) feature a faster born-again evolution than MLT models, quantitatively in agreement with observations. The physical interpretation is that rapid nuclear burning on the convective time scale releases energy and adds buoyancy to down-flowing convective bubbles, leading to additional breaking of the plumes. The temperature and time scales for nucleosynthesis during the HIF would be significantly different. This has not yet been applied to HIF models at $Z = 0$ or extremely low metallicity.

## 3.3 AGB nucleosynthesis at extremely low metallicity

By definition all yields of Pop III stars are primary, because the initial composition contains only Big Bang material. Nuclear production is secondary when nuclei heavier than H and He are already present in the initial abundance distribution and transformed in other species. POP III or EMP AGB stars can produce a large number of species in primary mode, including the CNO elements, Ne, Na, Mg and Al. The production of heavier elements depends on the availabilitity of neutrons. Some of these primary species, like oxygen, are usually not considered a product of AGB evolution at moderate metal-deficiency or solar metallicity. However, the initial envelope mass fraction of oxygen for models with $Z = 10^{-4}$ is $< 5 \cdot 10^{-5}$. $^{16}O$ in the intershell material (cf. Fig. 6) that is dredged-up is primary and even at this low metallicity of the order $1\%$ by mass. Dredge-up of such material will enhance the envelope abundance significantly.

A set of low- and intermediate mass AGB stellar evolution models with $Z = 10^{-4}$ ([Fe/H] = -2.3) with detailed structure, nucleosynthesis and yield predictions has been presented by Herwig (2004b). The oxygen overabundance in these models is $\log(X/X_{\odot}) = 0.5 \cdots 1.5$, depending on mass. The abundance evolution for C, N, O and Na is shown in Fig. 3 for two initial masses. The evolution of C and N is qualitatively different in the two cases. The lower mass sequence shows the increase of C and some O due to the repeated dredge-up events. No significant amount of $^{14}$N is dredged-up. The result is a large C/N ratio. The $5\,\mathrm{M}_{\odot}$ track shows the opposite behaviour. The much lower C/N ratio is the result of hot-bottom burning (Boothroyd et al. 1993). C and O are dredged-up after the thermal pulse as in the lower mass case, but during the following quiescent intperpulse phase the envelope convection reaches into the region hot enough for H-burning and the envelope C is transformed into N. The initial-mass transition between hot-bottom burning models with large N abdundances and C dredge-up models without hot-bottom burning is very sharp. This precludes the notion that EMP stars with simultaneously large N and C overabundances, like CS 29497-030 with $[\mathrm{C/Fe}] = 2.4$ and $[\mathrm{N/Fe}] = 1.9$ (Sivarani et al. 2004), may be poluted by such AGB stars in the transition regime where hot-bottom is only partially efficent (see § 5).

The signature of hot-bottom burning in Fig. 3 is a gradual change of abundance between the steps caused by dredge-up. From this it can be seen that Na has a different nuclear production site depending on initial mass. In the $2\,\mathrm{M}_{\odot}$ case Na is produced in the He-shell flash, by n-captures on $^{22}\mathrm{Ne}$. The neutrons come from the $^{22}\mathrm{Ne}(\alpha, \mathrm{n})$ reaction. $^{23}\mathrm{Na}$ is dredged-up but not produced during the interpulse phase. The neutron-heavy Mg isotope $^{26}\mathrm{Mg}$ is produced in a similar way. In the $5\,\mathrm{M}_{\odot}$ case Na is mainly produced in hot-bottom burning, by p-capture on dredged-up $^{22}\mathrm{Ne}$.

In the previous section it was argued that due to the large primary production of CNO and many other species the $Z = 0$ model predictions would also apply approximately to stars with ultra-low metallicity as well. Neutron capture reactions play an important role, and need to be included in yield-predictions, together with a neutron-sink approximation for species not explicitely included in the network. This has been done in a qualitatively similar way by both Herwig (2004b) for the $Z = 10^{-4}$ models and Siess et al. (2002) for their $Z = 0$ models, and the abundance evolution is quantitatively similar. Both model sets predict for non-hot-bottom burning cases large primary production of C and O, while the $^{14}$N abundance does not change. $^{22}\mathrm{Ne}$ is produced in both cases in the He-shell flashes, and in both cases the final $^{22}\mathrm{Ne}$ exceeds that of $^{14}$N. The evolution of $^{23}\mathrm{Na}$, $^{25}\mathrm{Mg}$ and $^{26}\mathrm{Mg}$ is qualitatively the same too. In particular the ratio of these three isotopes is quantitatively the same in the last computed model of the $Z = 0$ sequence and the final AGB model at $Z = 10^{-4}$. This is in particular interesting as these low-mass AGB models predict that $^{26}\mathrm{Mg} : {}^{25}\mathrm{Mg} > 1$, contrary to the signature of hot-bottom burning.

One of the largest uncertainties is the adopted mass loss. While mass loss in AGB stars of solar metallicity can be constrained observationally, this has not been possible for extremely low metallicity. Other uncertainties relate to mixing in EMP AGB stars. Herwig (2004a) has shown that convection induced extra-mixing, like exponential overshoot that is used to generate a $^{13}$C pocket for the $s$-process in higher-metallicity models, may lead to vigorous H-burning during the third dredge-up. Depending on the efficiency of such mixing the third dredge-up may turn into a flame-like burning front, leading to very deep core penetration. This may significantly impact the formation and effectiveness of a $^{13}$C pocket for the $s$ process (Goriely & Siess 2004).

# 4 Near-field cosmology application of stellar yield calculations

Accurate stellar yields for intermediate mass stars are requested by another emerging field, near-field cosmology (Bland-Hawthorn & Freeman 2000, Freeman & Bland-Hawthorn 2002). The baryon (stellar) halo of the Milky Way retains a fossil imprint of the merging history of the galaxy. Different merging components, like the infalling dwarf galaxies with a range of masses, have different star formation histories that translate into different abundance signatures of the member stars of these components. The current surge of spectroscopic multi-object capabilities at large telescopes will likely significantly enhance the importance of this approach.

An example is the recent work by Venn et al. (2004) in which abundances of halo stars are compared with abundances of stars in satellites, dwarf galaxies trapped in the potential well of the Galaxy. Here, the basic idea is that dwarfs are merging with a galaxy at different times at which point star formation and chemical evolution stops. Satellites that survive until the present day should show the signature of more evolved chemical evolution, for example including the $s$-process elements associated with the long-lived low-mass stars. In contrast, halo stars that are the dispersed members of dwarf galaxies that have merged at an earlier time in the evolution of the galaxy, have less evolved chemical evolution patterns, for example showing more clearly the patterns of nucleosynthesis in massive star evolution.

Potentially, the implications that can be derived from comparison of halo stars and satellite members may include some fundamental questions of nucleosynthesis itself. The data presented in Venn et al. (2004) show that $[\alpha/\mathrm{Fe}]$ is systematically smaller in stars belonging to satellites compared to their halo counterparts. This can be explained in the $\Lambda$ cold dark matter model that predicts that most halo stars have formed in rather massive dwarf galaxies, which merged with the galaxy a long time ago (Robertson et al. 2005). These stars show the signature of truncated chemical evolution, dominated by the yields of supernovae type II and their significant $\alpha$-element contribution and moderate Fe ejecta. Stars in present, less massive satellite galaxies show the signature of chemical evolution components that take more time, like the SN Ia. These events add Fe but little $\alpha$-elements, and their $[\alpha/\mathrm{Fe}]$ ratio is therefore smaller.

Abundances of other elements could be compared too. For example, Venn et al. (2004) compare among others the abundances of Y, Ba and Eu in halo stars and satellite dwarf galaxies. Ba is typically considered a main-component $s$-process element (at least at $[\mathrm{Fe}/\mathrm{H}] > -2$) because the elemental solar abundance has a $81\%$ $s$-process contribution (Arlandini et al. 1999). Only the smaller remaining fraction is made in the r process. As discussed above, the nuclear production site of the main-component of the s process are low- and intermediate mass AGB stars, in the initial mass range $1.5 < M_{\rm ini}/\,{\rm M}_\odot < 3.0$. This component of galaxy chemical evolution needs even more time to contribute than the SN Ia. And indeed, [Ba/Fe] behaves differently from $[\alpha/{\rm Fe}]$ in halo stars and satellite stars. [Ba/Fe] is on average higher in the satellite stars than in halo stars, with a considerable spread. This is consistent with the framework of truncated chemical evolution of systems that were early disrupted in merger events, and with the understanding of stellar evolution and nucleosynthesis that $\alpha$-elements are predominanty made in short-lived massive stars, while Ba originates in rather old populations. It is then extremely interesting to consider the r-process element Eu, which in the solar abundance distribution has a very small $s$-process contribution of only $5.8\%$. This element does not behave like $\alpha$-elements. Instead, [Eu/Fe] is on average the same in halo stars and dwarf galaxy stars, however with a larger spread in the latter. In order to remain in the proposed scenario of why abundances in the halo differ from those in the nearby dwarf galaxies one would then have to assume that at least an important fraction of the r-process elements does not originate in SN II, which eject their yields on a short time scale. Instead, one would have to assume that Eu in the satellite stars comes from a source that releases the ejecta on a time scale comparable to or longer than the Fe production in SN Ia. Such sources could be the accretion induced collapse (Fryer et al. 1999) or the collapse of a super-AGB star in the initial mass range $8 \ldots 10\,{\rm M}_\odot$, if the core mass grows to the Chandrasekhar limit (Nomoto 1984). It would certainly go to far at this point to draw any further conclusion, for example in conjunction with the two r process source model proposed by Qian & Wasserburg (2003).

It is clear, that the full potential of using the abundances patterns of stars to reconstruct the formation of galaxies in their cosmological context can only be reached with detailed yield predictions including all masses and reaching down to extremely low metallicity. Y is another example that reinforces this notion. In the solar abundance distribution it has a $92\%$ fraction from the main-component $s$ process originating in low-mass stars. However, it is also the termination point of the weak $s$ process from massive stars, and it does have some r-process contribution. In the data presented by (Venn et al. 2004) it behaves – with regard to the difference between halo and satellite stars – like the $\alpha$-elements, which in this context would imply an important contribution from massive stars.

# 5 The origin of nitrogen in the early universe

Nitrogen is the fifth most common element in the universe. In stellar evolution of very low metallicity the primary production is of particular importance. Strictly from a nucleosynthesis point of view nitrogen is always made in a well established sequence of events, involving first the triple-$\alpha$ process that makes $^{12}$C and then two subsequent proton captures that are part of the CN-cylce. In a stellar model sequence the problem is to identify the mixing processes that can account for this chain of events. How can $^{12}$C, that is made in the He-burning that requires higher temperatures, be brought into layers of lower temperature where H-burning still takes place. Or, how can protons be mixed down into the He-burning layers and be captured by

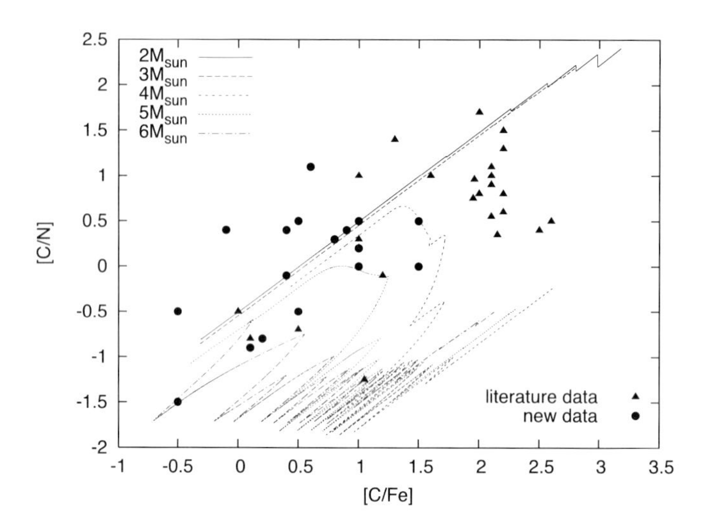


Figure 4: Observed nitrogen and carbon abundances (including literature data from Barbuy et al. 1997, Preston & Sneden 2001, Aoki et al. 2001, Lucatello et al. 2003) compared to the surface abundance evolution of AGB model tracks.

$^{12}$C to eventually form $^{14}$N, but then not be exposed to $\alpha$-captures that would convert $^{14}$N into $^{22}$Ne. Because the primary production of nitrogen depends so much on stellar mixing, this element assumes a key role in understanding stellar nucleosynthesis at extremely- and ultra-low metallicity.

Nitrogen is certainly present in the most metal-poor stars in the Milky Way (Laird 1985). According to standard stellar evolution and nucleosynthesis models primary N comes from intermediate mass (IM) AGB stars (Herwig 2004b), with perhaps a small contribution from rotating massive stars (Meynet & Maeder 2002). In AGB stars the repeating sequence of thermal pulses induce convective mixing events that lead to the prodution of N. C is made in the He-shell flash. Dredge-up after the flash mixes that C into the envelope, and in massive AGB stars the C in the envelope is then transformed into N. The low [N/$\alpha$] abundances observed in some damped Lyman-$\alpha$ systems (Pettini et al. 1999) have prompted suggestions that the IMF may be biased in favor of massive stars in some systems (Prochaska et al. 2002), or that these systems represent the earliest stages of chemical evolution of massive stars only (Centurión et al. 2003).

Observations of N and C abundances in C-rich EMP stars with $s$-process signature provide another way to study the N production at low metallicity. About 30% of all extremely metal-poor stars ([Fe/H] < -2) are strongly C- and to a lesser extent

N-enhanced (§ 2). In particular the overabundance of C and of N in addition to their s-process signature has lead to the assumption that the AGB star progenitor of the current white dwarf companion to the CEMP-s star is responsible for the observed abundance pattern. Because of the established binary nature and the $s$-process signature the CEMP-s stars are assumed to be poluted by the individual AGB stars that are the progenitors of their present white dwarfs companions.

In Fig. 4 the [C/N] and [C/Fe] ratios of literature data are shown, together with the AGB model prediction of 2 - 6 $M_\odot$ tracks. The literature data show systematically lower [C/N] ratios than what is expected by 2 or 3 $M_\odot$ models that dredge-up C, but do not produce N. This poses the question of the primary origin of the N in these stars.

Models of initial masses between 4 and 6 feature efficient hot-bottom burning (HBB) which turns most dredged-up primary carbon into nitrogen resulting in [C/N]<-1 and [C/Fe]<1.5. However, none of the literature data seem to show the very low [C/N] ratio that would be expected for an EMP star that happened to have an AGB companion in the $4 - 6\,M_\odot$ initial mass range. We have carried out observations of EMP targets with 0.5<[C/Fe]<1 using the CH and the NH band for abundance determination (Johnson et al. 2005, Herwig et al. 2004). Specifically we wanted to find out whether the paucity of EMP stars with [C/N]<-1 is a selection effect or a systematic observational bias imposed by the abundance indicators employed in previous studies. While C-strong stars can be identified by the G-Band at 4305Å, the strong NH band at 3360Å is never observed in the medium-resolution surveys that provide the targets for detailed abundance studies (e.g. Beers et al. 1992). Only the CN bands at 3883Å and 4215Å are included, and CN lines are not strong in N-rich stars unless C is also enhanced by a large amount. Therefore, studies of C and N abundances have favored C-rich stars that are easily identified, and may have missed stars that have $[C/Fe] \sim 0 - 1$, but are more rich in N. Thus the literature data in Fig. 4, drawn from high-resolution follow-up of the medium-resolution candidates, may be the result of a strong observational bias against finding N-rich stars.

In order to overcome this bias we obtained spectra for 18 new stars. In Fig. 4, we plot the preliminary estimates of the [C/N] ratios of 18 stars, based on observations of the NH bands during 6 nights at CTIO/KPNO. We did not find any stars with low [C/N], and the analysis of additional data is underway to put our findings on a more robust statistical basis. This leaves us with two important open questions: (1) Where are the EMP stars polluted by massive AGB stars, and (2) where does the N in the CEMP stars come from that we do observe? While we do not have any idea at this point of what the answer to the first question could be, there are a number of possibilities to address the absence of primary N. These include rotationally induced mixing before the AGB phase (§ 3.1), the H-ingestion flash (§ 3.2), and extra-mixing in AGB stars (Nollett et al. 2003) as envoked in models of RGB stars to account for abundance anomalies in globular cluster member stars (Denissenkov & VandenBerg 2003). In any of these cases the interpretation of N abundances in metal-poor systems of all kinds may have to be re-evaluated.

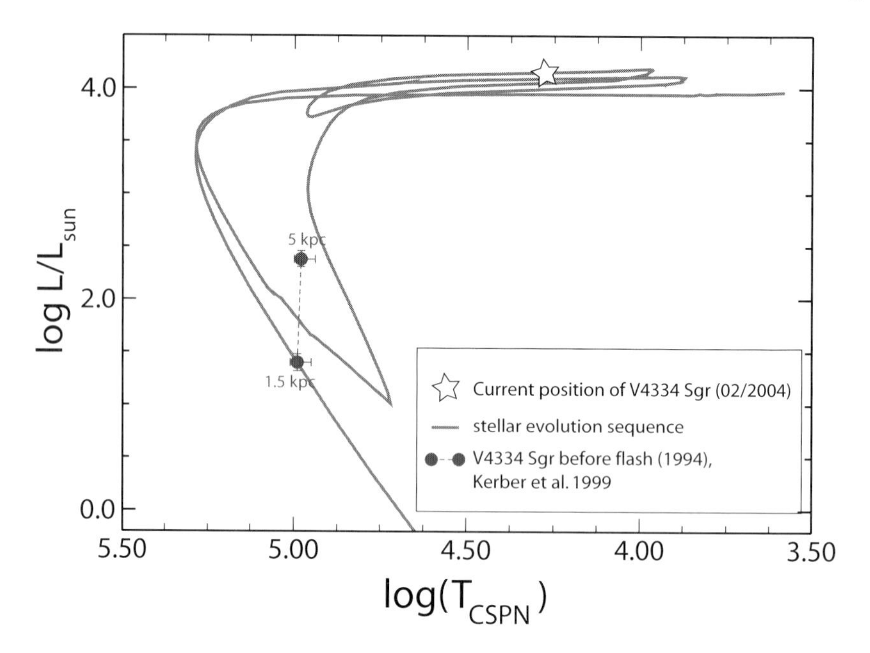


Figure 5: Observational position of Sakurai's object (V4334 Sgr) at different times, and evolutionary sequence of a $0.604\,\mathrm{M}_\odot$ central star of a planetary nebula track (Hajduk et al. 2005) with a very late thermal pulse (H-ingestion flash) with modified convective mixing velocities according to (Herwig 2001).

# 6 H-ingestion flash and born-again evolution

Nuclear production in the AGB stellar interior can become observable at a later time in the evolution, during the post-AGB phase, on the surface of those central stars of planetary nebulae that become H-deficient. Here H-deficient means in most cases that the mass fraction of H as observed in the stellar spectrum is less than $\approx 2\%$. This leads to additional important constraints on the evolution of AGB stars, which is in particular important to study AGB evolution at extremely low metallicitiy.

H-deficient central stars of planetary nebulae (CSPN) as well as very young white dwarfs have been analysed in detail (e.g. Koesterke 2001, Werner 2001, Hamann et al. 2003) and new important information about the abundances of these stars is yet to be revealed. Stellar evolution predicts that about 25% of all post-AGB stars will eventually loose all their small remaining H-rich envelope mass of the order $10^{-4}\,\mathrm{M}_\odot$ and expose the bare H-free cores. The carbon, oxygen and helium rich nature of these objects is evident from their Wolf-Rayet or PG 1159 type emission line spectra, and reflects nuclear processed material that has been built up during the stars progenitor evolution. Although there is considerable spread in the observed abundances of these stars, the pattern can be summarized in mass fraction as He $\sim$ C $\sim 0.4$ and O $= 0.08 \ldots 0.15$.

These H-deficient CSPN are important for AGB evolution modeling, because they provide a unique opportunity to study directly the nuclear processing shells in AGB stars. In order to explain the evolutionary origin Iben et al. (1983) introduced the born-again evolution scenario . The star evolves off the AGB, and becomes a hydrogen-rich central star and eventually a very hot, young white dwarf. However, the He-shell may still be capable to ignite a late He-shell flash, and in that case the star retraces its evolution in the HRD, back into the giant region (Fig. 5). As a result of hydrogen-ingestion into the He-convection and rapid burning, or by mixing from the emerging convective envelope (or because of both) the surface abundance of such born-again stars will be extremely H-deficient or even H-free.

Stellar models of this evolutionary origin scenario connect the surface abundance of the Wolf-Rayet central stars and the PG1159 stars with the intershell abundance of the progenitor AGB star. The intershell layer between the He- and the H-shell is well mixed during each He-shell flash. This zone contains the main nuclear production site of the progenitor AGB star, and the abundance of this zone reflects contributions from both He- and H-shell burning (Fig. 6). As a result of the born-again evolution these layers become visible at the surface of the resulting H-deficient CSPN.

The initial models of the H-deficient central stars by Iben and collaborators showed qualitatively that the born-again scenario could account for high He and C abundances, as these elements were abundant in the intershell of the AGB progenitor model they used. However, they could not account for the high observed oxygen abundance. Herwig et al. (1997) were the first to propose that the solution could be non-standard mixing during the AGB evolution. Models with overshooting at the bottom of the pulse-driven convection zone do not only feature higher temperatures at the bottom of that layer (§ 7.1), but they also show higher C and in particular O abundances in the PDCZ. Subsequently, Herwig (2000) has explored in detail how the various abundances depend on the overshoot efficiency and other details. In essence, overshooting brings AGB intershell abundances of He, C and O in very good quantitative agreement with the observed abundances of Wolf-Rayet type central stars and PG1159 stars, in the framework of the born-again evolution. More effort is needed to consolidate these constraints on the intershell abundance with the possibly tight upper limits on overshooting at the bottom of the PDCZ that the $s$ process may provide (§ 7.1).

The connection of the surface abundances of the hot PG1159 stars and the progenitor AGB intershell abundance has been reinforced recently by several new observational findings. These include a substantial overabundance of Ne (Werner et al. 2004), F abundances ranging from solar up to 250 times solar (Werner et al. 2005) in good agreement with AGB nucleosynthesis predictions by Lugaro et al. (2004), and Fe-deficiencies of at least 1 dex compared to solar (Miksa et al. 2002), which may reflect the depeletion of Fe in the AGB intershell due to $s$-process n-captures. In that case the Fe/Ni ratio should be low, as in fact observed in Sakurai's object. This born-again star is the smoking gun of the very late thermal pulse scenario and links this evolutionary scenario with the H-deficient PG 1159 and Wolf-Rayet type central stars (Duerbeck et al. 2000, Herwig 2001). Much of the observed abundance pattern of Sakurai's object as well as the rapid evolutionary time scale has been reproduced quantitatively with stellar models (Asplund et al. 1999, Herwig 2001). Similar to the

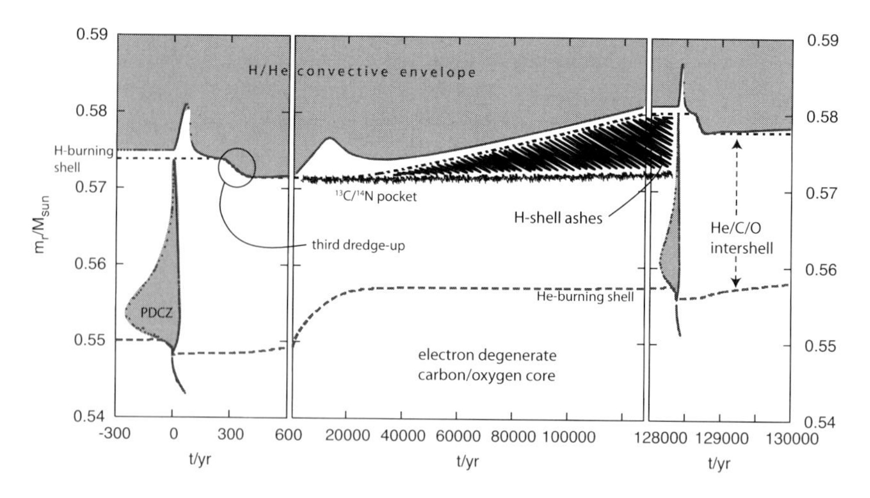


Figure 6: Evolution of convection zones (grey regions) and burning shells (dashed lines) in 1D stellar track including two He-shell flashes and the quiescent H-burning phase in between. The zig-zag indicates the location of the $^{13}$C-rich region that contains the bulk $s$-process production. The hashed region contains the ashes of the H-shell that are engulfed by the following He-shell flash convection zone.

PG 1159 stars, Sakurai's object shows Fe depletion of about 1dex compared to solar. The low ratio of $(\mathrm{Fe/Ni}) = 3$, shows that Sakurai's object shows surface material that has been directly irradiated with neutrons (Herwig et al. 2003b). New radio observations show that Sakurai's object has started reheating again, on its second evolution into the CSPN (Hajduk et al. 2005). This marks a new phase of the evolution of this star, and confirms the concept of convective mixing efficiency modified by nuclear burning (Herwig 2001).

# 7 $s$ process and AGB stars

The $s$ process is the origin of half of all elements heavier than iron (Arlandini et al. 1999). It is also important for isotopic ratios and in some cases elemental abundances of lighter elements - in particular at extremely low metallicity. The heavy elements are made by the $s$ process through neutron captures that are slow compared to the competing $\beta$-decay. Starting from the abundant iron group elements it follows closely the valley of stability in the chart of isotopes. It is characterized by neutron densities $N_\mathrm{n} < 10^{10}\mathrm{cm}^{-3}$. Observations and theory agree that the nuclear production site of the main and strong component of the $s$ process $(90 < A < 204)$ originates in low mass Asymptotic Giant Branch (AGB) stars. The weak component below the first $s$-process peak at $A = 90$ is produced during He and C burning in massive stars. Here we deal only with the $s$ process in AGB stars.

The $s$ process is important in the context of the second stars for two reasons. First, the observed elemental abundance distribution of the trans-iron elements is at any given metallicitiy a mix of s- and r-process contribution. In order to disentangle the two contributions in very-low metallicity stars, the $s$ process in this metallicity regime needs to be understood. This may than enable progress in identifying the conditions for the r process. Second, $s$-process branchings can be used to probe the physics for stellar mixing (§ 7.2), which is particularly usefull when modeling the evolution of EMP stars.

The $s$ process in low-mass AGB stars has two neutron sources. The main source is $^{13}$C, that forms via the $^{12}\mathrm{C}(\mathrm{p},\gamma)^{13}\mathrm{N}(\beta^{+})^{13}\mathrm{C}$ reaction. At the end of the dredge-up phase (Fig. 6) after the He-shell flash, when the bottom of the H-rich convective envelope has penetrated into the $^{12}$C rich intershell layer, partial mixing at this interface would create a thin layer providing simultaneously the required protons and $^{12}$C. The $s$ process in the $^{13}$C pocket is characterized by low neutron densities ($\log N_{\mathrm{n}} \sim 7$) that last for several thousand years under radiative, convectively stable conditions during the quiescent interpulse phase. The physics of mixing at the H/$^{12}$C interface at the end of the third dredge-up phase has not yet been clearly identified (see below). Most likely it is some type of convection induced mixing beyond the convection boundary.

The $^{22}\mathrm{Ne}(\alpha,\mathrm{n})^{25}\mathrm{Mg}$ reaction requires the high temperatures that can be found at the bottom of the pulse-driven convection zone (PDCZ) during the He-shell flash ($T > 2.5 \cdot 10^{8}$ K). The neutrons are released with high density ($\log N_{\mathrm{n}} \sim 9 \ldots 11$) in a short burst (Gallino et al. 1998). These peak neutron densities are realised for only about a year, followed by a neutron density tail that lasts a few years, depending on the stellar model assumptions. The current quantitative modeling of the $s$ process uses the thermodynamic output from a stellar evolution calculation including mass loss as input for nucleosynthesis calculations with a complete $s$-process network (Busso et al. 1999). The post-processing accounts for both the $^{13}$C neutron source as well as for the $^{22}$Ne source, and mixes the different contributions according to the information provided by the stellar evolution calculations. The free parameter of the model is the $^{13}$C abundance in the $^{13}$C pocket that is proportional to the neutron exposure that results from burning the $^{13}$C in the $(\alpha, n)$ reaction. Physical mixing processes which are responsible for bringing protons down from the envelope into the $^{12}$C-rich core to enable $^{13}$C formation, are not explicitely included in this model.

Observationally, models have to account for observed spread in observables that are related to the neutron exposure (e.g. Van Winckel & Reyniers 2000, Eck, S. van et al. 2003, Nicolussi et al. 1998). This spread is not only evident from stellar spectroscopy, but also from SiC grain data, that indicate that a spread by a factor of five is necessary for the neutron exposure for a given mass and metallicity (Lugaro et al. 2003a). Currently, this spread is accounted for by a range of different cases in each of which the $^{13}$C abundance in the pocket is assumed to be different (Busso et al. 2001). However, the physics of the mixing that is associated with the range of neutron exposures has yet to be identified.

## 7.1 Convection and rotationally induced mixing for the $s$ process

It is useful to distinguish $s$-process mixing for solar and moderately metal-poor stellar evolution, and the extremely and ultra metal-poor cases. In the first, the assumption that $s$-process mixing does not feedback strongly into the thermodynamic evolution is generally valid, in the latter it is generally not. Many more observational constraints exist for the solar and moderately metal-poor $s$ process, in particular the isotopic information from the pre-solar SiC grains.

Mixing for the solar and moderately metal-poor $s$ process has to satisfy two general constraints: (1) How is the partial mixing zone of H and $^{12}$C generated that eventually forms the neutron source species $^{13}$C, and (2) what is the origin of the observed spread in neutron exposures. Mixing for the $^{13}$C pocket is probably related to the penetrative evolution of the bottom of the convective envelope during the third dredge-up. Possible mechanisms include exponential diffusive overshooting (Herwig et al. 1997, Herwig 2000), mixing induced by rotation (Langer et al. 1999), and mixing by internal gravity waves (Denissenkov & Tout 2003). Each of these effectively leads to a continuously and quickly decreasing mixing efficiency from the H-rich convection zone into the radiative $^{12}$C-rich layer, and each of these will lead to the formation of two pockets which are overlapping (Lugaro et al. 2003b). At low H/$^{12}$C ratios H-burning is proton-limited, and protons will make $^{13}$C via the $^{12}\mathrm{C(p,\gamma)}$ reaction but no (or little) $^{14}$N. At larger H/$^{12}$C ratio a $^{14}$N pocket forms. The maximum abundance of both $^{13}$C and $^{14}$N depends on the $^{12}$C abundance in the intershell. Therefore, the conditions in the $^{13}$C pocket are not independent of, for example, the mixing at the bottom of the PDCZ. Lugaro et al. (2003b) derive the relationship $\tau_{\mathrm{max}} = 1.2X(^{12}\mathrm{C})_{\mathrm{IS}} + 0.4$, where $X(^{12}\mathrm{C})_{\mathrm{IS}}$ is the intershell $^{12}$C mass fraction and $\tau_{\mathrm{max}}$ is the maximum neutron exposure reached in the $^{13}$C pocket. The observations of H-deficient post-AGB stars described in § 6 require that the $^{12}$C abundance in the PDCZ of AGB stars is about $40\%$. This value is reproduced in AGB stellar evolution models with overshooting at the bottom of the PDCZ, and about twice as large as in models without overshooting.

Apart from the maximum amount of $^{13}$C in the pocket the mixing process must produce a partial mixing layer of the right mass $\Delta M_{\mathrm{spr}}$ which eventually contains $s$-process enriched material. In the exponential overshooting model of Herwig et al. (1997) the mixing coefficient is written as $D_{\mathrm{OV}} = D_0 \exp\left(\frac{-2z}{f_{\mathrm{ov}} \cdot H_{\mathrm{p}}}\right)$, where $D_0$ is the mixing-length theory mixing coefficient at the base of the convection zone, $z$ is the geometric distance to the convective boundary, $H_{\mathrm{p}}$ is the pressure scale height at the convective boundary, and $f_{\mathrm{ov}}$ is the overshooting parameter. If applied to core convection $f_{\mathrm{ov}} = 0.016$ reproduces the observed width of the main sequence. Lugaro et al. (2003b) find that $f_{\mathrm{ov}} = 0.128$ at the bottom of the convective envelope generates a large enough $^{13}$C pocket. However, the maximum neutron exposure in the $^{13}$C pocket of the overshooting model is $0.7 \cdots 0.8\mathrm{mbarn}^{-1}$, while in the non-overshooting models this value is $0.4\mathrm{mbarn}^{-1}$. Correspondingly Lugaro et al. (2003b) obtained only negative values for the logarithmic ratio [hs/ls][1], while the overshooting model with the larger neutron exposure predicts [hs/ls]$\sim 0$. This compares to an observed range of $-0.6 < [\mathrm{hs/ls}] < 0.0$ for stars of solar metallicity (see Busso et al. 2001 for a compilation of observational data). Models with overshooting at all convective boundaries can reproduce only the largest observed hs/ls

[1] hs and ls are the heavy and light $s$-process indices which are the average of the abundances of elements around the neutron magic numbers 50 and 80.

ratios, indicating that the neutron exposure in the $^{13}$C pocket in these models is at the maximum of the observationally bounded range.

Overshooting alone is not able to account for all features of $s$-process-mixing. In particular there is no mechanism to acount for the spread in neutron exposures within the overshooting framework. Rotation, however, may induce a range of mixing efficiencies for a sample of stars with otherwise identical parameters. Models of rotating AGB stars were presented by Langer et al. (1999), and the $s$ process was analyzed in detail by Herwig et al. (2003a) and Siess et al. (2004). The implementation of rotation for the AGB models was the same as the one that had been used previously to construct rotating models of massive stars (Heger et al. 2000). This implementation yields a $^{13}$C pocket generated by shear mixing below the envelope convection base, however an order of magnitude smaller than what is needed in the partial mixing zone of a non-rotating model to reproduce the observed overproduction in stars. The second important finding is that shear mixing which initially generates the small $^{13}$C pocket, prevails throughout the interpulse phase, even when the base of the convection is receding in mass after H-shell burning has resumed. When the dredge-up ends, the low-density slowly rotating convective envelope and the fast rotating compact radiative core are in contact and mixing is induced through shear at this location of large differential rotation. This radial velocity gradient remains as a source for shear mixing at exactly the mass coordinate of the $^{13}$C pocket with important consequence for the $s$ process. Shear mixing during the interpulse phase swamps the $^{13}$C pocket with $^{14}$N from the pocket just above. By the time the temperature has reached about $9 \cdot 10^7$ K and $^{13}$C starts to release neutrons via $^{13}\mathrm{C}(\alpha, \mathrm{n})^{16}\mathrm{O}$, $^{14}$N is in fact more abundant than $^{13}$C in all layers of the $^{13}$C pocket. $^{14}$N is a very efficient neutron poison. It has a very large $^{14}\mathrm{N}(\mathrm{n}, \mathrm{p})^{14}\mathrm{C}$ rate and simply steels neutrons from the iron seed. As a result, the neutron exposure is only $\sim 0.04\mathrm{mbarn}^{-1}$, about a factor of ten too small to generate the observed $s$-process abundance distribution. The detailed post-processing models of current rotating AGB stars showed that they are not capable to account for the observed $s$-process overabundances. Parametric models show that weaker shear mixing during the interpulse leads to a weaker poisoning effect. For very small poisoning effects the neutron exposures are still large enough to reproduce some observations.

While neither rotation or overshooting alone provide the right amount of $s$-process mixing it is interesting to consider a combination of these. In essence the idea is that overshooting would provide mixing for the formation for the $^{13}$C pocket and for a larger $^{12}$C abundance in the PDCZ to obtain a large neutron exposure in the $^{13}$C pocket. Then, shear mixing during the interpulse could add some poison and result in the observed spread. Magnetic fields have not yet been considered in AGB stellar evolution. However, one may assume that qualitatively magnetic fields will add coupling between the fast rotating core and the slowly rotating envelope and provide additional angular momentum transport. Models including this effect could result in smaller shear mixing than predict by the current rotating AGB stellar models.

It is interesting to note that such an effect of magnetic fields may also help to reconcile the predicted rotation rates of AGB cores of $\sim$ 30km/s with the rotation rate determinations of white dwarfs. Spectroscopic determinations of rotation

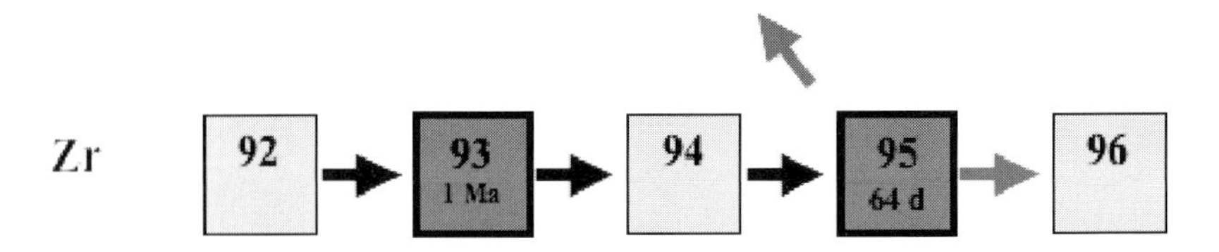


Figure 7: Small section of the chart of nuclides, showing the branching at $^{95}\mathrm{Zr}$. The $s$ process proceeds by adding neutrons to both stable (light grey) and unstable (dark grey) species. Radioactive species like $^{95}\mathrm{Zr}$ can either $\beta$-decay or capture a neutron.

rates of white dwarfs of spectral type DA can not rule out such values, but most are also consistent with zero or very low rotation (Koester et al. 1998, Heber et al. 1997). However, asterioseismological measurements of WD rotation rates clearly yield smaller values in the range $0.1 < v_{\mathrm{rot}}/(\mathrm{km/s}) < 1$ (see references in Kawaler 2003).

## 7.2 $s$ process as a diagnostic tool

The high precision information on the pre-solar meteoritic SiC grains provide isotope ratio measurements that allow to probe the conditions at the $s$-process nuclear production site in more detail. One example is the highly temperature dependent nucleosynthesis triggered by the release of neutrons from $^{22}\mathrm{Ne}$ at the bottom of the PDCZ. The temperature at the bottom of the PDCZ correlates with the efficiency of extra mixing like overshooting at the bottom of this convection zone (Herwig 2000). Neutrons in the PDCZ are generated by the $^{22}\mathrm{Ne}(\alpha,\mathrm{n})^{25}\mathrm{Mg}$ reaction. For larger temperatures the neutron density is higher. Isotopic ratios that enclose a branch point isotope in the $s$-process path will be more neutron heavy for higher neutron densities. An example is the $^{96}\mathrm{Zr}/^{94}\mathrm{Zr}$-ratio, that is set by the branching at the radioactive $^{95}\mathrm{Zr}$ (Fig. 7). For low neutron densities $^{95}\mathrm{Zr}$ decays. For $N_{\mathrm{n}} > 3{\cdot}10^{8}\mathrm{cm}^{-3}$ $^{95}\mathrm{Zr}(\mathrm{n},\gamma)^{96}\mathrm{Zr}$ becomes significant, hence $^{96}\mathrm{Zr}$ is produced. If the temperature is larger the neutron density is larger and the $^{96}\mathrm{Zr}/^{94}\mathrm{Zr}$ ratio, which can be measured in SiC grains, is larger as well.

Lugaro et al. (2003a) have studied the measured isotopic ratios of Mo and Zr as well as Sr and Ba from SiC grains. They evaluate the sensitivity of their results in terms of nuclear reaction rate uncertainties. All branchings that are activated by the $^{22}\mathrm{Ne}$ neutron source depend on the still uncertain $^{22}\mathrm{Ne}(\alpha,\mathrm{n})^{25}\mathrm{Mg}$ rate. In addition, the $^{96}\mathrm{Zr}/^{94}\mathrm{Zr}$ ratio depends on the neutron cross-section of the unstable isotope $^{95}\mathrm{Zr}$. Like for almost all radioactive nuclei the $(\mathrm{n},\gamma)$ rate of $^{95}\mathrm{Zr}$ is not measured. The theoretical estimates vary from Maxwellian-averaged cross sections of $20\mathrm{mb}$ (Beer et al. 1992) to $140\mathrm{mb}$ (JENDL-3.2). In order to make full use of the potent method of using the $s$ process as a diagnostic tool it is critical that n-capture rates of the radioactive $s$ process-branch point nuclides are measured. Different $s$ process-branchings are sensitiv to different mixing processes, including those that may be induced by rotation.

The $s$ process as a diagnostic tool can provide information on mixing processes that are potentially relevant for the evolution of stars of all masses, including the progenitors of supernovae. In particular the details of the initial model for a supernova calculations determines important properties of the explosion, like asymmetries, or the final fate as black hole or neutron star (Young et al. 2005).

# 8 Carbon star formation and nuclear reaction rate input

In order to understand C-rich extremely metal-poor (CEMP) stars the formation of C-rich AGB stars has to be understood in a quantitative way. AGB stars become C-rich because of the third dredge-up, which mixes C-rich material from the intershell into the envelope (Fig. 6). The evolution of C and O in AGB stars and the problems related to modelling the third dredge-up are discussed in Herwig (2005). In summary, the third dredge-up is now obtained in 1D stellar evolution calculations in sufficient amount and at sufficient low core mass. These calculations take into account convection-induced mixing into the stable layers in a time- and depth dependent way, and use high numerical resolution. Uncertainty is introduced by the choice of the mixing length parameter (Boothroyd & Sackmann 1988).

Understanding the dredge-up properties of AGB stars is important, because the dredge-up dependent yield predictions for low and intermediate mass stars enter models for galaxy chemical evolution. AGB stars serve as diagnostics for extra-galactic populations, and for this purpose the conditions of the O-rich to C-rich transitions needs to be known. C-rich giants are the brightest infrared population in extra-galactic systems. Finally, the envelope enrichment of AGB stars with the s-process elements is intimately related to the dredge-up properties of the models.

Recently, it has been shown that uncertainties in nuclear reaction rates propagate in a significant way into the dredge-up and thereby yield predictions. Herwig & Austin (2004) calculated an extensive grid of $2\,\mathrm{M}_{\odot}$, $Z = 0.01$ tracks for combinations of rates for the $^{14}\mathrm{N}(\mathrm{p},\gamma)^{14}\mathrm{O}$, the triple-$\alpha$ and the $^{12}\mathrm{C}(\alpha,\gamma)^{16}\mathrm{O}$ rate within the errors given in the NACRE compilation (Angulo et al. 1999). The main result was that dredge-up and C yields are larger for lower $^{14}\mathrm{N}(\mathrm{p},\gamma)$ rate and for larger triple-$\alpha$ rate. The $^{12}\mathrm{C}(\alpha,\gamma)$ rate plays a less important role. It was also found that nuclear physics work since the 1999 NACRE compilation require a downward revision of the $^{14}\mathrm{N}(p,\gamma)$ rate, by almost a factor of 2. Fig. 8 shows the revised range for this rate and the yields from the calculations assuming the uncertainty range of the nuclear reaction rates.

# 9 Conclusions

Extremely metal-poor stars are an emerging field of astrophysical and astronomical research, pushing the limits of observations, and theory and numerical simulation. Potentially, much can be learned about challenging questions of astrophysics. How do galaxies like our Milky Way form? How did the first stars and their cosmological

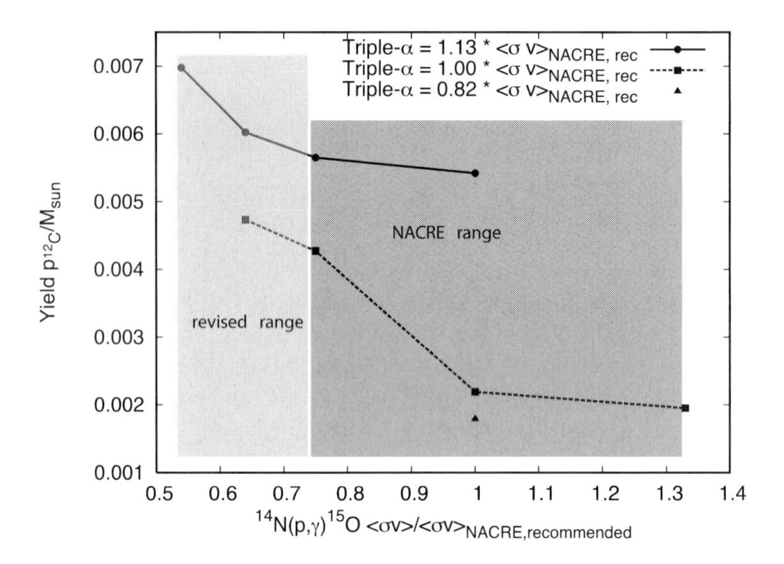


Figure 8: Carbon yield as a function of nuclear physics input. Each point gives the yield from one full $2\,\mathrm{M}_\odot$, $Z = 0.01$ TP-AGB stellar evolution sequence with the indicated choice for the $^{14}\mathrm{N}(\mathrm{p},\gamma)^{14}\mathrm{O}$ and the triple-$\alpha$ reaction. The yield of a species is the abundance in the ejected material minus the initial abundance integrated over the mass loss.

environment form and evolve? Some exciting clues about the evolution of stars come from the smallest astrophysical bodies, pre-solar stardust extracted from primitive meteorites. Thus stellar evolution and nucleosynthesis connects nearby phenomena of planetary system formation with star formation and evolution in the earliest time of the universe as recorded in the element distribution patterns of the most metal-poor stars.

In the future the quantity and quality of spectroscopic data of stars, in particular the valuable most metal-poor stars, will increase dramatically, due to multi-object spectroscopy and large spectroscopic surveys. It is already becoming clear that this data can be put to full use only with qualitatively and quantitatively improved simulations of nuclear production in stars, including low and intermediate mass and massive stars.

## Acknowledgements

I would like to thank D. Schoenberner for nominating me for the Ludwig-Biermann award, and for directing my interest to stellar evolution and nucleosynthesis as my *Doktorvater*. I am very grateful to N. Langer and D. VandenBerg for their continuing support. I would also like to thank my colleagues at the Theoretical Astrophysics group (T-6) at Los Alamos National Laboratory, in particular C. Fryer, A. Heger, and F. Timmes, as well as R. Reifarth at LANSCE-3, who contribute to a very stimulating atmosphere. Finally, I would like to thank A. Font and B. O'Shea for stimulating discussions, that have contributed to some views expressed in this article. This work was funded under the auspices of the U.S. Dept. of Energy, and supported by its contract W-7405-ENG-36 to Los Alamos National Laboratory.

## References

Abel, T., Bryan, G. L., Norman, M. L. 2002, Science, 295, 93

Angulo, C., Arnould, M., Rayet, M. et al. 1999, Nucl. Phys., A 656, 3, NACRE compilation

Aoki, W., Ando, H., Honda, S., Iye, M., Izumiura, H., Kajino, T., Kambe, E., Kawanomonoto, S., Noguchi, K., Okita, K., Sadakane, K., Sato, B., Shelton, I., Takada-Hidai, M., Takeda, Y., Watanabe, E., Yoshida, M. 2002, PASJ, 54, 427

Aoki, W., Ryan, S. G., Norris, J. E., Beers, T. C., Ando, H., Iwamoto, N., Kajino, T., Mathews, G. J., Fujimoto, M. Y. 2001, ApJ, 561, 346

Arlandini, C., Käppeler, F., Wisshak, K., Gallino, R., Lugaro, M., Busso, M., Straniero, O. 1999, ApJ, 525, 886

Asplund, M., Lambert, D. L., Kipper, T., Pollacco, D., Shetrone, M. D. 1999, A&A, 343, 507

Barbuy, B., Cayrel, R., Spite, M., Beers, T. C., Spite, F., Nordstroem, B., Nissen, P. E. 1997, A&A, 317, L63

Beer, H., Voss, F., Winters, R. R. 1992, ApJS, 80, 403

Beers, T. C., Christlieb, N. 2005, ARAA, 43, in press

Beers, T. C., Preston, G. W., Shectman, S. A. 1992, AJ, 103, 1987

Bessell, M. S., Christlieb, N., Gustafsson, B. 2004, ApJ Lett., 612, L61

Bland-Hawthorn, J., Freeman, K. 2000, Science, 287, 79

Böhm-Vitense, E. 1958, Z. Astrophys., 46, 108

Boothroyd, A. I., Sackmann, I.-J. 1988, ApJ, 328, 671

Boothroyd, A. I., Sackmann, I.-J., Ahern, S. C. 1993, ApJ, 416, 762

Busso, M., Gallino, R., Lambert, D. L., Travaglio, C., Smith, V. V. 2001, ApJ, 557, 802

Busso, M., Gallino, R., Wasserburg, G. J. 1999, ARA&A, 37, 239

Centurión, M., Molaro, P., Vladilo, G., Péroux, C., Levshakov, S. A., D'Odorico, V. 2003, A&A, 403, 55

Chieffi, A., Dominguez, I., Limongi, M., Straniero, O. 2001, ApJ, 554, 1159

Christlieb, N., Bessell, M. S., Beers, T. C., Gustafsson, B., Korn, A., Barklem, P. S., Karlsson, T., Mizuno-Wiedner, M., Rossi, S. 2002, Nature, 419

Christlieb, N., Green, P. J., Wisotzki, L., Reimers, D. 2001, A&A, 375, 366

Christlieb, N., Gustafsson, B., Korn, A. J., Barklem, P. S., Beers, T. C., Bessell, M. S., Karlsson, T., Mizuno-Wiedner, M. 2004, ApJ, 603, 708

Denissenkov, P. A., Tout, C. A. 2003, MNRAS, 340, 722

Denissenkov, P. A., VandenBerg, D. A. 2003, ApJ, 593, 509

Duerbeck, H. W., Liller, W., Sterken, C., Benetti, S., van Genderen, A. M., Arts, J., Kurk, J. D., Janson, M., Voskes, T., Brogt, E., Arentoft, T., van der Meer, A., Dijkstra, R. 2000, AJ, 119, 2360

Eck, S. van, Goriely, S.,Jorissen, A., Plez, B. 2003, A&A, 404, 291

Freeman, K., Bland-Hawthorn, J. 2002, ARA&A, 40, 487

Fryer, C., Benz, W., Herant, M., Colgate, S. A. 1999, ApJ, 516, 892

Fujimoto, M. Y., Ikeda, Y., Iben, I., J. 2000, ApJ Lett., 529, L25

Gallino, R., Arlandini, C., Busso, M., Lugaro, M., Travaglio, C., Straniero, O., Chieffi, A., Limongi, M. 1998, ApJ, 497, 388

Goriely, S., Siess, L. 2001, A&A, 378, L25

Goriely, S., Siess, L. 2004, A&A, 421, L25

Hajduk, M., Zijlstra, A. A., Herwig, F., et al. 2005, Science, 308, 231

Hamann, W.-R., Gräfener, G., Koesterke, L. 2003, in Planetary Nebulae. Their Evolution and Role in the Universe, ed. M. D. etal., PASP Conf. Ser., IAU Symp 209, 203

Heber, U., Napiwotzki, R., Reid, I. N. 1997, A&A, 323, 819

Heger, A., Langer, N., Woosley, S. E. 2000, ApJ, 528, 368

Heger, A., Woosley, S. E. 2002, ApJ, 567, 532

Herwig, F. 2000, A&A, 360, 952

Herwig, F. 2001, ApJ Lett., 554, L71

Herwig, F. 2003, in CNO in the Universe, ASP Conf. Ser. astro-ph/0212366

Herwig, F. 2004a, ApJ, 605, 425

Herwig, F. 2004b, ApJS, 155, 651

Herwig, F. 2005, ARAA, 43, in press

Herwig, F., Austin, S. M. 2004, ApJ Lett., 613, L73

Herwig, F., Blöcker, T., Langer, N., Driebe, T. 1999, A&A, 349, L5

Herwig, F., Blöcker, T., Schönberner, D., El Eid, M. F. 1997, A&A, 324, L81

Herwig, F., Johnson, J., Beers, T. C., Christlieb, N. 2004, BAAS 205

Herwig, F., Langer, N. 2001, Nucl. Phys. A, 688, 221, astro-ph/0010120

Herwig, F., Langer, N., Lugaro, M. 2003a, ApJ, 593, 1056

Herwig, F., Lugaro, M., Werner, K. 2003b, in Planetary Nebulae. Their Evolution and Role in the Universe, ed. M. D. et al., PASP Conf. Ser., IAU Symp 209, astro-ph/0202143

Iben, Jr., I., Kaler, J. B., Truran, J. W., Renzini, A. 1983, ApJ, 264, 605

Iwamoto, N., Kajino, T., Mathews, G. J., Fujimoto, M. Y., Aoki, W. 2004, ApJ, 602, 378

Johnson, J., Herwig, F., Beers, T. C., Christlieb, N. 2005, AJ, in prep.

Kawaler, S. D. 2003, in Stellar Rotation, ed. A. Maeder & P. Eenens, IAU Symp. 215, astro-ph/0301539

Koester, D., Dreizler, S., Weidemann, V., Allard, N. F. 1998, A&A, 338, 612

Koesterke, L. 2001, Ap&SS, 275, 41

Laird, J. B. 1985, ApJ, 289, 556

Langer, N., Heger, A., Wellstein, S., Herwig, F. 1999, A&A, 346, L37

Lucatello, S., Gratton, R., Cohen, J. G., Beers, T. C., Christlieb, N., Carretta, E., Ramirez, S. 2003, AJ, 125, 875

Lucatello, S., Tsangarides, S., Beers, T. C., Carretta, E., Gratton, R. G., Ryan, S. G. 2004, ArXiv Astrophysics e-prints, apJ, in press.

Lugaro, M., Davis, A. M., Gallino, R., Pellin, M. J., Straniero, O., Käppeler, F. 2003a, ApJ, 593, 486

Lugaro, M., Herwig, F., Lattanzio, J. C., Gallino, R., Straniero, O. 2003b, ApJ, 586, 1305

Lugaro, M., Ugalde, C., Karakas, A. I., Görres, J., Wiescher, M., Lattanzio, J. C., Cannon, R. C. 2004, ApJ, 615, 934

Meynet, G., Maeder, A. 2002, A&A, 390, 561

Miksa, S., Deetjen, J. L., Dreizler, S., Kruk, J. W., Rauch, T., Werner, K. 2002, A&A, 389, 953

Nicolussi, G. K., Pellin, M. J., Lewis, R. S., Davis, A. M., Clayton, R. N., Amari, S. 1998, Phys. Rev. Lett., 81, 3583

Nollett, K. M., Busso, M., Wasserburg, G. J. 2003, ApJ, 582, 1036

Nomoto, K. 1984, ApJ, 277, 791

O'Shea, B., Abel, T., Norman, M. L., Whalen, D. 2005, in prep.

Pettini, M., Ellison, S. L., Steidel, C. C., Bowen, D. V. 1999, ApJ, 510, 576

Preston, G. W., Sneden, C. 2001, AJ, 122, 1545

Prochaska, J. X., Henry, R. B. C., O'Meara, J. M., Tytler, D., Wolfe, A. M., Kirkman, D., Lubin, D., Suzuki, N. 2002, PASP, 114, 933

Qian, Y.-Z., Wasserburg, G. J. 2003, ApJ, 588, 1099

Robertson, B., Bullock, J. S., Font, A. S., Johnston, K. V., Hernquist, L. 2005, ArXiv Astrophysics e-prints

Siess, L., Goriely, S., Langer, N. 2004, A&A, 415, 1089

Siess, L., Livio, M., Lattanzio, J. 2002, ApJ, 570, 329

Sivarani, T., Bonifacio, P., Molaro, P., Cayrel, R., Spite, M., Spite, F., Plez, B., Andersen, J., Barbuy, B., Beers, T. C., Depagne, E., Hill, V., François, P., Nordström, B., Primas, F. 2004, A&A, 413, 1073

Suda, T., Aikawa, M., Machida, M. N., Fujimoto, M. Y., Iben, I. J. 2004, ApJ, 611, 476

Umeda, H., Nomoto, K. 2003, Nature, 422, 871

Van Winckel, H., Reyniers, M. 2000, A&A, 354, 135

Venn, K. A., Irwin, M., Shetrone, M. D., Tout, C. A., Hill, V., Tolstoy, E. 2004, AJ, 128, 1177

Weiss, A., Schlattl, H., Salaris, M., Cassisi, S. 2004, A&A, 422, 217

Werner, K. 2001, Ap&SS, 275, 27

Werner, K., Rauch, T., Kruk, J. W. 2005, A&A, astro-ph/0410690

Werner, K., Rauch, T., Reiff, E., Kruk, J. W., Napiwotzki, R. 2004, A&A, 427, 685

Woosley, S. E., Weaver, T. A. 1995, APJS, 101, 181

Young, P., Meakin, C., Fryer, C. L., Arnett, D. 2005, ApJ, in prep.

# Cosmological Structures behind the Milky Way

Renée C. Kraan-Korteweg

Departamento de Astronomía, Universidad de Guanajuato,
Apartado Postal 144, Guanajuato GTO 36000, México
Astronomy Department, University of Cape Town [1],
Rondebosch 7700, South Africa
kraan@circinus.ast.uct.ac.za,
http://mensa.ast.uct.ac.za/∼kraan

**Abstract**

*This paper provides an update to the review on extragalactic large-scale structures uncovered in the Zone of Avoidance (ZOA) by Kraan-Korteweg & Lahav 2000, in particular in the Great Attractor (GA) region. Emphasis is given to the penetration of the ZOA with the in 2003 released near–infrared (NIR) 2MASS Extended Source Catalog. A comparison with deep optical searches confirms that the distribution is little affected by the foreground dust. Galaxies can be identified to extinction levels of over* $A_B \gtrsim 10^{\rm m}$ *compared to about* $3\overset{\rm m}{.}0$ *in the optical. However, star density has been found to be a strong delimiting factor. In the wider Galactic Bulge region* ($\ell = 0^\circ \pm 90^\circ$) *this does not hold and optical surveys actually probe deeper (see Fig. 9). The shape of the NIR-ZOA is quite asymmetric due to Galactic features such as spiral arms and the Bulge, something that should not be ignored when using NIR samples for studies such as dipole determinations.*

*Various systematic surveys have been undertaken with radio telescopes to detect gas-rich galaxies in the optically and NIR impenetratable part of the ZOA. We present results from the recently finished deep blind HI ZOA survey performed with the Multibeam Receiver at the 64 m Parkes telescope* ($v \lesssim 12\,700$ km s$^{-1}$)*. The distribution of the roughly one thousand discovered spiral galaxies within* $|b| < 5^\circ$ *clearly depict the prominence of the Norma Supercluster. In combination with the optically identified galaxies in the ZOA, a picture emerges that bears a striking resemblance to the Coma cluster in the Great Wall in the first redshift slice of the CFA2 survey (de Lapparent, Geller & Huchra 1986): the rich Norma cluster (ACO 3627) lies within a great-wall like structure that can be traced at the redshift range of the cluster over* $\sim 90^\circ$ *on the sky, with two foreground filaments – reminiscent of the legs in the famous stick man – that merge in an overdensity at slightly lower redshifts around the radio galaxy PKS 1343−601 (see Figs. 14 & 16).*

[1] since January 2005

*Reviews in Modern Astronomy 18.* Edited by S. Röser
 ISBN 3-527-40608-5.

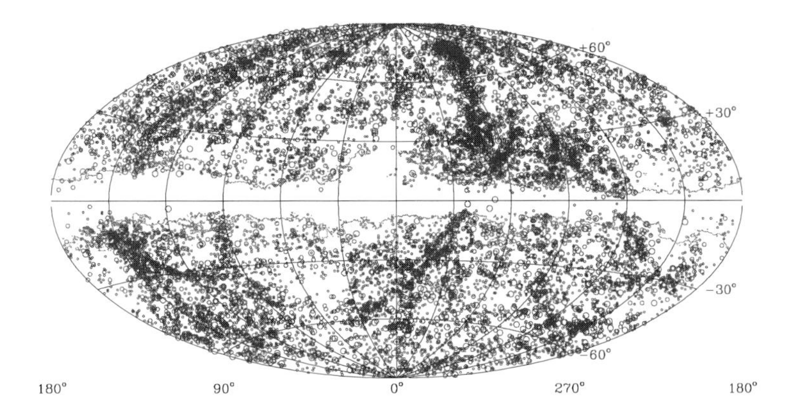


Figure 1: Aitoff equal-area projection in Galactic coordinates of galaxies with $D \geq 1\overset{\prime}{.}3$. The galaxies are diameter-coded. The contour marks absorption in the blue of $A_B = 1\overset{m}{.}0$ as determined from the Schlegel, Finkbeiner & Davis (1998) dust extinction maps. Figure from KK&L2000.

# 1 Introduction

The absorption of light due to dust particles and the increase in star density close to the Galactic Equator and around the Galactic Bulge creates a "Zone of Avoidance" in the distribution of galaxies, the size and shape of which depends on the wavelength at which galaxies are sampled. Figure 1 shows a complete sample of optically cataloged galaxies in an Aitoff projection in Galactic coordinates (see Kraan-Korteweg & Lahav 2000 for details; henceforth KK&L2000). The broad band void of galaxies takes up about 20% of the sky. Its form is like a near-perfect negative of the optical light distribution as depicted in the famous composite by Lundmark (1940), and is well traced by the dust (see contour in Fig. 1).

The ZOA has, with few exceptions, been avoided by astronomers studying the extragalactic sky because of the inherent difficulties in determining the physical parameters of galaxies lying behind the disk of our Galaxy – if they can be identified at all. The effect of absorption and star-crowding is illustrated in a simulation made by Nagayama (2004) which is reproduced in Fig. 2. The images are based on observations made with the Japanese 1.4 m Infrared Survey Facility (IRSF) at the Sutherland observing site of the South African Astronomical Observatory. The camera on the IRSF has the ability to simultaneously take $J$, $H$, and $K$ data with a field of view of $8' \times 8'$.

The left-hand panel shows a combined $JHK$ field in the Hydra cluster where absorption is negligible and star-crowding a minor problem. Subjecting this image to a foreground extinction of $A_B = 12^{\rm m}$ which in the NIR bands $JHK$ reduces to a mere $2\overset{m}{.}5$, $1\overset{m}{.}7$ and $1\overset{m}{.}1$ respectively (see Sect. 3 for details) results in the image of the middle panel. While the originally small and faint galaxies are lost completely, the larger galaxies are smaller in size, of lower surface-brightness, and redder. A further difficulty in identifying galaxies and determining their prop-

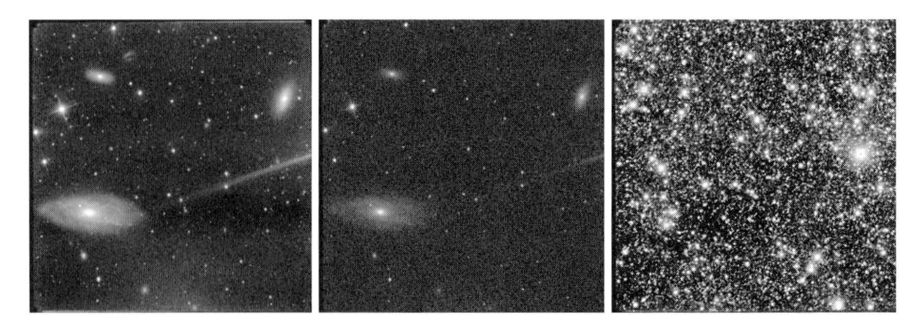

Figure 2: Left: A $JHK$ image of 8'×8' obtained with the Japanese IRSF of the Hydra galaxy cluster. Middle: simulation of same field seen through an obscuration layer of $12\overset{m}{.}0$, $2\overset{m}{.}5$, $1\overset{m}{.}7$, and $1\overset{m}{.}1$ of extinction in the $BJHK$ bands respectively. Right: previous field now positioned in the area of the low latitude radio galaxy PKS 1343−601. Figure adopted from Nagayama 2004.

erties is star-crowding. To illustrate this, Nagayama then combined this artificially absorbed field with a field in the surroundings of the radio galaxy PKS 1343−601 at low Galactic latitudes ($\ell = 309\overset{\circ}{.}7, b = +1\overset{\circ}{.}8$), also obtained with IRSF. This is reproduced in the right-hand panel of Fig. 2. Here, the recognition of even the intrinsically largest galaxies is hard when knowing where the galaxies are located (see panels to the left). Further examples of extinction and star-crowding effects can be found on http://www.z.phys.nagoya-u.ac.jp/~nagayama/hydra. This site also shows some examples of real galaxy candidates detected behind this thick obscuration and star layer.

The simulation clearly demonstrates why the incentive was low to map and investigate galaxies, their properties and their distribution in space in the ZOA. However, with the realization that galaxies are located predominantly in clusters, sheets and filaments, leaving large areas devoid of luminous matter, came the understanding that a concensus of galaxies of the "whole sky" is required when addressing various cosmological questions related to the dynamics of the local Universe.

The closest superclusters such as the Local Supercluster, the Centaurus Wall, the Perseus-Pisces chain, the Great Attractor – a large mass overdensity of about $5 \cdot 10^{16}\mathcal{M}_\odot$ that was predicted from the systematic infall pattern of 400 ellipiticals (Dressler et al. 1987) – all are bisected by the Milky Way, making a complete mapping and determination of their extent and dynamics impossible. Moreover, the irregular distribution of mass induces systematic flow patterns over and above the uniform Hubble expansion of the Universe. This effect is seen in the peculiar motion of the Local Group with respect to the Cosmic Microwave Background (CMB; e.g. Kogut et al. 1993). Such systematic flow patterns were first mapped within the Virgo Supercluster (Tonry & Davis 1981) and later on a much larger scale later in the Great Attractor region. It might even perturb the motions of galaxies in a volume all the way out to the Shapley Concentration, including the GA as a whole, though this still remains controversial (e.g. Kocevski, Mullis, & Ebeling 2004; Lucey, Radburn-Smith & Hudson, 2005; Hudson et al. 2004).

Kolatt, Dekel & Lahav (1995), have shown that the mass distribution of the inner ZOA ($b \pm 20°$) as derived from theoretical reconstructions of the density field is crucial to the derivation of the gravitational acceleration of the LG. This not only concerns hidden clusters, filaments, and voids. Nearby massive galaxies may also contribute significantly to the dipole and many of the nearby luminous galaxies do actually lie behind the Milky Way (e.g. Kraan-Korteweg et al. 1994). Moreover, a hidden Andromeda-like galaxy will influence the internal dynamics of the LG, its mass derivation and the present density determination of the Universe from timing arguments (Peebles 1994).

For our understanding of velocity flow fields, in particular the Great Attractor, the ZOA constitutes a severe barrier. The various 2 and 3-dimensional reconstruction methods find the flow towards this GA to be due to a quite extended region of moderately enhanced galaxy density centered on the Milky Way (about $\sim 40° \times 40°$, centered on $\ell, v, b \sim 320°, 0°, 4500\,\mathrm{km\,s^{-1}}$; see e.g. Fig. 1b in Kolatt et al. 1995). Although a considerable excess of galaxies is seen in that general region of the sky (Lynden-Bell & Lahav 1988; Fig. 1 here), no dominant cluster or central peak had been identified. Whether it existed and whether galaxies were fair tracers of the dynamically implicated mass distribution could not be answered, because a dominant fraction of the GA was hidden by the Milky Way.

For these reasons, various groups began projects in the last 10–15 years to try to unveil the galaxy distribution behind our Milky Way. Preliminary results, mainly based on optical, HI observations and follow-up of selected IRAS-PSC galaxy candidates, were presented at the first conference on this topic in 1994 (see Balkowski & Kraan-Korteweg (eds.) 1994, ASP Conf. Ser. 67). Meanwhile most of the ZOA has been probed in a "systematic" manner in "all" wavelength ranges of the electromagnetic spectrum (optical, near- and far-infrared, HI and X-ray), next to, and in comparison to, the reconstructed density fields in the ZOA. Many of these surveys and subsequent results are described in the proceedings of the second and third meeting on this topic (Kraan-Korteweg, Henning & Andernach (eds.) 2000, ASP Conf. Ser. 218; Fairall & Woudt (eds.) 2005, ASP Conf. Ser. 329).

A comprehensive overview on the then current status of all the ZOA projects was prepared in 2000 by Kraan-Korteweg & Lahav. It provides a detailed introduction on the motivation of ZOA studies, the status of and the results from the different survey methods, including a discussion on the limitations and selection effects of the various approaches, as well as how they complement each other. In this paper, I will build on the information given there, and concentrate mainly on new results, although a summary and an update on the results from deep optical galaxy searches and redshift follow-ups in the Great Attractor region is given in Sect. 2, as these are relevant to the discussions in the subsequent sections. Sect. 3 describes the enormous progress made in NIR-surveys with the release of the 2MASS Extended Source Catalog (2MASX) which contains 1.65 million galaxies or other extended sources over the whole-sky. It contains a discussion on the characteristics of this survey, in particular to penetrating the ZOA in comparison to optical surveys. Sect. 4 is dedicated to the results obtained from the systematic HI-survey of the southern ZOA that was performed between 1997 and 2002 with the Multibeam Receiver at the 64 m Parkes radio telescope. Next to the instrument and survey technique, the newly uncovered

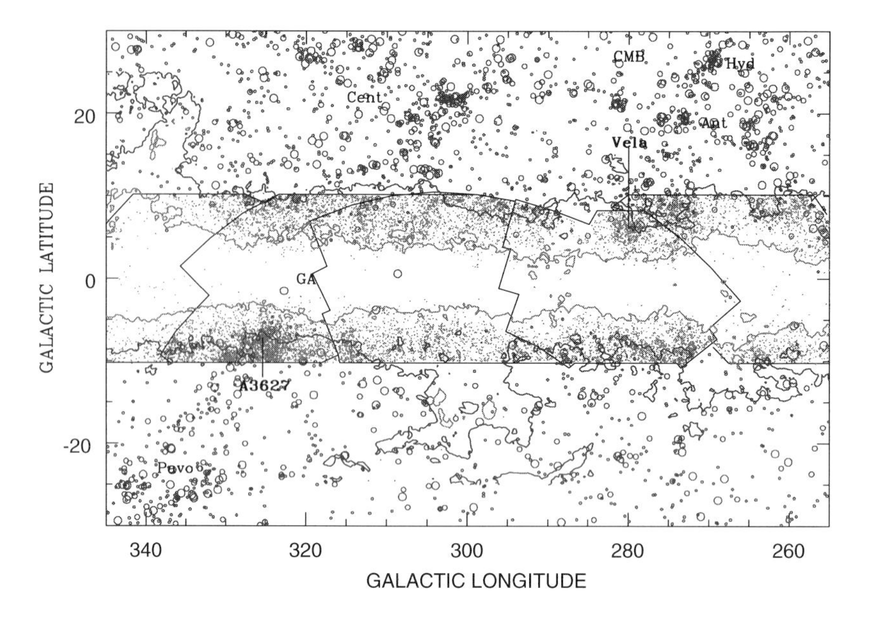


Figure 3: Distribution of Lauberts (1982) galaxies with $D \geq 1\overset{'}{.}3$ (open circles) and galaxies with $D \geq 12''$ (small dots) identified in the optical galaxy searches. The contours represent extinction levels of $A_B = 1\overset{m}{.}0$ and $3\overset{m}{.}0$.

galaxy distribution is discussed. The last section (Sect. 5) will then describe the emerging picture of the Great Attractor overdensity including the data obtained from the various ZOA survey methods.

## 2 Optical Galaxy Searches and the GA

Optical galaxy catalogs become increasingly incomplete for intrinsically large galaxies towards the Galactic Plane, because of the reduction in brightness and 'visible' extent by the thickening dust layer and the increase in star density (Fig. 2). However, this is a gradual effect. Deeper searches for partially obscured galaxies – fainter and smaller than existing catalogs – have been succesfully performed on existing sky survey plates. Over 50 000 unknown galaxies were uncovered, resulting in a considerable reduction of the ZOA.

Figure 3 gives an example of results obtained by our group in and around the Great Attractor region. This comprises (from right to left) the Vela region (Salem & Kraan-Korteweg, in prep.), Hydra/Antlia (Kraan-Korteweg 2000), Crux and Great Attractor (Woudt & Kraan-Korteweg 2001) and Scorpius (Fairall & Kraan-Korteweg 2000, 2005). Using a viewer with a 50 times magnification on IIIaJ film copies of the ESO/SRC survey resulted in the identification of over 17 000 galaxies down to a diameter limit of $D = 0\overset{'}{.}2$ within Galactic latitudes of $|b| \lesssim 10°$ over a longitude range of $250° \lesssim \ell \lesssim 350°$. 97% were previously unknown. Their distribution

Figure 4: A $BRI$ composite of the central $37' \times 37'$ of the cluster ACO 3627 obtained with the WFI of the ESO 2.2 m telescope (left). Note the 2 dominant cD galaxies at the center of this cluster and the richness of the dwarf population in the close-up (4x) on the right. Figure from Woudt et al. 2000.

is displayed in Fig. 3 together with earlier known galaxies. Note how the ZOA could be filled from $A_B = 1\overset{\mathrm{m}}{.}0$, the approximate completeness limit of previous catalogs, to $A_B = 3\overset{\mathrm{m}}{.}0$, with a few galaxies still recognizable up to extinction levels of $A_B = 5\overset{\mathrm{m}}{.}0$.

Distinct large-scale structures uncorrelated with the foreground obscuration can be discerned. The most extreme overdensity $(\ell, b) \sim (325°, -7°)$ is found close to the core of the GA. It is centered on the cluster ACO 3627 (Abell, Corwin & Olowin 1989) and is at least a factor 10 denser compared to regions at similar extinction levels, and contains a significant larger fraction of brighter galaxies as well as elliptical galaxies. The prominance of this cluster had never been realized because of its location in the Milky Way. Within the Abell radius (defined as 3 $h_{50}^{-1}$ Mpc) of this cluster, a total of 603 galaxies with $D \geq 0\overset{\prime}{.}2$ have been identified, of which only 31 were cataloged before by Lauberts (1982).

Figure 4, a deeper $BRI$ composite image obtained later by Woudt with the Wide Field Imager at the ESO 2.2 m telescope in la Silla (Woudt, Kraan-Korteweg & Fairall 2000), displays the center of this cluster in its full glory, despite the approximate 200 000 foreground stars. On this $37' \times 37'$ image (left panel), even more galaxies can be identified than the 74 galaxies found in the same area on the IIIaJ fields. The cluster has, like the Coma cluster, two cD galaxies at its center. The richness of this cluster can be fully appreciated in the close-up in the right-hand panel with the there apparent large dwarf galaxy population (image centered on the right of center cD of the left panel).

A quantification of the relevance of the newly uncovered galaxy distribution in context to known structures can be made when studying the completeness limits of the ZOA galaxy catalogs and correcting the observed parameters of the galaxies for the diminishing effects due to absorption applying the inverse Cameron (1990) laws.

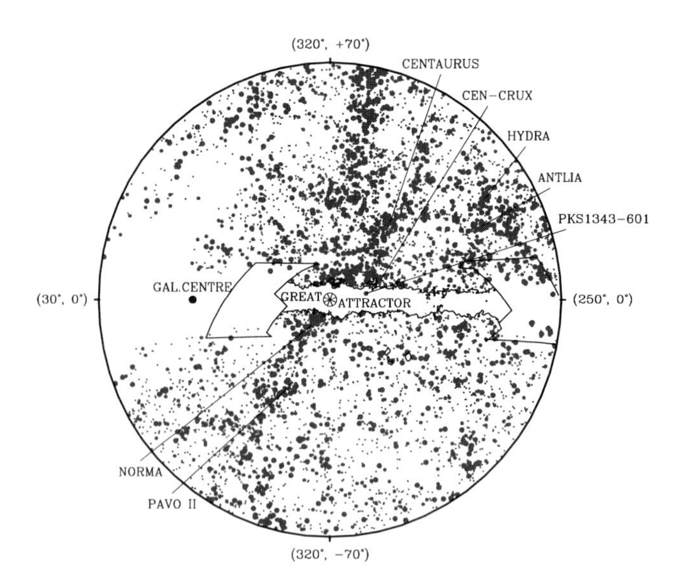


Figure 5: Equal area projection of galaxies with $D^0 \geq 1\farcm3$ and $A_B \leq 3^{\rm m}$ centered on the GA at $(\ell, b) = (320°, 0°)$ within a radius of 70°. The galaxies are taken from the ESO, the UGC and the MGC, complemented by our ZOA catalogs. The extinction contour of $A_B = 3^{\rm m}$ is superimposed. Search areas in progress are indicated. Figure from Woudt & Kraan-Korteweg 2001.

The latter provides equations to correct the observed magnitudes and diameters for their reduction as a function of extinction at their positon behind the Milky Way. Kraan-Korteweg (2000) and Woudt & Kraan-Korteweg (2001) have shown that their catalogs are complete to $D = 14''$ down to extinction levels of $A_B \leq 3\fm0$. Corrections for the apparently smallest galaxy at maximum extinction indicates that these surveys are complete for galaxies with extinction-corrected diameters of $D^o = 1\farcm0$ down to $A_B \leq 3\fm0$.

Knowing that the combined ESO Lauberts (1982) catalog, the Uppsala General Catalog UGC (Nilson 1973), and the Morphological Catalog of Galaxies MGC (Vorontsov-Velyaminov & Archipova 1962-74) displayed in Fig. 1 are complete to $D = 1\fdg3$ (Hudson & Lynden-Bell 1991), we can complement this merged whole-sky catalog down to $A_B \leq 3\fm0$ with galaxies that would appear in these catalogs were they not lying behind the Milky Way, i.e. with $D^o = 1\farcm3$. This has been done in Fig. 5, in an equal area projection centered on the Great Attractor at $(\ell, b) = (320°, 0°)$ and includes all galaxies with extinction-corrected diameters larger than $D^0 \geq 1\farcm3$ and $A_B \leq 3^{\rm m}$ from the Hydra/Antlia, Crux and GA catalogs, next to the previously known galaxies from the ESO, UGC and MGC catalogs.

Figure 5 provides the most complete view of the optical galaxy distribution in the Great Attractor region to date. Comparing Fig. 5 with Fig. 1 demonstrates the reduction of the optical ZOA (over 50%). The final galaxy distribution not only shows the dominance of the ACO 3627 cluster, henceforth called the Norma cluster for the constellation in which it is located, but also displays a high galaxy density in the wider GA region on both sides of the Galactic Plane. Though the reduction of the ZOA is significant, optical approaches clearly do not fully succeed in penetrating the ZOA.

## 2.1 Velocity Distribution

Redshift coverage of the nearby galaxy population in the ZOA is essential in determining their impact on the local dynamics and also to identify the features that form part of the GA overdensity. We have aimed to obtain a fairly homogeneous and complete coverage of the (extinction-corrected) brighter and larger newly uncovered galaxies in the ZOA. Three distinctly different observational approaches were used: (i) optical spectroscopy for individual galaxies of high central surface brightness at the 1.9 m telescope of the SAAO (Kraan-Korteweg, Fairall, & Balkowski 1995; Fairall, Woudt, & Kraan-Korteweg 1998; Woudt, Kraan-Korteweg, & Fairall 1999), (ii) HI-line observations with the 64 m Parkes radio telescope for low surface brightness gas-rich spiral galaxies (Kraan-Korteweg, Henning & Schröder 2002; Schröder, Kraan-Korteweg & Henning, in prep.), (iii) low resolution, multi-fiber spectroscopy for the high-density regions with Optopus and MEFOS at the 3.6 m telescope of ESO, LaSilla (see Woudt et al. 2004 for MEFOS results). With the above observations, we typically obtain redshifts of 15% of the galaxies and can trace large-scale structures fairly well out to recession velocities of about 20 000 $\mathrm{km\,s^{-1}}$.

The resulting velocity histograms are plotted in Fig. 6, separately for the three search areas. One glance immediately reveals the striking difference between the Hydra/Antlia, Crux and GA region, although all three have been sampled to approximately the same depth. The velocity distribution in Hydra/Antlia is overall quite shallow, with a peak at $v \sim 2750\,\mathrm{km\,s^{-1}}$, which corresponds to the extension of the Hydra/Antlia filament into the ZOA, and a broader overdensity at about $6000\,\mathrm{km\,s^{-1}}$, associated with the Vela overdensity $(280^\circ, +6^\circ)$, next to some higher velocity peaks.

In the Crux region, a broad concentration of galaxies is present from about 3500 to $8500\,\mathrm{km\,s^{-1}}$. This feature is already influenced by the GA overdensity. It is due to a wall-like structure that seems to connect the Norma cluster across the Centaurus-Crux, respectively the CIZA J 1324.7$-$5736 cluster at $(\ell, b, v) \sim (307^\circ, 5^\circ, 6200\,\mathrm{km\,s^{-1}})$ (Woudt 1998; Ebeling, Mullis & Tully 2002) to the Vela overdensity at $(280^\circ, +6^\circ, 6000\,\mathrm{km\,s^{-1}})$ (see also Fig. 15 and 16).

The GA histogram, in comparison, is strongly dominated by the very high peak associated with the Norma cluster at $4848\,\mathrm{km\,s^{-1}}$ (Kraan-Korteweg et al. 1996; Woudt 1998) and the surrounding great wall-like structure in which it is embedded. The peak corresponds exactly to the predicted mean redshift of the Great Attractor. The Norma cluster, with a velocity dispersion of $896\,\mathrm{km\,s^{-1}}$, has been found to have a virial mass of the same order as the Coma cluster (Kraan-Korteweg et al. 1996; Woudt 1998). This is confirmed independently by the X-ray observations with ROSAT (Böhringer et al. 1996) which finds the Norma cluster to be the 6th brightest ROSAT cluster in the sky. Simulations have furthermore shown that the well-known Coma cluster would appear the same as the Norma cluster if Coma were

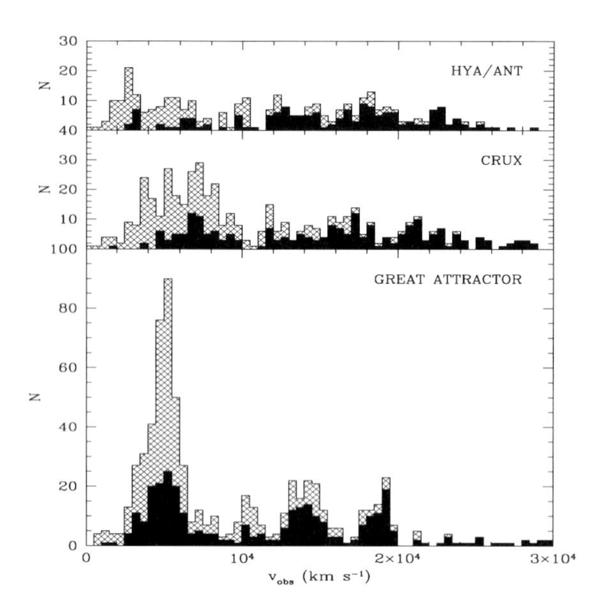


Figure 6: Velocity histograms in the Hydra/Antlia, Crux and GA regions. Dark shaded area correspond to the heliocentric velocities obtained with MEFOS, cross-hatched region includes the SAAO and Parkes observations as well as velocities from the literature. Figure from Woudt et al. 2004.

located behind the Milky Way at the location of the Norma cluster (Woudt 1998). The Norma cluster does not only seem the most likely candidate to define the core of the GA, it also seems to have superseded the well-known Coma cluster as being the nearest rich cluster to the LG, and therefore is an interesting cluster on its own, irrespective of its central position in the GA region.

A detailed dynamical analysis of the Norma cluster is in preparation (see Woudt 1998, Woudt et al. 2000, for preliminary results), as well as a precise determination of its distance (Woudt et al. 2005, Woudt et al., in prep.), in order to determine whether the Norma cluster, and therefore the GA as a whole, is at rest with respect to the CMB, or partakes in the earlier mentioned, still controversial flow towards to Shapley Concentration.

Besides the dominant peak due to the GA, the peak at 14 000 km s$^{-1}$ in the lower panel of Fig. 6 is noteworthy. It is due to the Ara cluster (Woudt 1998; Ebeling et al. 2002) which together with the adjacent Triangulum-Australis cluster (McHardy et al. 1981) forms a larger overdensity referred to as a 'Greater Attractor behind the Great Attractor' by Saunders et al. (2000). They find a signature of this overdensity in the reconstructed IRAS galaxy density field.

## 2.2 Does the Norma Cluster Define the Core of the GA?

The emerging optical picture of the Great Attractor so far is that of a confluence of superclusters (the Centaurus Wall and the Norma supercluster) with the Norma cluster being the most likely candidate for the Great Attractor's previously unseen center. However, the potential well of the GA might be rather shallow and extended. Seen that the ZOA has not been completely reduced by deep optical surveys it is not inconceivable that further prospective galaxy clusters might be located at the bottom of the GA's potential well at extinction levels $A_B \geq 3^{\rm m}$ (Fig. 5).

Detecting clusters at higher extinction levels is not straightforward. The X-ray band is potentially an excellent window for studies of large-scale structure in the ZOA, because the Milky Way is transparent to the hard X-ray emission above a few keV, and because rich clusters are strong X-ray emitters. But although dust extinction and stellar confusion are unimportant in the X-ray band, photoelectric absorption by the Galactic hydrogen atoms – the X-ray absorbing equivalent hydrogen column density – does also limit detections close to the Galactic Plane. A systematic X-ray search for clusters in the ZOA ($|b| \leq 20°$) has been performed by Ebeling et al. (2002). They found only the above-mentioned Centaurus-Crux or CIZA J 1324.7$-$5736 cluster, as a previously unknown component, that might form part of of the GA (see also Kocevski et al. 2004, Mullis et al. 2005), though it is by no means as centrally located in the GA, nor as massive as the Norma cluster.

Alternatively, a strong central radio source, such as PKS 1610$-$608 in the Norma cluster, could also point to unidentified clusters. Exactly such a source lies in the deepest layers of the Galactic foreground extinction ($A_B = 12^{\rm m}$) at $(\ell, b, v) = (309.7°, +1.7°, 3872$ km s$^{-1}$). This strong radio source was suspected for a long time by Kraan-Korteweg & Woudt (1999) of being the principal member of an unknown rich galaxy cluster. An overdensity of galaxies around this massive galaxy had indeed been seen in blind HI-surveys, which are uneffected by extinction (see Sect. 4). Moreover, dedicated infrared studies in the surroundings of PKS 1343$-$601 also found an excess of galaxies (e.g. Kraan-Korteweg et al. 2005a in the $I$-band; Schröder et al. 2005 on DENIS $IHK$- images; and Nagayama et al. 2004, 2005, in $JHK$). The data are, however, not supportive of this being a rich massive cluster. This is consistent with the upper limit of X-ray emission determined from ROSAT data by Ebeling et al. 2002.

ACO 3627 thus remains the most likely candidate of constituting the central density peak of the potential well of the Great Attractor overdensity.

# 3 2MASS Galaxies and the ZOA

Observations in the near infrared (NIR) can provide important complementary data to other surveys. With extinction decreasing as a function of wavelength, NIR photons are much less affected by absorption compared to optical surveys. The $I$, $J$, $H$ and $K$ band extinction is only 45%, 21%, 14% and 9% compared to the optical $B$ band – hence, as the mean longitude of the surveys in the respective wavebands increases, a progressively deeper search into thicker obscuration layers at lower Galactic latitudes is possible.

The NIR is sensitive to early-type galaxies – tracers of massive groups and clusters – which are missed in far infrared (FIR) surveys not discussed in this paper (but see KK&L2000) and HI surveys (Sect. 4). Moreover, recent star formation contributes only little to the NIR flux of galaxies (in contrast to optical and FIR emission) and therefore provides a better estimate of the stellar mass content of galaxies.

Two systematic near infrared surveys have been performed: DENIS, the DEep Near Infrared Southern Sky Survey, has imaged the southern sky from $-88° < \delta < +2°$ in the $I_c$ (0.8μm), $J$ (1.25μm) and $K_s$ (2.15μm) bands with magnitude completeness limits for stars of $I = 18\overset{m}{.}5$, $J = 16\overset{m}{.}5$ and $K = 14\overset{m}{.}0$ leading to a prediction of the detection of 100 million stars (see http://www-denis.iap.fr for further details). The DENIS completeness limits (total magnitudes) for highly reliable automated galaxy extraction away from the ZOA ($|b| > 10°$) was determined as $I = 16\overset{m}{.}5$, $J = 14\overset{m}{.}8$, $K_s = 12\overset{m}{.}0$ by Mamon (1998), leading to a predicted extraction of roughly 250 000 galaxies. A provisional DENIS $I$-band catalog of galaxies with $I \leq 14\overset{m}{.}5$ for 67% of the southern sky has been released by Paturel, Rousseau & Vauglin (2003), and over 2000 serendipitous DENIS detections of galaxies behind the Milky Way by Vauglin et al. 2002 (see also Rousseau et al. 2000 for the description of some noteworthy DENIS galaxies uncovered in the ZOA). Results of pilot studies in probing the ZOA using DENIS data concentrated on the Great Attractor region, in particular in the surroundings of the cluster ACO 3627 and the radio source PKS 1343−601, have been given in Schröder et al. (1997, 1999, 2000, 2005) and Kraan-Korteweg et al. (1998), and KK&L2000.

2MASS, the 2 Micron All Sky Survey, covers the whole sky in the $J$ (1.25μm), $H$ (1.65μm) and $K_s$ (2.15μm) bands. Its point source sensitivity limits are $J = 15\overset{m}{.}8$, $H = 15\overset{m}{.}1$ and $K = 14\overset{m}{.}3$, whereas for galaxies and other spatially resolved objects, the survey should be complete away from the Galactic Plane to $J = 15\overset{m}{.}0$, $H = 14\overset{m}{.}3$ and $K = 13\overset{m}{.}5$ over a wide range of surface brightnesses (Jarrett et al. 2000a,b). Source extraction resulted in a Point Source Catalog (PSC) containing close to half a billion objects, most of which will be Milky Way stars (next to an estimated 3 to 5 million unresolved galaxies), as well as the 2MASS Extended Source Catalog (2MASX) which contains 1.65 million galaxies or other extended sources.

In the following, we will discuss the effectiveness of 2MASX with regard to Zone of Avoidance penetration and compare this to the reduced optical ZOA including the results from the deep optical surveys. This not only to determine the most effective method of uncovering galaxies hidden by the Milky Way, but by studying the magnitude completeness limits, colors and surface brightness as a function of extinction or star density, we hope, amongst others, to also optimize redshift follow-up observations in the ZOA.

## 3.1 The 2MASX Zone of Avoidance

The final release of the 2MASX by the Two Micron All Sky Survey Team in 2003 (Jarrett et al. 2000b) now allows a detailed study of the performance of 2MASS in mapping the extragalactic large-scale structures across the ZOA, as well as compare and cross-correlate the 2MASS galaxy distribution with the deep optical ZOA catalogs (see also Kraan-Korteweg & Jarrett 2005).

Figure 7: Distribution of 2MASX sources with $K < 14^{\rm m}\!\!.0$ in an equal area Aitoff projection in Galactic coordinates centered on the Galactic Bulge. Stars from the PSC area also displayed. Figure adopted from Jarrett 2004.

Figure 7 (adopted from Jarrett 2004) shows an Aitoff projection in Galactic coordinates centered on the Galactic Bulge of all the 1.65 million resolved MASX sources with magnitudes brighter than $K < 14^{\rm m}\!\!.0$, in addition to the nearly 0.5 billion Milky Way stars. A description of the large-scale structure of these galaxies based on redshifts estimated from their NIR colors is given in Jarrett 2004. Here we will concentrate on the penetration of the ZOA.

Compared to the optical, the 2MASX sources provide a much deeper and uniform view of the whole extragalactic sky. The galaxy distribution can be traced without hardly any hindrance in the Galactic Anticenter. However, the wider Galactic Bulge region – represented here by the half billion Galactic stars from the PSC – continues to hide a non-negligible part of the extragalactic sky. Despite the fact that NIR surveys should in principle be able to uncover galaxies to extinction levels of about $A_B = 10^{\rm m}$ compared to $3^{\rm m}$ in the optical, the NIR ZOA does not appear narrower here compared to the reduced optical ZOA which on average has an approximate width of $b \lesssim \pm 5^\circ$ around the Galactic Plane (see Fig. 4 in KK&L2000 which shows a whole-sky distribution of galaxies with $D \geq 1\!\!'.3$ that has been complemented with all published ZOA galaxies that also meet that criterium).

Thus, there clearly also exists a NIR ZOA, but its form is quite distinct from the optical one. To understand the differences between these two ZOAs we compared in detail the results from our optical survey in the Scorpius region with 2MASX detections. The Scorpius region lies to the left of the GA region (most left search area in Fig. 3) and to the right of the Galactic Center, where confusion from the foreground Milky Way is extreme.

The optical and 2MASS galaxy distributions are displayed jointly in Fig. 8. The large circles represent optically detected galaxies in Scorpius (Fairall & Kraan-Korteweg 2000, 2005), the small dots 2MASX objects with $K \leq 14^{\rm m}\!\!.0$. It should be noted that extremely blue objects ($(J - H) < 0^{\rm m}\!\!.0$) are excluded, as well as 2MASX sources that were rejected as likely galaxy candidates upon visual examination by

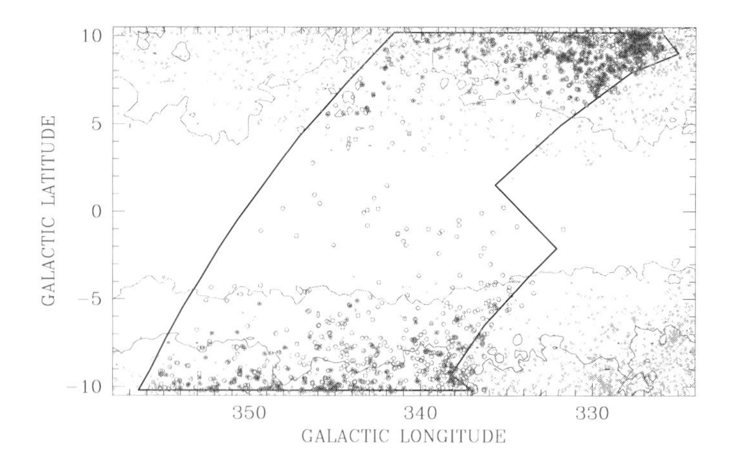


Figure 8: Distribution of optically detected galaxies (circles) with $D > 12''$ in the Scorpius regions, and 2MASS galaxies (small dots) with $K \leq 14\overset{m}{.}0$ and $(J - H) > 0\overset{m}{.}0$ in the Scorpius search region and surroundings. The contours mark extinction levels of $A_B = 1\overset{m}{.}0$ and $3\overset{m}{.}0$.

Jarrett (priv. comm.). These generally are Galactic objects, such as HII regions and Planetary Nebulae, the prime contaminant at $|b| < 2°$ (optical catalogs also contain a small fraction of them). The $1\overset{m}{.}0$ and $3\overset{m}{.}0$ optical extinction contours are also drawn.

Figure 8 confirms that even this close to the Galactic Bulge, where star densities are high, deep optical searches are fairly faithful tracers of the galaxy distribution to $A_B \sim 3^{m}$, with only a few – mostly uncertain – galaxy candidates peaking through higher dust levels. Though the extinction in the $K$-band is only 9% of that in the $B$-band, only about one-third of the optically identified galaxies in the Scorpius region have a counterpart in 2MASS. Although the 2MASX sources seem to probe deeper into the Milky Way above the Galactic Plane, this trend is not seen at negative latitudes. There, clearly optical galaxies dominate and they also probe the galaxy distribution deeper into the plane.

A direct correlation between dust absorption and 2MASX source density is not seen here. NIR surveys become progressively less succesful compared to optical surveys when approaching the Galactic Center. In fact, this progressive loss of 2MASX galaxies is already noticeable in the fraction of optical galaxies that have counterparts in the 2MASX catalog. This fraction is 47% in the Hydra/Antlia region ($\ell \approx 280°$; see Fig. 3 for orientation), and decreases to 39% in the Crux region ($\ell \approx 310°$), 37% in the GA region ($\ell \approx 325°$), 33% in the Scorpius region ($\ell \approx 340°$), and to a mere 16% in the by Wakamatsu et al. (2000, 2005) explored Ophiuchus cluster region ($\ell = 0\overset{\circ}{.}5, b = +9\overset{\circ}{.}5$).

It should be maintained though, that there also are 2MASX galaxies in the optically surveyed regions that have no optical counterpart. This fraction increases inversely, i.e. the farther away from the Galactic Bulge the higher this fraction. Not unsurprisingly, the majority of ZOA galaxies that are seen in 2MASS galaxies but not in the optical are on average quite red and faint (mostly with $K \gtrsim 12^{m}$). This is partly a ZOA effect with NIR surveys finding red galaxies more readily at higher absorption level. But it is also due to the inherent characteristics of the two surveys. Whereas the NIR is better at detecting old galaxies and ellipticals, the optical surveys – although susceptible to all galaxy types – are best at finding spirals and late-type galaxies (especially low surface-brightness galaxies and dwarfs). The surveys are in fact complementary.

The success rate of galaxy identification in the NIR actually depends much more strongly on *star density* than dust extinction. This effect is corroborated by Fig. 9, which shows 2MASX sources with $K \leq 14\overset{\mathrm{m}}{.}0$ within $\pm 15°$ of the Galactic equator. In the left panel, DIRBE/IRAS extinction contours of $A_B = 1\overset{\mathrm{m}}{.}0, 3\overset{\mathrm{m}}{.}0$ and $5\overset{\mathrm{m}}{.}0$ are superimposed. The right panel emphasizes the locations of 2MASX sources in regions where the density of stars in the PSC with $K \leq 14\overset{\mathrm{m}}{.}0$ per square degree is log$N$ = 3.50, 3.75, and 4.00.

An examination of the dust extinction contours with the area in which 2MASX does not find extended sources does not suggest a correlation between them. In the Galactic Anticenter, roughly defined here as $\ell \sim 180° \pm 90°$, 2MASX objects seem to cross the Plane without any hindrance. For this half of the ZOA, plots of NIR magnitude or diameter versus extinction (not shown here) confirm that galaxies can be easily identified up to extinction levels equivalent to $A_B \gtrsim 10^{\mathrm{m}}$, and extinction-corrected $K^o$-band magnitudes versus extinction diagrams imply that 2MASS remains quite complete up to $K^o \lesssim 13\overset{\mathrm{m}}{.}0$ for $A_B \sim 10\overset{\mathrm{m}}{.}5$.

This is not at all true for the wider Galactic Bulge region ($\ell \sim 0° \pm 90°$), where 2MASX detects objects to lower extinction levels only and where the completeness limit is at least one magnitude lower compared to the Anticenter. And although Galactic dust does reduce the completeness limit of 2MASX sources at low latitudes – although to a much lower extent than in the optical – the origin of the NIR ZOA is mainly due to source confusion in regions of high *star density*.

This region in which NIR surveys fail completely has a very well-defined shape. It is traced by the star density isopleth (stars per square degree with $K < 14\overset{\mathrm{m}}{.}0$) of log $(N) = 4.00$ , i.e. the innermost contour in the right panel of Fig. 9. At this level, the completeness has already dropped significantly. Above this limit, the point sources are packed so densely that extended sources can not be extracted anymore. The hoped-for improvement of uncovering the galaxy overdensity with NIR surveys to lower latitudes compared to the optical, as for instance in the Great Attractor region (compare Fig. 3 or 5 to Fig. 7), has not been achieved.

It should be maintained, however, that for the ZOA away from the Bulge – as well as for the rest of the sky – 2MASS as a homogeneous whole-sky survey obviously is far superior. An optimal approach in revealing the galaxy distribution behind the Milky Way will actually result from a combination of deep optical and near-infrared surveys. Not only because of the different susceptibility to morphology, but because the former are sensitive to galaxies located behind regions of high source confusion, and the latter to galaxies located behind thick dust walls.

A reduction of the ZOA common to both the optical and NIR might be achieved by a combination of a deep $R$-band survey with a spatially higher resolved NIR survey (higher than the current NIR surveys 2MASS and DENIS) which should diminish the source confusion in high star density regions.

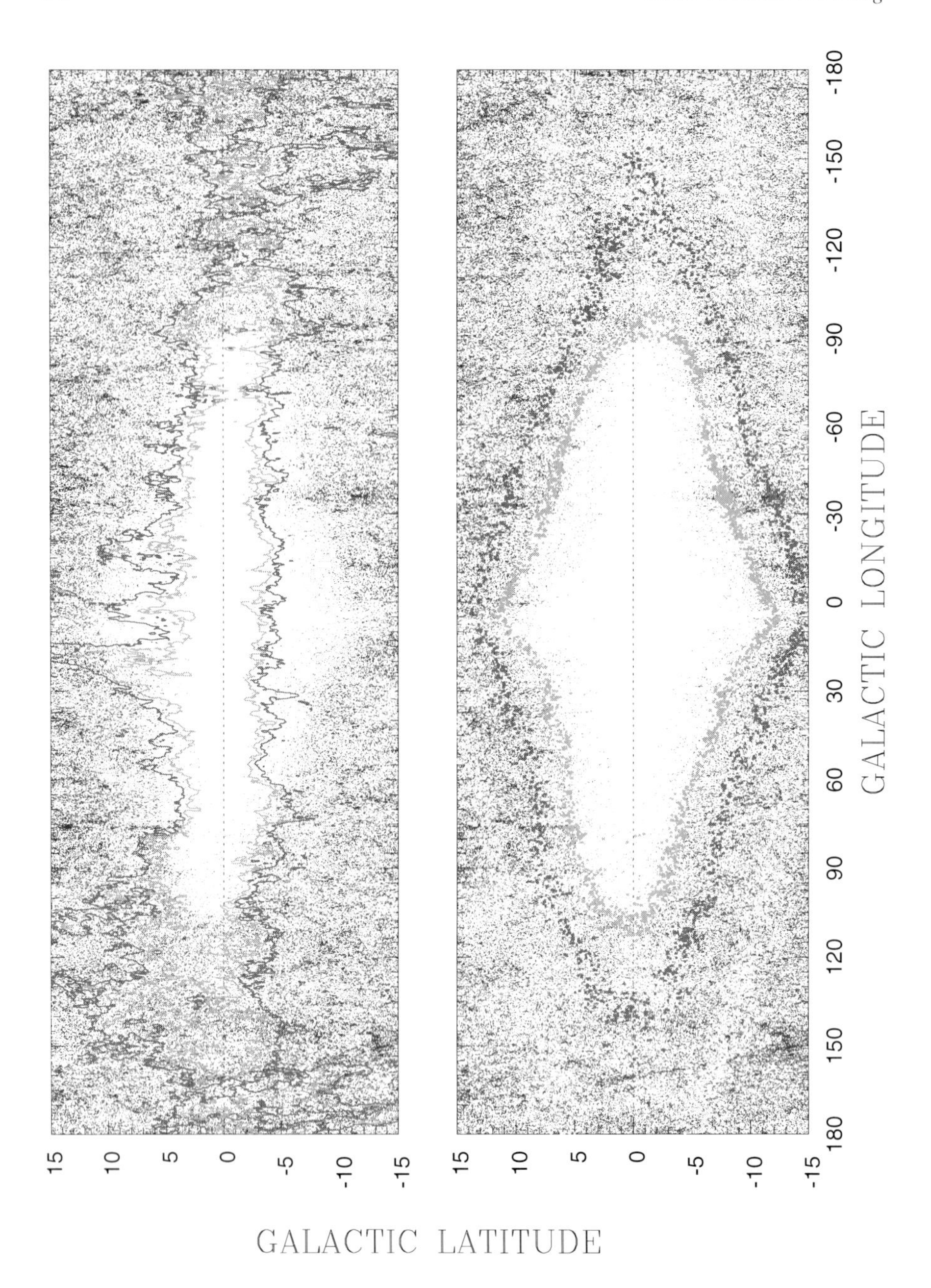


Figure 9: Distribution of 2MASX sources with $K \leq 14\overset{\mathrm{m}}{.}0$ along the Galactic equator within $b \leq \pm 15°$. In the left panel, DIRBE/IRAS extinction contours of $A_B = 1\overset{\mathrm{m}}{.}0, 3\overset{\mathrm{m}}{.}0$ and $5\overset{\mathrm{m}}{.}0$ are superimposed. In the right panel, galaxies found in regions of star densities of log$N = 3.50$, 3.75, and 4.00 per square degree ($K \leq 14\overset{\mathrm{m}}{.}0$) are enhanced.

## 3.2 The Shape of the NIR ZOA

For the discussion in this section on the shape of the NIR ZOA, we define it more or less ad hoc as the nearly completely empty region as outlined by the isopleth given with the 2MASS PSC star density of log$N = 4.00$ per square degree for stars with $K \leq 14\overset{m}{.}0$ (inner contour of left panel of Fig. 9). A closer look at the NIR ZOA reveals some interesting asymmetries that are independent of extragalactic large-scale structure but are actually due to Galactic structure.

The bulge and disk as outlined by the lack of galaxies seem to be inclined with respect to the Galactic equator. Its mean latitude is offset to positive latitudes for $\ell \sim 90°$, and to negative latitudes for $\ell \sim 270°$ $(-90°)$. This lopsidedness has been known for a long time. It was first established from the Galactic hydrogen column densities by Kerr & Westerhout in 1965 for longitudes between $\ell = 220° - 330°$. The same inclination is evidenced also in the dust contours (see left panel of Fig. 9).

Another asymmetry becomes obvious when regarding the width of the NIR ZOA with respect to the Galactic Center. Whereas the NIR ZOA can be followed to about $\ell \sim 120°$ on the one side, it stretches over 'only' $90°$ on the opposite side of the Galactic Center (to $\ell \sim 270°$). The explanation for this asymmetry lies in the relative location of the Sun with respect to the Galactic spiral arms. When looking towards the local Orion arm, our line of sight is hit directly with a high density of nearby stars, blocking a larger fraction of the extragalactic sky from our view, whereas the opposite line of sight is nearly free of stars for quite a distance until it hits the more distant spiral arm, allowing us to identify galaxies more easily between the fainter and smaller stars of this star population.

Furthermore, when regarding the location where the Galactic Bulge is highest, one notes that it does not peak at $\ell = 0°$ as expected, but is centered on $\ell \sim +5°$. This offset probably is due to the bar of our Galaxy. Its near side points towards us (positive longitudes), reducing our view of the extragalactic sky stronger compared to our line of sight towards the far side of the Galactic bar.

As these asymmetries have more to do with Galactic structure than extragalactic structure, one might be tempted to ignore them. However, these asymmetries in the distribution of galaxies should be taken into account when using the 2MASX catalog – or redshift surveys based on 2MASX subsamples – for dipole determinations. These structures might have a significant effect on the results if not properly corrected for.

## 3.3 A Redshift Zone of Avoidance

A reduction of the ZOA on the sky does not imply at all that a similar reduction can also be attained in redshift space. This is seen most clearly in Fig. 10 which displays the distribution of 2MASX galaxies that have a redshift listed in NED, the NASA/IPAC Extragalactic Database (Jarrett 2004, priv. comm.). The ZOA in this figure is quite distinct from the previously defined NIR ZOA. Its form is actually much more reminiscent of the optical ZOA delimited by the extinction contour of $A_B = 3\overset{m}{.}0$ (see Fig. 4 in KK&L2000). Hardly any redshifts are available for latitudes of $b \lesssim 5°$.

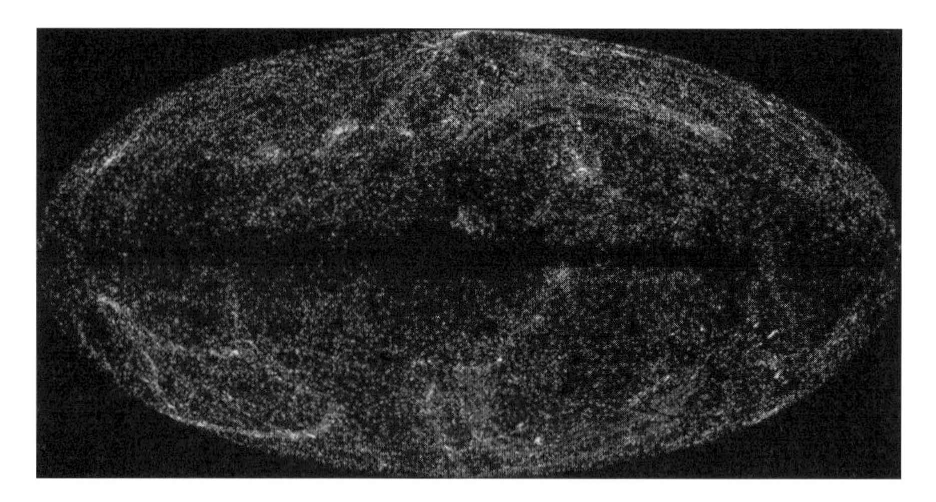

Figure 10: Distribution of 2MASX sources in an equal area Aitoff projection in Galactic coordinates that have redshifts listed in NED (Jarrett 2004, priv. comm.).

One might argue that this is an artifact because it is based mainly on optically selected targets. 2MASX has been released only fairly recently and systematic follow-up redshifts observations of the newly uncovered galaxies at low Galactic latitudes, in particular in the Galactic Anticenter ZOA half, have not yet been made and/or published. This is however not entirely true. In whatever kind of waveband a ZOA galaxy is identified, it remains inherently difficult to obtain a reliable optical redshift when the extinction in the optical towards this galaxy exceeds 3 magnitudes, i.e. the delimiting factor of optical surveys in general.

This is seen quite clearly in Fig. 11 which shows 2MASS galaxies in Puppis – a filament is crossing the Plane there (see Fig. 7) – on a sequence of 6-degree fields (6dF) centered on Dec $= -25^\circ$. This strip has been observed with the multifibre spectroscope at the UK Schmidt Telescope as part of a pilot project aimed at extending the 6dF Galaxy Survey towards lower latitudes. It now is restricted to the southern sky with $|b| \geq 10^\circ$ (see http://www.mso.anu.edu.au/6dFGS for further details).

The crosses mark galaxies for which a reliable redshift could be measured, the filled dots galaxies for which this could not be realized. At first glance, the plot seems to indicate a fantastic success rate for obtaining redshifts all the way across the Milky Way. However, it should be noted that the concerned ZOA strip lies in a region renowned for its low dust content (it hardly exceeds 3 magnitudes), and a careful inspection indicates that redshifts have generally not been obtained for galaxies that lie in pockets where the extinction is higher than $A_B \gtrsim 3\overset{\text{m}}{.}0$ (thick contour). So even when galaxies are identifiable deep in the Plane, reducing the *redshift ZOA* will remain hard, and an optimization of targeted ZOA galaxies is crucial for redshift follow-ups.

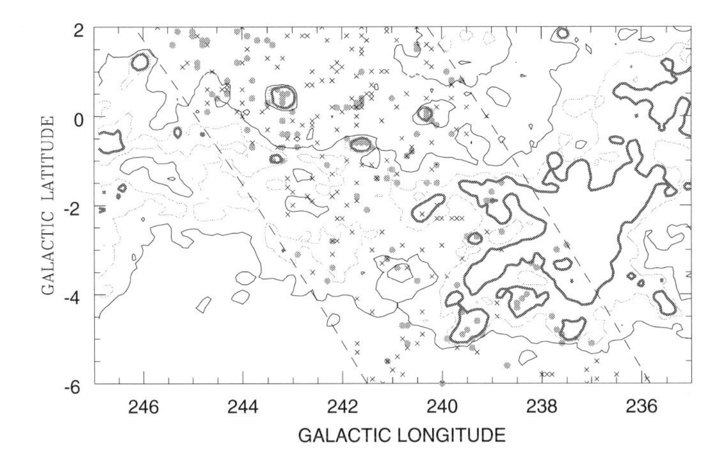


Figure 11: Distribution of 2MASS galaxies observed with 6dF in a strip crossing the Galactic Plane in the Puppis region. Circles represent galaxies with redshifts, crosses those without a reliable redshift determination. The contours indicate an optical extinction of $A_B = 1\overset{m}{.}0, 2\overset{m}{.}0$ and $3\overset{m}{.}0$ (thick contour).

# 4 Dedicated HI Galaxy Searches in the ZOA

Because the Galaxy is fully transparent to the 21cm line radiation of neutral hydrogen, HI-rich galaxies can readily be found through the detection of their redshifted 21cm emission in the regions of the highest obscuration and infrared confusion. Furthermore, with the detection of an HI signal, the redshift and rotational properties of an external galaxy are immediately known, providing insight not only on its location in redshift space but also on the intrinsic properties of such obscured galaxies. This makes systematic blind HI surveys powerful tools in mapping large-scale structures behind the Milky Way.

Early-type galaxies – tracers of massive groups and clusters – are gas-poor and will, however, not be identified in these surveys. Furthermore, low-velocity extragalactic sources that fall within the velocity range of the strong emission of the Galactic gas ($v \lesssim \pm 250\,\mathrm{km\,s^{-1}}$) will be missed. Galaxies that lie close in position to radio continuum sources may also be missed because of the baseline ripples they produce over the whole observed requency range.

Two systematic blind HI searches for galaxies behind the Milky Way have been made. The first used the 25 m Dwingeloo radio to survey the whole northern Galactic Plane for galaxies out to 4000 $\mathrm{km\,s^{-1}}$ with a sensitivity of $rms$ of 40 mJy for a 1 hr integration (see KK&L2000 for a summary of the results). A more sensitive survey ($rms$ of typically 6 mJy beam$^{-1}$), probing a considerably larger volume (out to 12 700 $\mathrm{km\,s^{-1}}$), has been performed with the Multibeam Receiver at the Parkes 64 m radio telescope in the southern sky. In the following, the most recent results of this survey are given.

## 4.1 The Parkes Multibeam HI ZOA Survey

The Multibeam receiver at the 64 m Parkes telescope was specifically constructed to efficiently search for galaxies of low optical surface brightness, or galaxies at high optical extinction, over large areas of the sky. It has 13 beams, each with a beamwidth of $14\rlap{.}'4$, arranged in a hexagonal grid in the focal plane array (Staveley-Smith et al. 1996) which allows – with its large footprint of $2\rlap{.}^\circ5$ on the sky -- rapid sampling of large areas.

In March 1997, this instrument was mounted on the telescope and various surveys were started, one being a systematic blind HI survey within $b < \pm5^\circ$ of the ZOA. The observations were performed in scanning mode. Fields of length $\Delta\ell = 8^\circ$ centered on the Galactic Plane were surveyed along constant Galactic latitudes where each scan was offset by $35'$ in latitude until the final width of $\Delta b = \pm5^\circ$ had been attained (17 passages back and forth). The final goal was 25 repetitions per field. With an effective integration time of 25 min/beam, a $3\,\sigma$ detection limit of 25 mJy was obtained. The correlator bandwidth of 64 MHz was set to cover a velocity range of $-1200 \lesssim v \lesssim 12700$ km s$^{-1}$. The survey therewith is sensitive to normal spiral galaxies well beyond the Great Attractor region (e.g. $5 \cdot 10^9$ M$_\odot$ at 60 Mpc for a galaxy with a linewidth of 200 km s$^{-1}$), next to the lowest mass dwarf galaxies in the local neighborhood ($10^6 - 10^7$ M$_\odot$), or extremely massive galaxies beyond 10 000 km s$^{-1}$ such as the extraordinarily massive galaxy HIZOA J0836-43 with a HI mass of $7 \cdot 10^{10}$ M$_\odot$ found in one of the ZOA data cubes (Kraan-Korteweg et al. 2005b; Donley et al. in prep.).

The data are in the form of three-dimensional data cubes (position-position-velocity, with pixel and beam sizes of $4' \times 4'$, and $15\rlap{.}'5$, respectively). Experimentation with automatic galaxy detection algorithms indicated that visual inspection of the data cubes is more efficient for the ZOA, where the noise due to continuum sources and Galactic HI is high and variable. The ZOA cubes were inspected by at least two, sometimes three, individual researchers, with the subsequent neutral evaluation of inconsistent cases in the detection lists by a third party.

An first analysis covering the *southern* Milky Way ($212^\circ \leq \ell \leq 36^\circ$) based on 2 out of the foreseen 25 passages (the HI ZOA Shallow Survey; henceforth HIZSS) with and $rms$ noise of 13 mJy beam$^{-1}$ led to the discovery of 110 galaxies, two thirds of which were previously unknown (Henning et al. 2000). A final catalog of the 23 central cubes of the full-sensitivity survey is in preparation (Henning et al., in prep.).

The data and plots of the full sensitivity survey are presented in the next section. They are based on a provisional version of this catalog which might still contain a few galaxy candidates that will be rejected for inclusion in the final catalog. They furthermore include detections from an extension to the north (Dec$> 0^\circ$), which was done at a later stage, resulting in 2 further cubes on both sides of the southern ZOA (Donley et al. 2005). The data set regarded here thus consists of 27 data cubes that cover the ZOA between $196^\circ \leq \ell \leq 52^\circ$ for $|b| \leq 5^\circ$. A total of slightly over one thousand galaxies were identified in these data cubes.

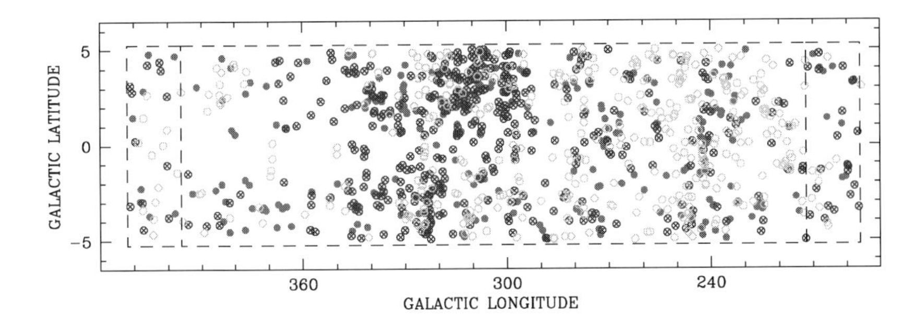


Figure 12: Distribution in Galactic coordinates of the galaxies detected in the deep HI ZOA survey. Open circles: $v_{\rm hel} < 3500$; circled crosses: $3500 < v_{\rm hel} < 6500$; filled circles: $v_{\rm hel} > 9500\,{\rm km\,s^{-1}}$.

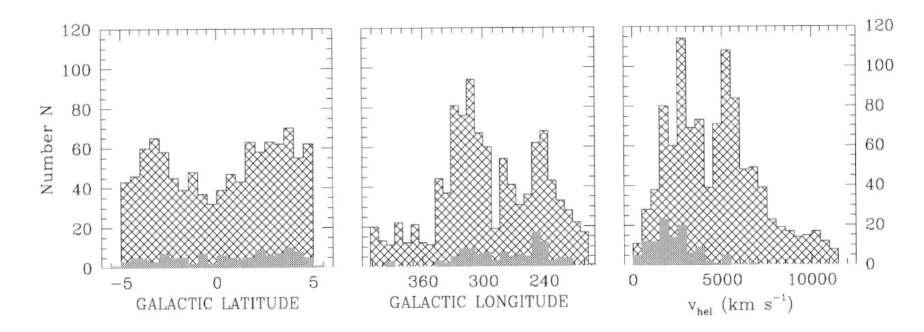


Figure 13: Distributions as a function of the Galactic latitude, longitude and the heliocentric velocity of the in HI detected galaxies. The lower histograms represent the results from the HIZSS.

## 4.2 The Detected Galaxies

Figure 12 displays the distribution along the Milky Way of the in HI detected galaxies. An inspection of this distribution shows that the HI survey nearly fully penetrates the ZOA with hardly any dependence on Galactic latitude.

This is confirmed by the left panel in Fig. 13, which shows the detection rate as a function of Galactic latitude. The small dip in the detection rate between $-2° \lesssim b \lesssim +1°$ stems mainly from the Galactic Bulge region and to a lesser extent from the GA region ($\ell \approx 300° - 340°$). The former is due to the high number of continuum sources at low latitudes in the Galactic Bulge region. In the GA region the gap is possibly related to the high galaxian density for the on average higher velocity, hence fainter, galaxies. There, a moderate number of continuum sources may already result in a detection-loss. This explanation is supported by the fact that this dip is not noticeable in the shallower HIZSS data (lower histogram).

A much stronger variation is apparent in the number density as a function of Galactic longitude (see also middle panel of Fig. 13). This can be explained entirely with large-scale structures such as the nearby (and therefore prominent in HIZSS) Puppis filament ($\ell \approx 240°$), the Hydra-Antlia filament ($\ell \approx 280°$), the very dense

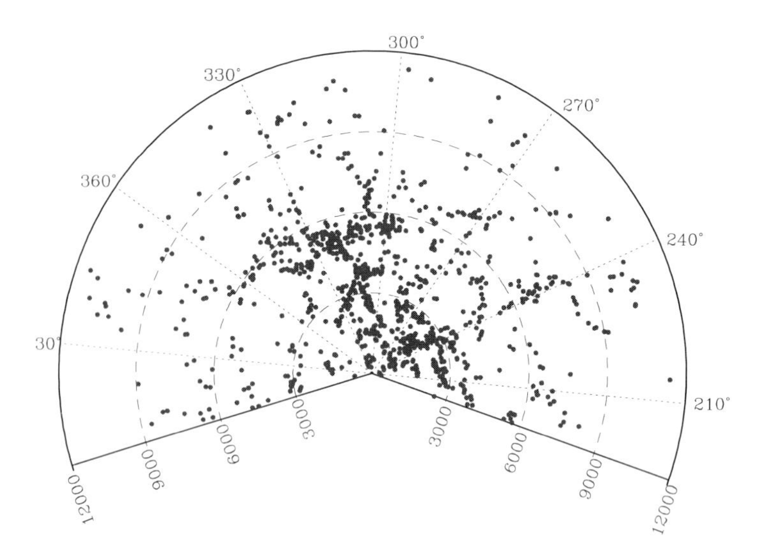


Figure 14: Galactic latitude slice with $|b| \leq 5°$ out to 12000 km s$^{-1}$ of the slightly over 1000 in HI detected galaxies. Circles mark intervals of 3000 km s$^{-1}$.

GA region ($\ell \approx 300 - 340°$), followed by an underdense region ($52° \gtrsim \ell \gtrsim 350°$) due to the Local and Sagittarius Void.

These large-scale structures have left their imprint also on the velocity diagram (right panel of Fig. 13), which shows two conspicuous broad peaks. The low-velocity one is due to a blend of various structures in or crossing the Galactic Plane while the second around 5000 km s$^{-1}$ clearly is due to the GA overdensity (see also Fig. 14 – 16). The velocity histogram moreover shows that galaxies are found all the way out to the velocity limit of the survey of $\sim$ 12 000 km s$^{-1}$, hence probe the galaxy distribution considerably deeper than either the shallow ZOA survey HIZSS (lower histogram), or the southern sky HI surveys also made with the Multibeam instrument, the HI Bright Galaxy Catalog (BGC; Koribalski et al. 2003) and the HI Parkes All Sky Survey (HIPASS; Meyer et al. 2004).

## 4.3 Uncovered Large-Scale Structures in the GA

Figure 14 shows a Galactic latitude slice with $|b| \leq 5°$ out to 12 000 km s$^{-1}$ of the galaxies detected in the deep Parkes HI ZOA survey (henceforth HIZOA) for the longitude range $196° \leq \ell \leq 52°$. The clear conclusion when inspecting this figure is that the HI survey really permits the tracing of large-scale structures in the most opaque part of the ZOA; and this in a homogenous way, unbiased by the clumpiness of the foreground dust contamination.

In the following, some of the most interesting features revealed in Fig. 14 will be discussed. It is suggested to simultaneously consult Fig. 15, which shows the HI-ZOA data with data extracted from LEDA surrounding the ZOA in sky projections for three velocity shells of thickness 3000 km s$^{-1}$. This helps to show the newly discovered features in context to known structures. Viewing the distribution of galaxies within $\Delta b \leq 5°$ in this figure illustrates quite clearly how the HI survey has managed to fill in that part of the ZOA, tracing various contiguous structures across the plane of the Milky Way.

The most prominent large-scale structure in Fig. 14 certainly is the Norma Supercluster which seems to stretch from $360°$ to $290°$ in this plot, lying always just below the $6000\,\mathrm{km\,s^{-1}}$ circle, with a weakly visible extension towards Vela ($\sim 270°$). The latter is more pronounced at higher latitudes (see panel 2 in Fig. 15, and Fig. 16). This wall-like feature seems to be formed of various agglomerations. The first one around $340°$ is seen for the first time with new HI data. Because of the high extinction there, it cannot be assessed whether this overdensity continues for $|b| > 5°$. The clump at $325°$ is due to the outer boundaries at lower latitudes side of the Norma cluster A3627 ($\ell, b, v = 325°, -7°, 4880\,\mathrm{km\,s^{-1}}$; Kraan-Korteweg et al. 1996). The next two are previously unrecognized due to groups (or small clusters) at $310°$ and $300°$, both at $|b| \sim +4°$. They are very distinct in the middle panel of Fig. 15. Between these two clusters at slightly higher latitude we see a small finger of God, which belongs to the Centaurus-Crux/CIZA J 1324.7$-$5736 cluster.

This is not the only overdensity in the GA region. A significant agglomeration of galaxies is evident closer by at $(\ell, v) = (311°, 3900\,\mathrm{km\,s^{-1}})$ with two filaments merging into it. Although no finger of God is visible (never very notable in HI redshift slices), this concentration forms part of the previously discussed galaxy concentration around the strong radio source PKS 1343$-$601 which is consistent with an intermediate size cluster residing there.

Next to the Norma cluster, only the Centaurus-Crux cluster has an appreciable X-ray emission (Ebeling et al. 2002; Mullis et al. 2005). It seems therefore unlikely that any of the newly apparent galaxy concentrations are a signature of a further massive cluster that conforms part of the GA overdensity. But it should be kept in mind that X-ray photons are subjected to photoelectric absorption by the Galactic hydrogen atoms – the X-ray absorbing equivalent hydrogen column density – which does limit detections close to the Galactic Plane. This effect is particularly severe for the softest X-ray emission, as observed by ROSAT (0.1-2.4 keV), and is seen in the CIZA cluster distribution (CIZA standing for Clusters in the Zone of Avoidance, a systematic search for X-ray clusters with Galactic latitiudes $|b| \leq 20°$). Hardly any clusters are found for Galactic HI column density over $N_{\mathrm{HI}} > 5 \times 10^{21}\,\mathrm{cm^{-2}}$, which creates an X-ray ZOA of about $\Delta b \lesssim 5°$ on average (see Fig. 14 in KK&L2000; Fig. 7 in Ebeling et al. 2002).

Although the overdensity in the GA region looks quite impressive here, a preliminary quantitative analysis of the 4 cubes covering $300° \leq \ell \leq 332°$ by Staveley-Smith et al. (2000) find a mass excess of 'only' $\sim 2 \cdot 10^{15} \Omega_0 \mathrm{M}_\odot$ over the background, thus considerably lower than the predictions from the infall pattern. Then again, the signatures of overdensities and clusters are overall much shallower in HI surveys compared to, e.g., optical surveys (e.g. Fig. 22 versus Fig. 23 in Koribalski et al. 2004).

To the left of the GA overdensity a few other features are worthwile describing. An underdense region comprised of the Local Void and the Sagittarius Void is seen around $\ell = 360°$ at central velocities of $\sim 1500$ and $4500\,\mathrm{km\,s^{-1}}$ (see

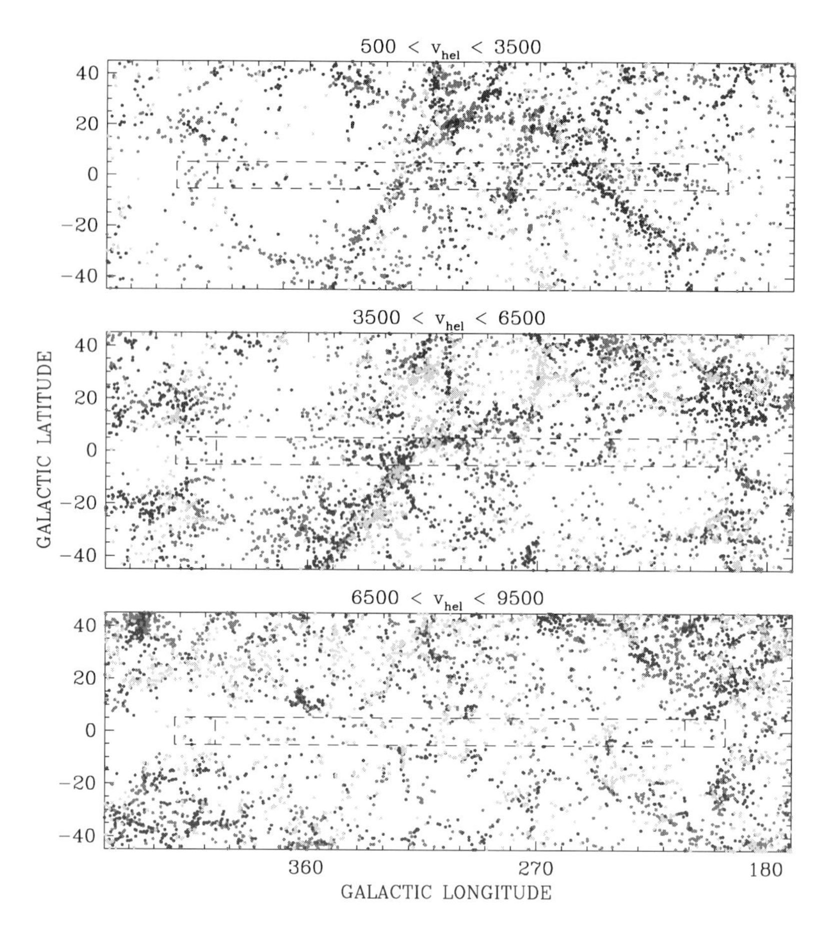


Figure 15: Sky projections of three redshift slices of depth $\Delta v = 3000\,\mathrm{km\,s^{-1}}$ showing the HIZOA data in combination with data from LEDA. The HIZOA survey area is outlined.

also the first two panels of Fig. 15). Except for the tiny group of galaxies at about $(350^\circ, 3000\,\mathrm{km\,s^{-1}})$, the distribution here and in Fig. 14 suggest one big void rather than two seperate ones. In contrast, the righthand side of Fig. 15 is quite crowded: the Puppis region ($\ell \sim 240^\circ$) with its two nearby groups (800 and $1500\,\mathrm{km\,s^{-1}}$) followed by the Hydra Wall at about $3000\,\mathrm{km\,s^{-1}}$ that extends from the Monocerus group ($210^\circ$) to the concentration at $280^\circ$. The latter is not the signature of a group but due to a filament emerging out of the Antlia cluster ($273^\circ, 19^\circ$; see Fig. 15).

With this systematic HI survey, we could map for the first time large-scale structures without any hindrance across the Milky Way (Figs. 14 and 15). It is the only approach that easily uncovers galaxies in the ZOA - and records their redshift. For this reason we are currently extending the Parkes HI ZOA survey to higher Galactic latitudes in the Galactic Bulge region ($332^\circ < \ell < 36^\circ$) where the optical and NIR ZOAs are wider and knowledge about the structures very poor. They will improve the knowledge on the borders of the Local and Sagittarius Void, as well as the Ophiuchus cluster studied optically by Wakamatsu et al. (2000; 2005).

# 5 Discussion

In the last decade, enormous progress has been made in unveiling the extragalactic sky behind the Milky Way. At optical wavebands, the entire ZOA has been systematically surveyed, reducing the optical ZOA by about a factor of $2 - 2.5$, i.e. from $A_B = 1\overset{\mathrm{m}}{.}0$ to $A_B = 3\overset{\mathrm{m}}{.}0$. Its average width is about $\pm 5°$, except in the low-extinction Puppis area, where galaxies have been found at all latitudes, in contrast to the Galactic Bulge region where the $3\overset{\mathrm{m}}{.}0$ contour rises to higher latitudes for positive latitudes.

2MASS, as a homogeneous NIR survey, obviously is far superior to optical surveys, particularly considering that the optical ZOA surveys were not only performed on different plate material, but also by different searchers using different search techniques. Nevertheless, in the regions of highest star densities (over 10 000 stars with $K \leq 14^{\mathrm{m}}$ per square degree), i.e. around the Galactic Bulge ($\ell \lesssim \pm 90°$), the identification of galaxies fails for latitudes between $\pm 5°$ up to $\pm 10°$ (see innermost contour in right panel of Fig. 9) and the hoped-for improvement of uncovering further galaxy overdensities in, for instance, the Great Attractor, could not be realized.

Even when a galaxy can be identified at high extinction levels, such as is possible in the NIR and FIR, this will, however, not automatically reduce the ZOA in redshift space. As discussed in Sect. 3.3 (see Fig. 10) it remains nearly impossible to obtain optical redshifts for galaxies at extinction levels $A_B \gtrsim 3^{\mathrm{m}}$. At these levels only HI observations prevail – if the galaxies are gas-rich and not too distant. But even with a HI detection, cross-identification with its optical, 2MASS and/or IRAS counterpart (Donley et al. 2005) often remains ambiguous because of positional uncertainty due to the large beams of single-dish radio telescopes.

As seen in this paper, the mapping of the galaxy distribution behind the Milky Way requires considerable efforts. Still, combining data obtained from the various multi-wavelength approaches will reveal this hidden part of the Universe, as clearly illustrated with Fig. 16 for the Great Attractor region – though a quantification of the structures will remain difficult due to the different biases and selection effects in the different methods.

Figure 16 shows a redshift slice of width $|b| \leq 10°$ out to $12\,000\,\mathrm{km\,s^{-1}}$ compared to the $\pm 5°$ of the HI data alone. All galaxies with a redshift in the LEDA data base are included next to the data from the HIZOA. Note that this is not a homogenous data set. It clearly is deeper sampled from $270° \lesssim \ell \lesssim 340°$ because of the intensive redshift follow-up programs of the ZOA catalogs by Kraan-Korteweg, Woudt and collaborators in that longitude range (see Sect. 2.1).

Before the ZOA research programs, this slice only had a few points in them – mainly in the low extinction Puppis area ($\ell \sim 240°$) – and certainly did not allow any reliable description of large-scale structures. It now has been filled to a depth comparable to unobscured regions in the sky and the above diagram reveals various clusters, filamentary and wall-like structures, next to some sharply outlined

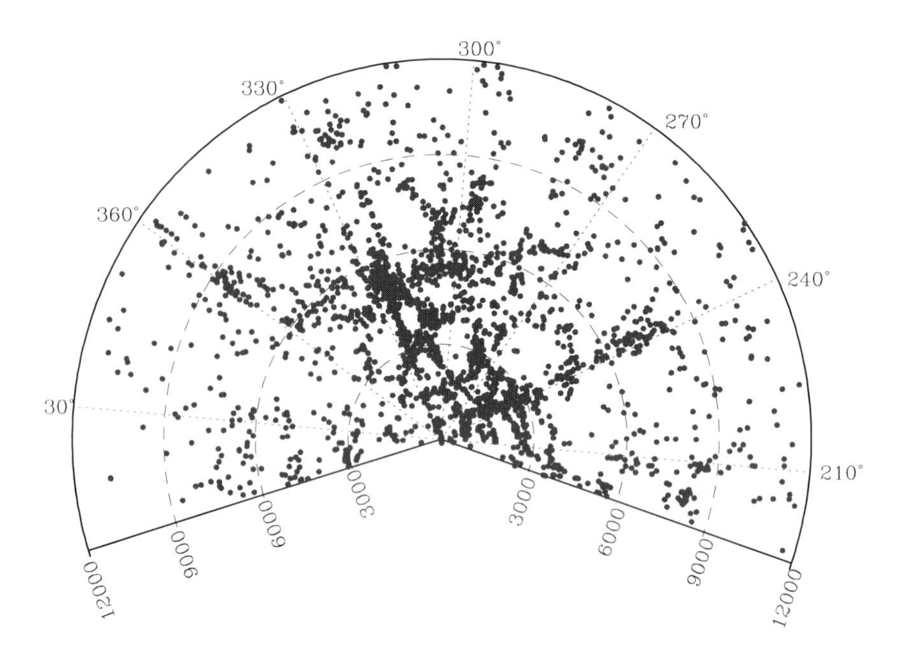


Figure 16: Galactic latitude slice within $|b| \leq 10°$ out to $12000\,\mathrm{km\,s^{-1}}$ with all the galaxies extracted from LEDA, including the HI Multibeam data displayed in Fig. 14.

voids which finally allow a fairly profound glimpse at the previously hidden core of the Great Attractor. Figure 16 clearly shows the prominence of the Norma cluster ($\ell = 325°$) as well as its central location in the great-wall like structure that can be followed from $270°$ to $360°$ within the redshift range $4000-6000\,\mathrm{km\,s^{-1}}$. In front of this wall, we see further filaments that merge in the galaxy concentration around PKS 1343−601. The combined structures in the vicinity of the Norma cluster show a strong similarity to the Coma cluster in the Great Wall from the CfA redshift slices by Huchra, Geller and collaborators. The flow field that pointed to a so-called Great Attractor does seem explained by what we now call the Norma Supercluster, together with the galaxy concentration around PKS 1343−601, as well as the Centaurus Wall, that stretches from the Pavo cluster ($332°, -23°$) across the Galactic Plane to the Centaurus cluster ($320°, +22°$) at slightly lower velocities than the Norma Supercluster.

# Acknowledgements

Discussions with P.A. Woudt and T. Jarrett have been invaluable in the preparation of this review. The contributions of the ZOA team L. Staveley-Smith (PI), B. Koribalski, P.A. Henning, A.J. Green, R.D. Ekers, R.F. Haynes, R.M. Price, E.M. Sadler, I. Stewart and A. Schröder and other participants in the Multibeam HI survey are gratefully acknowledged. This research used the Lyon-Meudon Extragalactic Database (LEDA), supplied by the LEDA team at the Centre de Recherche Astronomique de Lyon, Obs. de Lyon, and also the NASA/IPAC Infrared Science Archive (2MASS) and the NASA/IPAC Extragalactic Database (NED), which are operated by the Jet Propulsion Laboratory, California Institute of Technology, under contract with the National Aeronautics and Space Administration. RCKK thanks CONACyT for their support (research grant 40094F) and the Australian Telescope National Facility (CSIRO) for their support and hospitality during her sabbatical.

## References

Abell, G.O., Corwin, H.G.Jr., & Olowin, R.P. 1989, ApJS 70, 1

Balkowski, C., & Kraan-Korteweg, R.C. (eds.) 1994, "Unveiling Large-Scale Structures behind the Milky Way", ASP Conf. Ser. 67, (San Francisco: ASP)

Böhringer, H., Neumann, D.M., Schindler, S., & Kraan-Korteweg R.C. 1996, ApJ 467, 168

Cameron, L.M. 1990, A&A 233, 16

Donley, J.L., Staveley-Smith, L., Kraan-Korteweg, R.C., et al. 2005, AJ 129, 220

Dressler, A., Faber, S.M., Burstein, D., et al. 1987, ApJ 313, 37

Ebeling, H., Mullis, C.R., & Tully, B.R. 2002, ApJ 580, 774

Fairall, A.P., & Kraan-Korteweg, R.C. 2000, in "Mapping the Hidden Universe: The Universe behind the Milky Way - The Universe in HI", ASP Conf. Ser. 218, eds. R.C. Kraan-Korteweg, P.A. Henning & H. Andernach, (San Francisco: ASP), 35

Fairall, A.P., & Kraan-Korteweg, R.C. 2005, in prep.

Fairall, A.P., & Woudt, P.A. (eds.) 2005, in "Nearby Large-Scale Structures and the Zone of Avoidance", ASP Conf. Ser. 329 (San Francisco: ASP)

Fairall, A.P., Woudt, P.A., & Kraan-Korteweg, R.C. 1998, A&ASS 127, 463

Henning, P.A., Staveley-Smith, L., Ekers, R.D., et al. 2000, AJ 119, 2686

Hudson, M.J., & Lynden-Bell, D. 1991, MNRAS 252, 219

Hudson, M.J., Smith, R.J., Lucey, J.R., & Branchini, E. 2004, MNRAS 352, 61

De Lapparent, V., Geller, M.J., & Huchra, J.P. 1986, ApJ 302, L1

Jarrett, T. 2004, PASA 21, 396

Jarrett, T.-H., Chester, T., Cutri, R., Schneider, S., Rosenberg, J., Huchra, J.P., & Mader, J. 2000a, AJ 120, 298

Jarrett, T.-H., Chester, T., Cutri, R., Schneider, S., Skrutskie, M., & Huchra, J.P. 2000b, AJ 119, 2498

Joint IRAS Science Working Group 1988, *IRAS* Point Source Catalog, Version 2 (Washington: US Govt. Printing Office)

Kerr, F.J., & Westerhout, G. 1965, Galactic Structure (Chicago: University of Chicago), 186

Kocevski, D.D., Mullis, C.R., & Ebeling, H. 2004, ApJ 608, 721

Kogut, A., Lineweaver, C., Smoot, G.F., et al. 1993, ApJ 419, 1

Kolatt, T., Dekel, A., & Lahav, O. 1995, MNRAS 275, 797

Koribalski, B.S., Staveley-Smith, L., Kilborn, V.A., et al. 2004, AJ 128, 16

Kraan-Korteweg, R.C. 2000, A&ASS 141, 123

Kraan-Korteweg, R.C., & Jarrett, T. 2005, in "Nearby large-Scale Structures and the Zone of Avoidance", ASP Conf. Ser. 329, eds. A.P. Fairall & P.A. Woudt, (San Francisco: ASP), 119 (astro-ph/0409391)

Kraan-Korteweg, R.C., & Lahav, O. 2000, A&ARv 10, 211

Kraan-Korteweg, R.C., & Woudt, P.A. 1999, PASA 16, 53

Kraan-Korteweg, R.C., Loan, A.J., Burton, W.B., Lahav, O., Ferguson, H.C., Henning, P.A., & Lynden-Bell, D. 1994, Nature 372, 77

Kraan-Korteweg, R.C., Fairall, A.P., & Balkowski, C. 1995, A&A 297, 617

Kraan-Korteweg, R.C., Woudt, P.A., Cayatte, V., Fairall, A.P., Balkowski, C., & Henning, P.A. 1996, Nature 379, 519

Kraan-Korteweg, R.C., Schröder, A., Mamon, G., & Ruphy S. 1998, in "The Impact of Near-Infrared Surveys on Galactic and Extragalactic Astronomy", ed. N. Epchtein (Kluwer: Dordrecht), 205

Kraan-Korteweg R.C., Henning P.A., & Andernach H. (eds.) 2000, in "Mapping the Hidden Universe: The Universe Behind the Milky Way – The Universe in HI", ASP Conf. Ser. 218 (San Francisco: ASP)

Kraan-Korteweg R.C., Henning P.A., & Schröder, A.C. 2002, A&A 391, 887

Kraan-Korteweg, R.C., Ochoa, M., Woudt, P.A., & Andernach, H. 2005a, in "Nearby Large-Scale Structures and the Zone of Avoidance", ASP Conf. Ser. 329, eds. A.P. Fairall & P.A. Woudt, (San Francisco: ASP), 159 (astro-ph/0406044)

Kraan-Korteweg, R.C., Staveley-Smith, L., Donley, J., Koribalski, B., & Henning, P.A. 2005b, in "Maps of the Cosmos", IAU Symp. 216, eds. M. Colless & L. Staveley-Smith, (San Francisco: ASP), in press (astro-ph/0311129)

Lauberts, A. 1982, The ESO/Uppsala Survey of the ESO (B), (Atlas: ESO, Garching)

Lucey, J.R., Radburn-Smith, D.J., & Hudson, M.J. 2005, in "Nearby Large-Scale Structures and the Zone of Avoidance", ASP Conf. Ser. 329, eds. A.P. Fairall & P.A. Woudt, (San Francisco: ASP), 21 (astro-ph/0412329)

Lundmark, K. 1940, Lundmark Observatory

Lynden-Bell, D., & Lahav, O. 1988, in "Large-scale motions in the universe", eds. V.C. Rubin & G.V. Coyne (Princeton, NJ: Princeton University Press), 199

McHardy, I.M., Lawrence, A., Pye, J.P., et al. 1981, MNRAS 197, 893

Mamon, G.A. 1998, in "Wide Field Surveys in Cosmology", eds. Y. Mellier & S. Colombi, (Gif-sur-Yvette: Editions Frontières), 323

Meyer, M.J., Zwaan, M.A., Webster, R.L., et al. 2004, MNRAS 350, 1195

Mullis, C.R., Ebeling, H., Kocevski, D.D., & Tully, R.B. 2005, in "Nearby Large-Scale Structures and the Zone of Avoidance", ASP Conf. Ser. 329, eds. A.P. Fairall & P.A. Woudt (San Francisco: ASP), 183

Nagayama, T., Ph.D. thesis, Nagoya University, 2004

Nagayama, T., Woudt, P.A., Nagashima, C., et al. 2004, MNRAS 354, 980

Nagayama, T., Nagata, T., Sato, S., Woudt, P.A., & IRSF/SIRIUS team 2005, in "Nearby Large-Scale Structures and the Zone of Avoidance", ASP Conf. Ser. 329, eds. A.P. Fairall & P.A. Woudt (San Francisco: ASP), 177

Nilson, P. 1973, Uppsala General Catalog of Galaxies, (Uppsala: University of Uppsala)

Paturel, G., Petit, C., Rousseau, J., & Vauglin, I. 2003, A&A 405, 1

Peebles, P.J.E. 1994, ApJ 429, 43

Rousseau, J., Paturel, G., Vauglin, I., Schröder, A., et al. 2000, A&A 363, 62

Salem, C., & Kraan-Korteweg, R.C., in prep.

Saunders, W., D'Mellow, K.J., Valentine H., et al. 2000, in "Mapping the Hidden Universe: The Universe behind the Milky Way - The Universe in HI", ASP Conf. Ser. 218, eds. R.C. Kraan-Korteweg, P.A. Henning & H. Andernach, (San Francisco: ASP), 141

Schlegel, D.J., Finkbeiner, D.P., & Davis M. 1998, ApJ 500, 525

Schröder, A., Kraan-Korteweg, R.C., Mamon, G.A., & Ruphy S. 1997, in "Extragalactic Astronomy in the Infrared", eds. G.A. Mamon, T.X. Thuan & J. Tran Thanh Van (Editions Frontières: Gif-sur-Yvette), 381

Schröder, A., Kraan-Korteweg, R.C., & Mamon G.A. 1999, PASA 16, 42

Schröder, A., Kraan-Korteweg, R.C., & Mamon, G.A. 2000, in "Mapping the Hidden Universe: The Universe behind the Milky Way - The Universe in HI", ASP Conf. Ser. 218, eds. R.C. Kraan-Korteweg, P.A. Henning & H. Andernach, (San Francisco: ASP), 119

Schröder, A., Kraan-Korteweg, R.C., Mamon, G.A., & Woudt, P.A. 2005, in "Nearby Large-Scale Structures and the Zone of Avoidance", ASP Conf. Ser. 329, eds. A.P. Fairall & P.A. Woudt (San Francisco: ASP), 167 (astro-ph/0407019)

Schröder, A.C., Kraan-Korteweg, R.C., & Henning, P.A., in prep.

Staveley-Smith, L., Wilson, W.E., Bird, T.S., et al. 1996, PASA 13, 243

Tonry, J.L., & Davis, M. 1981, ApJ 246, 680

Vauglin, I., Rousseau, J., Paturel, G., et al. 2002, A&A 387, 1

Vorontsov-Velyaminov, B., Archipova, V.P., & Krasnogorskaja, A. 1962-1974, Morphological Catalogue of Galaxies, Vol. I-V (Moscow: Moscow State University)

Wakamatsu K., Parker Q.E., Malkan M., & Karoji H. 2000, in "Mapping the Hidden Universe: The Universe behind the Milky Way - The Universe in HI", ASP Conf. Ser. 218, eds. R.C. Kraan-Korteweg, P.A. Henning & H. Andernach, (San Francisco: ASP), 187

Wakamatsu, K., Malkan, M.A., Nishida, M.T., Parker, Q.A., Saunders, W., & Watson, F.G. 2005, in "Nearby Large-Scale Structures and the Zone of Avoidance", ASP Conf. Ser. 329, eds. A.P. Fairall & P.A. Woudt, (San Francisco: ASP), 149

Woudt, P.A. 1998, Ph.D. thesis, Univ. of Cape Town

Woudt, P.A., & Kraan-Korteweg, R.C. 2001, A&A 380, 441

Woudt P.A., Kraan-Korteweg, R.C., & Fairall, A.P. 1999, A&A 352, 39

Woudt, P.A., Kraan-Korteweg, R.C., & Fairall, A.P. 2000, in "Mapping the Hidden Universe: The Universe behind the Milky Way - The Universe in HI", ASP Conf. Ser. 218, eds. R.C. Kraan-Korteweg, P.A. Henning & H. Andernach, 203

Woudt, P.A., Kraan-Korteweg, R.C., Cayatte, V., Balkowski, C., & Felenbok, P. 2004, A&A 415, 9

# New Cosmology with Clusters of Galaxies

Peter Schuecker

Max-Planck-Institut für extraterrestrische Physik
Giessenbachstraße, Postfach 1312, D-85741 Garching
peters@mpe.mpg.de

**Abstract**

*The review summarizes present and future applications of galaxy clusters to cosmology with emphasis on nearby X-ray clusters. The discussion includes the density of dark matter, the normalization of the matter power spectrum, neutrino masses, and especially the equation of state of the dark energy, the interaction between dark energy and ordinary matter, gravitational holography, and the effects of extra-dimensions.*

# 1 Basic cosmological framework

The general framework for present cosmological work is set by three observational results. The perfect Planckian shape of the cosmic microwave background (CMB) spectrum as observed with the COBE satellite (Mather et al. 1990) clearly shows that the Universe must have evolved – from a hot, dense, and opaque phase. The very good correspondence of the observed abundance of light elements and the results of Big Bang Nucleosynthesis (BBN, e.g. Burles, Nollett & Turner 2001) shows that the cosmic expansion can be traced back to cosmological redshifts up to $z = 10^{10}$. Steigman (2002) pointed out that if these analyses would have been performed with Newton gravity and not with Einstein gravity, then the observed abundances could not be reconciled with the BBN predictions. One can take this as one of the few indications than Einstein gravity can in fact be applied within a cosmological context and underlines the importance of the BBN benchmark for any gravitational theory. Finally, the consistency of the ages of the oldest stars in globular clusters (e.g. Chaboyer & Krauss 2002) and the age of the Universe as obtained from cosmological observations can be regarded as the long-waited 'unification' of the theory of stellar structure and the theory of cosmic spacetime (Peebles & Ratra 2004). Traditionally, Friedmann-Lemaître (FL) world models as derived from Einstein's field equations for spatially homogeneous and isotropic systems, are assumed, characterized by the Hubble constant $H_0$ in units of $h = H_0/(100\,\mathrm{km\,s^{-1}\,Mpc^{-1}})$, the normalized density of cosmic matter $\Omega_\mathrm{m}$ (e.g., baryonic and Cold Dark Matter CDM),

*Reviews in Modern Astronomy 18.* Edited by S. Röser
 ISBN 3-527-40608-5

the normalized cosmological constant $\Omega_\Lambda$, and its equation of state $w$. Within this general framework, clusters of galaxies are traditionally used as cosmological probes on Gigaparsec scales. However, a precise test that one can apply Einstein gravity on such large scales is still missing.

In Sect. 2, a summary of the basic properties of nearby galaxy clusters is given. The hierarchical structure formation paradigm is tested with nearby galaxy clusters in Sect. 3. Constraints on the density of dark matter (DM), the normalization of the matter power spectrum, and neutrino masses are presented in Sect. 4. Observational effects of the equation of state of the dark energy (DE), and a first test of a non-gravitational interaction between DE and DM are presented in Sect. 5. The problem of the cosmological constant and its discussion in terms of the gravitional Holographic Principle as well as the effect of an extra-dimension of brane-world gravity are discussed in Sect. 6. Sect. 7 draws some conclusions. A general review on clusters is given in Bahcall (1999), whereas Edge (2004) focuses on nearby X-ray cluster surveys, Borgani & Guzzo (2001) on their spatial distribution, Rosati, Borgani & Norman (2002) and Voit (2004) on their evolution.

# 2 Galaxy clusters

Galaxy clusters are the largest virialized structures in the Universe. Only 5% of the bright galaxies ($> L_*$) are found in rich clusters, but more than 50% in groups and poor clusters. The number of cluster galaxies brighter than $m_3 + 2^m$ where $m_3$ is the magnitude of the third-brightest cluster galaxy, and located within $1.5\,h^{-1}$ Mpc radius from the cluster center, range for rich clusters from 30 to 300 galaxies, and for groups and poor clusters from 3 to 30. For cosmological tests, rich clusters will turn out to be of more importance so that the following considerations will mainly focus on the properties of this type. Rich clusters have typical radii of $1-2\,h^{-1}$ Mpc where the surface galaxy density drops to $\sim 1\%$ of the central density.

Baryonic gas, falling into the cluster potential well, is shock-heated up to temperatures of $T_e = 10^{7-8}$ K. The acceleration of the electrons in the hot plasma (intracluster medium ICM) gives thermal Bremsstrahlung with a maximum emissivity at $k_B T_e = 2-14$ keV so that they can be observed in X-rays together with some line emission. Typical X-ray luminosities range between $L_x = 10^{42-45}\,h^{-2}\,\mathrm{erg\,s^{-1}}$ in the energy interval $0.1 - 2.4$ keV. With X-ray satellites like ROSAT, Chandra, or XMM-Newton, these clusters can thus be detected up to cosmological interesting redshifts. However, only a few clusters are detected at redshifts beyond $z = 1$ (Rosati et al. 2002; Mullis et al. 2005).

Galaxy clusters are rare objects with number densities of $10^{-5}\,h^3\,\mathrm{Mpc}^{-3}$, strongly decreasing with X-ray luminosity or cluster mass (Böhringer et al. 2002). Current structure formation models predict of the order of $10^6$ rich galaxy clusters in the visible Universe, the majority with redshifts below $z = 2$. More than 5 000 nearby galaxy clusters are already identified in the optical as local concentrations of galaxies, and 2 000 by their (extended) X-ray emission. Surveys planned for the next few years like the Dark Universe Observatory DUO (Griffiths, Petre, Hasinger et al. 2004) could yield about $10^4$ clusters possibly up to $z = 2$, that is, already 1% of the total cluster population. It appears thus not completely illusory to finally get an almost complete census of all rich galaxy clusters in the visible Universe.

X-ray clusters get their importance for cosmology because of the tide correlations between observables like X-ray temperature or X-ray luminosity and total gravitating cluster mass which allow a precise reconstruction of the cosmic mass distribution on large scales.

Knowledge of the total gravitating mass of a cluster within a well-defined radius, is of crucial importance. The masses are summarized in cluster mass functions which depend on structure formation models through certain values of the cosmological parameters. However, cosmic mass function appear to be independent of cosmology when they are written in terms of natural "mass" and "time" variables (Lacey & Cole 1994). Model mass functions can either be predicted from semi-analytic models (e.g, Sheth & Tormen 2002, Schuecker et al. 2001a, Amossov & Schuecker 2004) or from N-body simulations, the latter with errors between 10 to 30% (Jenkins et al. 2001; Springel et al. 2005).

Cluster masses can be determined in the optical by the velocity dispersion of cluster galaxies or in X-rays from, e.g., the gas temperature and density profiles, assuming virial and hydrostatic equilibrium, respectively (and spherical symmetry). Gravitational lensing uses the distortion of background galaxies and determines the projected cluster mass without any specific assumption (e.g., Kaiser & Squires 1993). For regular clusters, the masses of galaxy clusters are consistently determined with the three methods and range between $10^{14} - 10^{15}\,h^{-1}\,M_{\odot}$ (e.g., Wu et al. 1998). Several projects are currently under way to compare the mass estimates obtained with the different methods in more detail. The baryonic mass in clusters comes from the ICM and the stars in the cluster galaxies. The ratio between the baryonic and total gravitating mass (baryon fraction) in a cluster is about $0.07h^{-1.5} + 0.05$.

Systematic X-ray studies of large samples of galaxy clusters have revealed that about half of the clusters have significant substructure in their surface brightness distributions, i. e., some deviations from a perfect regular shape (e.g. Schuecker et al. 2001b). For the detection of substructure, different methods as summarized in Feretti, Giovannini & Gioa (2002) give substructure occurence rates ranging from 20 to 80%. The large range clearly shows that the definition of a well-defined mass threshold for substructure and the measurement of the masses of the different subclumps is difficult and has not yet been regorously applied. Further interesting ambiguities arise because clusters appear more regular in X-ray pseudo pressure maps (product of projected gas mass density and gas temperature) whereas contact discontinuities and shock fronts caused by merging events appear more pronounced in pseudo entropy and temperature maps (Briel, Finoguenov & Henry 2004).

Substructering is taken as a signature of the dynamical youth of a galaxy cluster. The most dramatic distortions occure when two big equal mass clumps collide (major merger) to form a larger cluster. With the ROSAT satellite, merging events could be studied for the first time in X-rays in more detail (e.g. Briel, Henry & Böhringer 1992). A typical time scale of a merger event is $10^{9}$ yr where the increased gas density and X-ray temperature can boost X-ray luminosities up to factors of five (Randall, Sarazin & Ricker 2002). The XMM-Newton and especially the Chandra X-ray satellite allows more detailed studies of substructures down to arcsec scales. Sub-

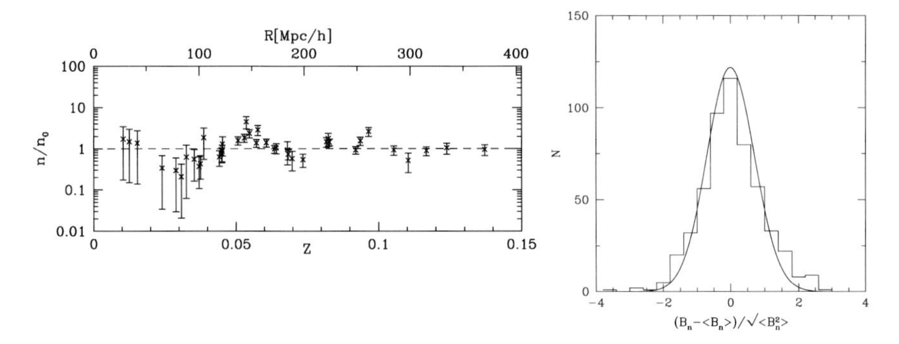


Figure 1: **Left:** Normalized comoving REFLEX cluster number densities as a function of redshift, and comoving radial distance $R$. Vertical error bars represent the formal $1\sigma$ Poisson errors. **Right:** Histogram of the normalized KL coefficients of the REFLEX sample and superposed Gaussian profile. The Kolmogorov-Smirnov probability for Gaussianity is 93%.

structures in form of cavities and bubbles (Böhringer et al. 1993, Fabian et al. 2000), cold fronts (Vikhlinin, Markevitch & Murray 2001), weak shocks and sound waves (Fabian et al. 2003), strong shocks (Forman et al. 2003), and turbulence (Schuecker et al. 2004) were discovered, possibly triggered by merging events and/or AGN activity. With the ASTRO-E2 satellite planned to be launched in 2005, the line-of-sight kinematics of the ICM will be studied for the first time to get more information about the dynamical state of the ICM. The majority of the abovementioned substructures have low amplitudes which do not much disturb radially-averaged cluster profiles (after masking) and thus cluster mass estimates. In fact, the hydrostatic equation relates the observed smooth pressure gradients to the total gravitating cluster mass, which makes the robustness of X-ray cluster mass estimates from numerical simulations plausible (Evrard, Metzler & Navarro 1996, but see Sect. 7). Present cosmological tests based on galaxy clusters assume that the diversity of regular and substructured clusters contribute only to the intrinsic scatter of the observed X-ray luminosity-mass relation or similar diagnostics, while keeping the shape and normalization of the original relation almost unaltered.

The remaining about 50% of the clusters appear quite regular - a significant fraction of these clusters have very bright X-ray cores, where the dense gas could significantly cool. Such cooling core clusters are expected to be in a very relaxed dynamical state since several Gigayears. Numerical simulations suggest that the baryon fraction in these clusters is close to the universal value and can be used after some corrections as a cosmic 'standard candle' (e.g. White et al. 1993).

For nearby ($z < 0.3$) rich systems, evolutionary effects on core radius and entropy input are found to be negligible (Rosati et al. 2002). Detailed XMM studies at $z \sim 0.3$ can be found in Zhang et al. (2004). Therefore, cosmological tests based on massive nearby clusters with gas temperatures $k_{\rm B}T_{\rm e} > 3\,{\rm keV}$ are expected to give reliable results. For these systems, the observed X-ray luminosity can be transformed into the theory-related cluster mass with empirical luminosity-mass or similar relations characterized by their shape, intrinsic scatter, and normalization (e.g., Reiprich & Böhringer 2002). It will be shown that with such methods, cosmological tests can be performed presently on the 20-30% accuracy level.

Further improvements on cluster scaling relations are thus necessary to reach (if possible) the few-percent level of 'precision cosmology'. Large and systematic observational programms based on Chandra and XMM-Newton observations are now under way which are expected to significantly improve the relations within the next few years (e.g., XMM Large Programme, Böhringer, H. et al., in prep., and a large XMM/Chandra project of Reiprich, T. H. et al., in prep.). For cosmological tests with distant rich clusters, additional work is necessary. Gravitationally-induced evolutionary effects due to structure growth, and non-gravitationally-induced evolutionary effects like ICM heating through galactic winds caused by supernovae (SNe), and heating by AGN cause systematic deviations from simple self-similarity expectations (Kaiser 1986; Ponman, Cannon & Navarro 1999). For cosmological tests, such evolutionary effects add further degrees of freedom to be determined simultaneously with the cosmological parameters (e.g., Borgani et al. 1999).

# 3 Hierarchical structure formation paradigm

Structure formation on the largest scales as probed by galaxy clusters is mainly driven by gravity and should thus be understandable in a simple manner. However, reconciling the tiny CMB anisotropies at $z \approx 1100$ with the very large inhomogeneities of the local galaxy distribution has shown that the majority of cosmic matter must come in nonbaryonic form (e.g., CDM). A direct consequence of such scenarios is that clusters should grow from Gaussian initial conditions in a quasi hierarchical manner, i.e., less rich clusters and groups tend to form first and later merge to build more massive clusters. The merging of galaxy clusters as seen in X-rays (Sect. 2) is a direct indication that such processes are still at work in the local universe.

A further argument for hierarchical structure growth comes from the spatial distribution of galaxy clusters on $10^2\,h^{-1}\,\mathrm{Mpc}$ scales. Less then 1/10 of this distance can be covered by cluster peculiar velocities within a Hubble time, keeping in this linear regime the Gaussianity of the cosmic matter field as generated by the chaotic processes in the early Universe almost intact. This Gaussianity formally stems from the random-phase superposition of plane waves and the central limit theorem (superposition approximation). The peaks of this random field will eventually collapse to form virialized clusters. The relation between the spatial fluctuations of the clusters and the underlying matter field is called 'biasing'. For Gaussian random fields, the biasing tend to concentrate the clusters in regions with the highest global matter density in a manner that their correlation strengths $r_0$ increase with cluster mass (Kaiser 1984) - otherwise they would immediately distroy Gaussianity (e.g. if we would put a very massive cluster into a void of galaxies). Peculiar velocities of the clusters induced by the resulting inhomogeneities modify the $r_0$-mass relation, but without disturbing the general trend (peak-background split of Efstathiou, Frank & White 1988, & Mo & White 1996). In the linear regime, we thus expect a Gaussian distribution of the amplitudes of cluster number fluctuations which increase with mass in a manner as predicted by the specific hierarchical scenario.

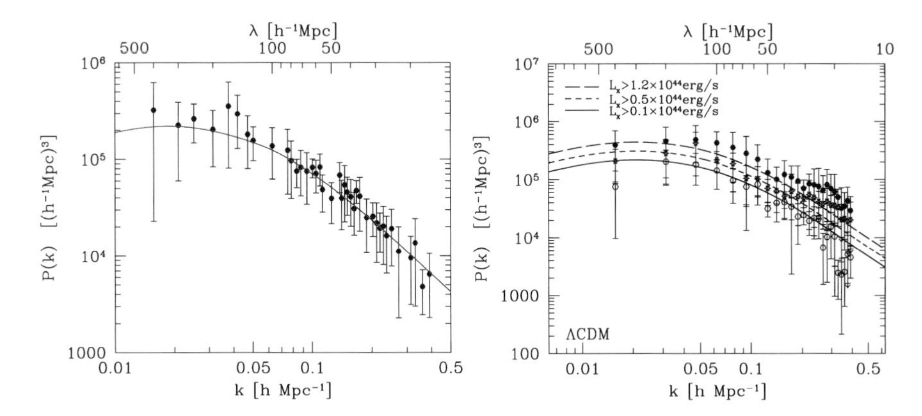


Figure 2: **Left:** Comparison of the observed REFLEX power spectrum (points with error bars) with the prediction of a spatially flat ΛCDM model with a matter density of $\Omega_{\rm m} = 0.34$ and $\sigma_8 = 0.71$. Errors include cosmic variance and are estimated with numerical simulations. **Right:** Comparison of observed power spectral densities and predictions of a low-density CDM semi-analytic model as a function of lower X-ray luminosity, i.e., lower X-ray cluster mass (Schuecker et al. 2001c). The errors include cosmic variance and are obtained from N-body simulations.

The REFLEX catalogue (Böhringer et al. 2004)[1] provides the largest homogeneously selected sample of X-ray clusters and is ideally suited for testing specific hierarchical structure formation models. The sample comprises 447 southern clusters with redshifts $z \leq 0.45$ (median at $z = 0.08$) down to X-ray fluxes of $3.0 \times 10^{-12}\,{\rm erg\,s^{-1}\,cm^{-2}}$ in the energy range $(0.1 - 2.4)\,{\rm keV}$, selected from the ROSAT All-Sky Survey (Böhringer et al. 2001). Several tests show that the sample cannot be seriously affected by unknown selection effects. An illustration is given by the normalized, radially-averaged comoving number densities along the redshift direction (Fig. 1 left). The densities fluctuate around a $z$-independent mean as expected when no unknown selection or evolutionary effects are present. Further tests can be found in Böhringer et al. (2001, 2004); Collins et al. (2000) and Schuecker et al. (2001c).

Tests of the Gaussianity of the cosmic matter field refer to the superposition approximation mentioned above. They may divide the survey volume into a set of large non-overlapping cells, count the clusters in each cell, decompose the fluctuation field of the cluster counts into plane waves via Fourier transformation, and check whether the frequency distribution of the amplitudes of the plane waves (Fourier modes with wavenumber $k$) follow a Gaussian distribution. However, the survey volume provides only a truncated view of the cosmic matter field which will result in an erroneous Fourier transform (the result obtained will be the convolution of the true Fourier transform with the survey window function). The truncation effect comprises both the reduction of fine details in the Fourier transform and the correlation

[1] http://www.xray.mpe.mpg.de/theorie/REFLEX/

of Fourier modes so that fluctuation power migrates between the modes. This leakage effect increases when the symmetry of the survey volume deviates from a perfect cubic shape. Uncorrelated amplitudes can be obtained, when the fluctuations are decomposed into modes which follow to some extent the shape of the survey volume. The Karhunen-Loèwe (KL) decomposition determines such eigenmodes under the constraint that the resulting KL fluctuation amplitudes are statistically uncorrelated. This construction is quite optimal for testing cosmic Gaussianity. The KL eigenmodes are the eigenvectors of the sample correlation matrix, i.e., the matrix giving the expected correlations between the number of clusters obtained in pairs of count cells as obtained with a fiducial (e.g. concordance) cosmological model. KL modes were first applied to CMB data by Bond (1995), to galaxy data by Vogeley & Szalay (1996), and to cluster data by Schuecker et al. (2002). The linearity of the KL transform and the direct biasing scheme expected for galaxy clusters suggest that the statistics of the KL coefficients should directly reflect the statistics of the underlying cosmic matter field.

Figure 1 (right) compares a standard Gaussian with the frequency distribution of the observed KL-transformed and normalized cluster counts obtained with REFLEX. The cell sizes are larger than $100\,h^{-1}$ Mpc and thus probe Gaussianity in the linear regime. The observed Gaussianity of the REFLEX data suggests Gaussianity of the underlying cosmic matter field on such large scales. This is a remarkable finding, taking into account the difficulties one has to test Gaussianity even with current CMB data (Komatsu et al. 2003, Cruz et al. 2004).

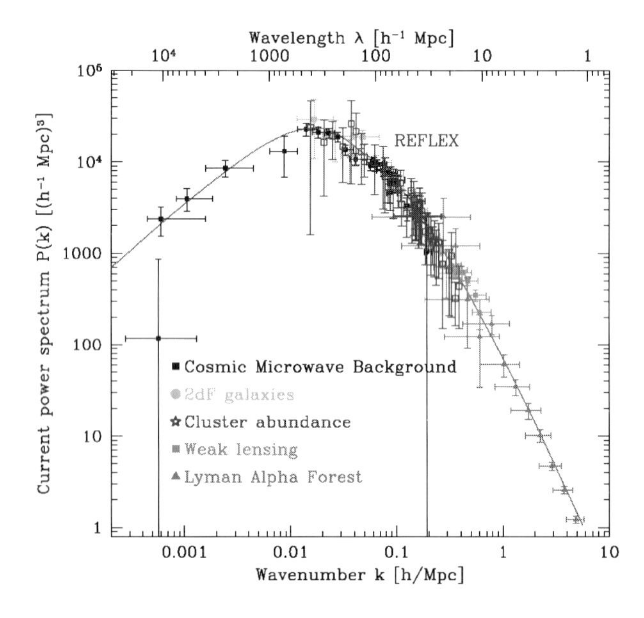


Figure 3: Compilation of fluctuation power spectra of various cosmological objects as complited by Tegmark & Zaldarriaga (2002) with the added REFLEX power spectrum. The continuous line represents the concordance ΛCDM structure formation model.

As mentioned above, hierarchical structure formation predicts that the amplitudes of the fluctuations should increase in a certain manner with mass. On scales small compared to the maximum extent of the survey volume, the fluctuation field roughly follows the superposition approximation. In this scale range, it is very convinient to test the mass-dependent amplitude effect with a simple plane wave decomposition as summarized by the power spectrum $P(k)$[2]. Fig. 2 (left) shows that the observed REFLEX power spectrum of the complete sample is well fit by a low-density ΛCDM model. Comparisons with other hierarchical scenarios are found to be less convincing (Schuecker et al. 2001c). In contrast to the 'standard CDM' model with $\Omega_{\rm m} = 1$, in low-density (open) CDM models, the epoch of equality of matter and radiation occure rather late and the growth of structure proceeds over a somewhat smaller range of redshift, until $(1+z) = \Omega_{\rm m}^{-1}$. Consequently, the turnover in $P(k)$ is at larger scales, leaving less power on small scales. The nonzero cosmological constant of a (flat) ΛCDM scenario stretches out the time scales of the model until $(1+z) = \Omega_{\rm m}^{-1/3}$. The differences in the dynamics of structure growth are thus not very large compared to an open CDM model and become only important at late stages. Note, however, that when all models are normalized to the local Universe, the opposite conclusion is true. The behaviour of the cluster fluctuation amplitude with mass (X-ray luminosity) for a low-density CDM model is shown in Fig. 2 (right). The predictions are shown as continuous and dashed lines which nicely follow the observed trends. The model includes an empirical relation to convert cluster mass to X-ray luminosity (Reiprich & Böhringer 2002), a model for quasi-nonlinear and linear structure growth (Peebles 1980), a biasing model (Mo & White 1996, Matarrese et al. 1997), and a model for the transformation of the power spectrum from real space to redshift space (Kaiser 1987).

However, one could still argue that clusters constitute only a small population of all cosmological objects visible over a limited redshift interval, and could therefore not give a representative view of the goodness of hierarchical structure formation models. Fig. 3 summarizes power spectra obtained with various cosmological tracer objects as compiled by Tegmark & Zaldarriaga (2002) including the REFLEX power spectrum. All spectra are normalized by their respective biasing parameters (if necessary). The combined power spectrum covers a spatial scale range of more than four orders of magnitude and redshifts between $z = 1100$ (CMB) and $z = 0$. The good fit of the ΛCDM model shows that this hierarchical structure formation model is really very successful in describing the clustering properties of cosmological objects. The following cosmological tests thus assume the validity of this structure formation model.

# 4 Ordinary matter

The observed cosmic density fluctuations are very well summarized by a low matter density ΛCDM model (Sect. 3). Therefore, many cosmological tests refer to this structure formation scenario. In general, baryonic matter, Cold Dark Matter

[2]The KL method would need many modes to test small scales which is presently too computer-intensive

(CDM), primeval thermal remnants (electromagnetic radiation, neutrinos), and an energy corresponding to the cosmological constant give the total (normalized) density of the present Universe, $\Omega_{\rm tot} = \Omega_{\rm b} + \Omega_{\rm CDM} + \Omega_{\rm r} + \Omega_{\Lambda}$. The normalized density of ordinary matter comprises the first three components. Recent CMB data suggest $\Omega_{\rm tot} = 1.02 \pm 0.02$ (Spergel et al. 2003), i.e., an effectively flat universe with a negligible spatial curvature. The same data suggest a baryon density of $\Omega_{\rm b}h^2 = 0.024 \pm 0.001$ and $h = 0.72 \pm 0.05$. For our purposes, the energy density of thermal remnants, $\Omega_{\rm r} = 0.0010 \pm 0.0005$ (Fukugita & Peebles 2004), can be neglected, yielding the present cosmic matter density $\Omega_{\rm m} = \Omega_{\rm b} + \Omega_{\rm CDM}$. At the end of this section, an estimate of $\Omega_{\rm r}$ including only the neutrinos is given.

Within this context of the hierarchical structure formation, the occurence rate of substructure seems to be a useful diagnostic to test different cosmological parameters because a high merger rate implies a high $\Omega_{\rm m}$ (e.g., Richstone, Loeb & Turner 1992, Lacey & Cole 1993). However, as mentioned in Sect. 2, the effects of substructure are difficult to measure and to quantify in terms of mass so that presently less stringent constraints are attainable (for a recent discussion see, e.g., Suwa et al. 2003).

A simple though $h$-dependent estimate of $\Omega_{\rm m}$ can be obtained from the comoving wavenumber of the turnover of the power spectrum because it corresponds to the horizon length at the epoch of matter-radiation equality $k_{\rm eq} = 0.195\Omega_{\rm m}h^2$ Mpc (e.g. Peebles 1993) below which most structure is smoothed-out by free-streaming CDM particles. A small $\Omega_{\rm m}$ or a small Hubble constant thus shifts the maximum of $P(k)$ towards larger scales. The product $\Gamma = \Omega_{\rm m}h$ is referred to as the shape parameter of the power spectrum. For the REFLEX power spectrum, the turnover is at $k_{\rm eq} = 0.025 \pm 0.005$ (Fig. 2), so that for $h = 0.72$ a matter density of $\Omega_{\rm m} = 0.25 \pm 0.05$ is obtained. In this case, the shape parameter is $\Gamma = 0.18 \pm 0.03$ which is typical for $\Lambda$CDM.

Cluster abundance measurements are a classical application of galaxy clusters in cosmology to determine the present density of cosmic matter, $\Omega_{\rm m}$, either assuming a negligible effect of $\Omega_{\Lambda}$ or not. The effective importance of $\Omega_{\Lambda}$ on geometry and structure growth cannot be neglected for clusters with $z > 0.5$. A related quantity is the variance of the matter fluctuations in spherical cells with radius $R$ and Fourier transform $W(kR)$: $\sigma^2(R) = \frac{1}{2\pi^2}\int_0^\infty dk\, k^2\, P(k)\,|W(kR)|^2$. The specific value $\sigma_8$ at $R = 8\,h^{-1}$ Mpc characterizes the normalization of the matter power spectrum $P(k)$. Recent CMB data suggest $\sigma_8 = 0.9 \pm 0.1$ (Spergel et al. 2003).

In the following, the abundance of galaxy clusters is used to determine simultaneously the values of $\Omega_{\rm m}$ and $\sigma_8$. Early applications of the method can be found in, e.g., White, Efstathiou & Frenk (1993), Eke, Cole & Frenk (1996), and Viana & Liddle (1996) suggesting a stronge degeneracy between $\Omega_{\rm m}$ and $\sigma_8$ of the form $\sigma_8 = (0.5 - 0.6)\Omega_{\rm m}^{-0.6}$. To understand this degeneracy and the high sensitivity of cluster counts on the values of the cosmological parameters, consider the expected number of clusters observed at a certain redshift and flux limit,

$$dN(z, f_{\rm lim}) = dV(z) \int_{M_{\rm lim}(z,f_{\rm lim})}^{\infty} dM \, \frac{dn(M, z, \sigma^2(M))}{dM} \,. \tag{1}$$

For optically selected samples, the flux limit has to be replaced by a richness (or optical luminosity) limit. The cosmology-dependency of $dN$ stems from the comoving volume element $dV$, the mass limit $M_{\rm lim}$ at a certain redshift, and the shape of the cosmic mass function $dn/dM$. Three basic cosmological tests are thus applied simultaneously, which explains the high sensitivity of cluster counts on cosmology, although sometimes effects related to structure growth and geometric volume can work against each other (Sect. 5).

The summation in (1) is over cluster mass whereas observations yield quantities like X-ray luminosity, gas temperature, richness etc. The conversion of such observables into mass is the most crucial step where most of the systematic errors can occure. For more massive systems, likely contributors to systematic errors are effects related to cluster merging, substructures, and cooling cores. Cluster merging increases the gas density and temperature and thus the X-ray luminosity which increases the detection probablity in X-rays. The overall statistical effect is difficult to quantify, but systematic errors in the cosmological parameters on the 20% level can be reached (Randall et al. 2002). For less massive systems, further effects related to additional heat input by AGN, star formation, galactic winds driven by SNe, etc. lead to deviations from self-similar expectations (Sect. 2), and increase the scatter in scaling relations. Such effects are quite difficult to simulate (e.g., Borgani et al. 2004, Ettori et al. 2004).

Equation (1) can directly be applied to flux-selected cluster samples as obtained in X-rays or millimeter wavelengths. The latter surveys detect clusters via the Sunyaev-Zel'dovich (SZ) effects (e.g., Birkinshaw, Gull & Hardebeck 1984, Carlstrom, Holder & Reese 2002). Here, energy of the ICM electrons is locally transferred through inverse Compton (Thomson) scattering to the CMB photons so that the number of photons on the long wavelength side of the Planck spectrum is depleted. After this blue-shift, each cluster is detected at wavelengths beyond 1.4 mm as decrements against the average CMB background, and at shorther wavelengths as increments. This process thus measures deviations relative to the actual CMB background and is thus redshift-independent so that cluster detection does not has to work against the $(1+z)^4$ Tolman's surface brightness dimming which is especially important for very distant clusters. Certain blind SZ surveys are now in preparation (SZ-Array starting 2004; AMI 2004, APEX-SZ 2005, ACT 2007, SPT 2007 and Planck 2007). The flux limits in X-rays and submm allow after some standard corrections a very accurate determination of the volume accessable by a cluster with certain X-ray or submm properties.

The detection of clusters in the optical is more complicated (e.g., red-sequence method in Gladders, Yee & Howard 2004, matched filter method in Postman et al. 1996, Schuecker & Böhringer 1998, Schuecker, Böhringer & Voges 2004). For the application of Eq. (1) to optically selected cluster samples, the mass limit $M_{\rm min}(z)$ has to be obtained with numerical simulations in a more model-dependent manner (e.g., Goto et al. 2002, Kim et al. 2002).

For cosmological tests, the values of the parameters are changed until observed and predicted numbers of clusters agree. In order to avoid the evaluation of 3rd and 4th-order statistics in the error determination, the parameter matrices should be as diagonal as possible. This can be achieved, when the cluster cell counts are transformed into the orthonormal base generated by the KL eigenvectors of the sample

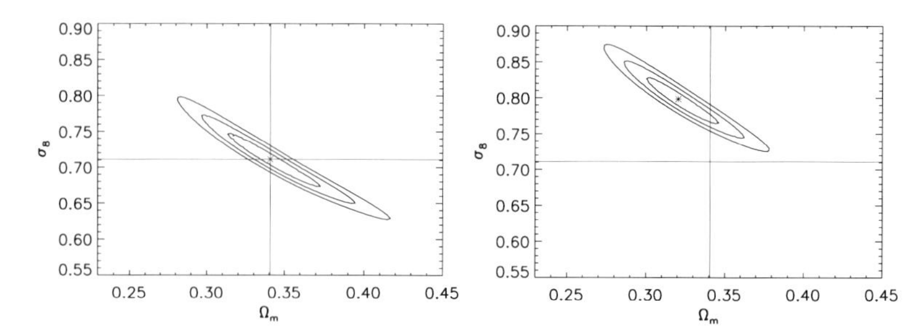


Figure 4: **Left:** Likelihood contours ($1-3\sigma$ level for two degrees of freedom) as obtained with the REFLEX sample. **Right:** Same likelihood contours as left for a different empirical mass/X-ray luminosity relation.

correlation matrix (Sect. 3). With the REFLEX sample, the classical $\Omega_{\rm m}$-$\sigma_8$ test was performed with the KL base (Schuecker et al. 2002, 2003a). The observed Gaussianity of the matter field directly translates into a multi-variant Gaussian likelihood function, and includes in a natural manner a weighting of the squared differences between KL-transformed observed and modeled cluster counts with the variances of the transformed counts. Not only the mean counts in the cells but also their variances from cell to cell depend on the cosmological model. The KL method thus simultaneously tests both mean counts and their fluctuations which increases the sensitivity of the method even more. The method was extensively tested with clusters selected from the Hubble Volume Simulation. Note that for the application of the KL method to galaxies of the Sloan Digital Sky Survey (SDSS, Szalay et al. 2003, Pope et al. 2004) only the fluctuations could be used and were in fact enough to provide constraints on the 10-percent level.

A typical result of a cosmological test of $\Omega_{\rm m}$ and $\sigma_8$ with REFLEX clusters is shown in Fig. 4. Note the small parameter range covered by the likelihood contours and the residual $\Omega_{\rm m}$-$\sigma_8$ degeneracy: For (flat) $\Lambda$CDM and low $z$, structure growth is negligible, and the $\Omega_{\rm m}$-$\sigma_8$ degeneracy is related to the fact that a small $\sigma_8$ (corresponding to a low-amplitude power spectrum) yields a small comoving cluster number density, whereas a large $\Omega_{\rm m}$ (corresponding to a low mass limit $M_{\rm min}$) yields a large comoving number density. For (flat) $\Lambda$CDM and high $z$, structure growth and comoving volume do again not strongly depend on $\Omega_{\rm m}$, but the number of high-$z$ clusters increases with decreasing $\Omega_{\rm m}$ because for a fixed cluster number density at $z=0$ the normalization $\sigma_8$ has to be increased when $\Omega_{\rm m}$ is decreased as shown above. However, the sensitivity on structure growth becomes apparent once open and flat models are compared (Bahcall & Fan 1998).

For the test, further cosmological parameters like the Hubble constant, the primordial slope of the power spectrum, the baryon density, the biasing model, and the empirical mass/X-ray luminosity relation had fixed prior values. The final REFLEX result is obtained by marginalizing over these parameters and yields the $1\sigma$ corridors $0.28 \le \Omega_m \le 0.37$ and $0.56 \le \sigma_8 \le 0.80$.

As mentioned above, the largest uncertainty in these estimates comes from the empirical mass/X-ray luminosity relation obtained for REFLEX from mainly ROSAT and ASCA pointed observations by Reiprich & Böhringer (2002) - compare Fig. 4 left and right. Tests are in preparation with a four-times larger X-ray cluster sample of 1 500 clusters combining a deeper version of REFLEX with an extended version of the cluster catalogue of Böhringer et al. (2000) of the northern hemisphere, plus a more precise M/L-relation obtained over a larger mass range with the XMM-Newton satellite. Errors below the 10-percent level are expected.

Variants of the cluster abundance method use the X-ray luminosity or the gas temperature function. For the transition from observables to mass, often the relations mass-temperature and luminosity-temperature are used. As an example, Borgani et al. (2001) obtained comparatively strong constraints using a sample of clusters up to $z = 1.27$ yielding the $1\sigma$ corridors $0.25 \leq \Omega_{\rm m} \leq 0.38$ and $0.61 \leq \sigma_8 \leq 0.72$.

White et al. (1993) pointed out that the matter content in rich nearby clusters provides a fair sample of the matter content of the Universe. The ratio of the baryonic to total mass in clusters should thus give a good estimate of $\Omega_{\rm b}/\Omega_{\rm m}$. The combination with determinations of $\Omega_{\rm b}$ from BBN (constrained by the observed abundances of light elements at high $z$) can thus be used to determine $\Omega_{\rm m}$ (David, Jones & Forman 1995; White & Fabian 1995; Evrard 1997). Extending the universality assumption on the gas mass fraction to distant clusters, Ettori & Fabian (1999) and later Allen et al. (2002) could show that at a certain distance from the center (density contrast) of quite relaxed distant clusters, the observed X-ray gas mass fraction tends to converge to a universal value. To illustrate the potential power of the method note that after further corrections, the results obtained by Allen et al. with only seven apparently relaxed clusters up to $z = 0.5$ were already sensitive enough to constrain the cosmic matter density, $\Omega_{\rm m} = 0.30^{+0.04}_{-0.03}$. Later work includes more clusters up to $z = 0.9$ and cluster abundances from the REFLEX-sample (Böhringer et al. 2004) and the BCS sample (Ebeling et al. 1998), and yields the $1\sigma$ error corridors $0.25 \leq \Omega_{\rm m} \leq 0.33$ and $0.66 \leq \sigma_8 \leq 0.74$ (Allen et al. 2003). However, the method shares some similarity with the type-Ia SNe method in the sense that the validity of the gas mass fraction as a cosmic standard candle especially at high $z$ is mainly based on observational arguments, partially supported by numerical simulations. The overlap of the error corridors of the less-degenerated results of Borgani et al. (2001), Schuecker et al. (2003a), and Allen et al. (2003) yields our final result

$$\Omega_{\rm m} = 0.31 \pm 0.03 \,. \tag{2}$$

Other measurements show the $\Omega_{\rm m}$-$\sigma_8$ degeneracy more pronounced over a larger range. When all measurements are evaluated at $\Omega_{\rm m} = 0.3$, the values of $\sigma_8$ appear quite consistent at a comparatively low normalization of

$$\sigma_8 = 0.76 \pm 0.10 \,, \tag{3}$$

within the total range $0.5 < \sigma_8 < 1.0$ (data compiled in Henry 2004 from Bahcall et al. 2003; Henry 2004; Pierpaoli et al. 2003; Ikebe et al. 2002; Reiprich & Böhringer

2002; Rosati et al. 2002; including Allen et al. 2003 and Schuecker et al. 2003a with small degeneracies)[3].

Recent neutrino experiments are based on atmospheric, solar, reactor, and accelerator neutrinos. All experiments suggest that neutrinos change flavour as they travel from the source to the detector. These experiments give strong arguments for neutrino oscillations and thus nonzero neutrino rest masses $m_\nu$ (e.g. Ashie et al. 2004 and references given therein). Further information can be obtained from astronomical data on cosmological scales. The basic idea is to measure the normalization of the matter CDM spectrum with CMB anisotropies on several hundred Mpc scales. This normalization is transformed with structure growth functions to $8\,h^{-1}$ Mpc at $z = 0$ assuming various neutrino contributions. This normalization should match the $\sigma_8$ normalization from cluster counts (e.g., Fukugita, Liu & Sugiyama 2000). Recent estimates are obtained by combining CMB-WMAP data with the 2dFGRS galaxy power spectrum, X-ray cluster gas mass fractions, and X-ray cluster luminosity functions (Allen, Schmidt & Bridle 2003). For a flat universe and three degenerate neutrino species, they measured the contribution of neutrinos to the energy density of the Universe, and a species-summed neutrino mass, and their respective $1\sigma$ errors,

$$\Omega_\nu = 0.006 \pm 0.003\,, \quad \sum_i m_i = 0.6 \pm 0.3\,\mathrm{eV}\,, \tag{4}$$

which formally corresponds to $m_\nu \approx 0.2\,\mathrm{eV}$ per neutrino. Their combined analysis yields a normalization of $\sigma_8 = 0.74^{+0.12}_{-0.07}$, which is consistent with the recent measurements with galaxy clusters mentioned above. From CMB, 2dFGRS and Ly-$\alpha$ forest data, Spergel et al. (2003) obtained the $2\sigma$ constraint $m_\nu < 0.23\,\mathrm{eV}$ per neutrino. In a similar analysis including also SDSS galaxy clustering, Seljak et al. (2004) found $m_\nu < 0.13\,\mathrm{eV}$ for the lightest neutrino (at $2\sigma$). Estimates from neutrino oscillations suggest $m_\nu \approx 0.05\,\mathrm{eV}$ for at least one of two neutrino species, consistent with the Fukugita & Peebles (2004) estimate given above.

# 5 Dark energy

The present state of the cosmological tests is illustrated in Fig. 5 (left). The combination of the likelihood contours obtained with three different observational approaches (type-Ia SNe: Riess et al. 2004; CMB: Spergel et al. 2003; galaxy clusters: Schuecker et al. 2003b) shows that the cosmic matter density is close to $\Omega_\mathrm{m} = 0.3$, and that the normalized cosmological constant is around $\Omega_\Lambda = 0.7$. This sums up to unit total cosmic energy density and suggests a spatially flat universe. However, the density of cosmic matter growths with redshift like $(1 + z)^3$ whereas the density $\rho_\Lambda$ related to the cosmological constant $\Lambda$ is independent of $z$. The ratio

[3] Vauclair et al. (2003) could find a consistent solution between local and high redshift X-ray temperature distribution functions and the redshift distributions of distant X-ray cluster surveys using mass-temperature and luminosity-temperature relations. Their best model has $\Omega_\mathrm{m} > 0.85$ and $\sigma_8 = 0.455$, and the shape parameter, $\Gamma = \Omega_\mathrm{m}\,h \approx 0.1$. For a 'standard' $\Lambda$CDM model this implies $h < 0.12$, in conflict with many observations. However, they use a different family of power spectra and thus work outside the standard $\Lambda$CDM paradigm.

$\Omega_\Lambda/\Omega_{\rm m}$ today is close to unity and must thus be a finely-tuned infinitesimal constant $\Omega_\Lambda/(\Omega_{\rm m}(1+z_\infty)^3)$ set in the very early Universe (cosmic coincidence problem). An alternative hypothesis is to consider a time-evolving 'dark energy' (DE), where

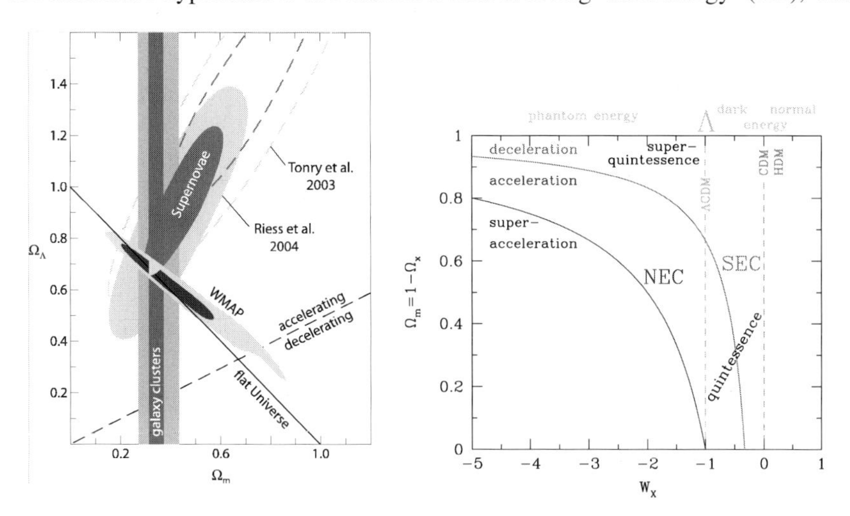


Figure 5: **Left:** Present situation of cosmological tests of the matter density $\Omega_{\rm m}$ and the normalized cosmological constant $\Omega_\Lambda$ from different observational approaches (Böhringer, priv. com.). **Right:** Null Energy Condition (NEC) and Strong Energy Condition (SEC) for a flat FL spacetime at redshift $z=0$ with negligible contributions from relativistic particles in the parameter space of the normalized cosmic matter density $\Omega_{\rm m}$ and the equation of state parameter of the dark energy $w_{\rm x}$. More details are given in the main text.

in Einstein's field equations the time-independent energy density $\rho_\Lambda$ of the cosmological constant is replaced by a time-dependent DE density $\rho_{\rm x}(t)$,

$$G_{\mu\nu} = -\frac{8\pi G}{c^4}\left[T_{\mu\nu} + \rho_{\Lambda\to{\rm x}}(t)\,c^2\,g_{\mu\nu}\right], \tag{5}$$

while assuming that the 'true' cosmological constant is either zero or negligible. Here, $G_{\mu\nu}$ is the Einstein tensor, $T_{\mu\nu}$ the energy-momentum tensor of ordinary matter, and $g_{\mu\nu}$ the metric tensor. For a time-evolving field (see, e.g., Ratra & Peebles 1988, Wetterich 1988, Caldwell et al. 1998, Zlatev, Wang & Steinhardt 1999, Caldwell 2002, recent review in Peebles & Ratra 2004) the aim is to understand the coincidence in terms of dynamics. A central rôle in these studies is assumed by the phenomenological ratio

$$w_{\rm x} = \frac{p_{\rm x}}{\rho_{\rm x}c^2} \tag{6}$$

(equation of state) between the pressure $p_{\rm x}$ of the unknown energy component and its rest energy density $\rho_{\rm x}$. Note that $w_{\rm x}=-1$ for Einstein's cosmological constant. The resulting phase space diagram of DE (Fig. 5, right) distinguishes different physical states of the two-component cosmic fluid – separated by two energy conditions of general relativity (Schuecker et al. 2003b).

Generally, assumptions on energy conditions form the basis for the well-known singularity theorems (Hawking & Ellis 1973), censorship theorems (e.g. Friedman et al. 1993) and no-hair theorems (e.g. Mayo & Bekenstein 1996). Quantized fields violate all local point-wise energy conditions (Epstein et al. 1965). In the present investigation we are, however, concerned with observational studies on macroscopic scales relevant for cosmology where $\rho_{\rm x}$ and $p_{\rm x}$ are expected to behave classically. Cosmic matter in the form of baryons and non-baryons, or relativistic particles like photons and neutrinos satisfy all standard energy conditions. The two energy conditions discussed below are given in a simplified form (see Wald 1984 and Barceló & Visser 2001).

The *Strong Energy Condition* (SEC): $\rho + 3p/c^2 \geq 0$ *and* $\rho + p/c^2 \geq 0$, derived from the more general condition $R_{\mu\nu}v^\mu v^\nu \geq 0$, where $R_{\mu\nu}$ is the Ricci tensor for the geometry and $v^\mu$ a timelike vector. The simplified condition is valid for diagonalizable energy-momentum tensors which describe all observed fields with non-zero rest mass and all zero rest mass fields except some special cases (see Hawking & Ellis 1973). The SEC ensures that gravity is always attractive. Certain singularity theorems (e.g., Hawking & Penrose 1970) relevant for proving the existence of an initial singularity in the Universe need an attracting gravitational force and thus assume SEC. Violations of this condition as discussed in Visser (1997) allow phenomena like inflationary processes expected to take place in the very early Universe or a moderate late-time accelerated cosmic expansion as suggested by the combination of recent astronomical observations (Fig. 5 left). Likewise, phenomena related to $\Lambda > 0$ and an effective version of $\Lambda$ whose energy and spatial distribution evolve with time (*quintessence*: Ratra & Peebles 1988, Wetterich 1988, Caldwell et al. 1998 etc.) are allowed consequences of the breaking of SEC – but not a prediction. However, a failure of SEC seems to have no severe consequences because the theoretical description of the relevant physical processes can still be provided in a canonical manner. Phenomenologically, violation of SEC means $w_{\rm x} < -1/3$ for a *single* energy component with density $\rho_{\rm x} > 0$. For $w_{\rm x} \geq -1/3$, SEC is not violated and we have a decelerated cosmic expansion.

The *Null Energy Condition* (NEC): $\rho + p/c^2 \geq 0$, derived from the more general condition $G_{\mu\nu}k^\mu k^\nu \geq 0$, where $G_{\mu\nu}$ is the geometry-dependent Einstein tensor and $k^\mu$ a null vector (energy-momentum tensors as for SEC). Violations of this condition are recently studied theoretically in the context of macroscopic traversable wormholes (see averaged NEC: Flanagan & Wald 1996, Barceló & Visser 2001) and the Holographic Principle (Sect. 6). The breaking of this criterion in a finite local region would have subtle consequences like the possibility for the creation of "time machines" (e.g. Morris, Thorne & Yurtsever 1988). Violating the energy condition in the cosmological case is not as dangerous (no threat to causality, no need to involve chronology protection, etc.), since one cannot isolate a chunk of the energy to power such exotic objects. Nevertheless, violation of NEC on cosmological scales could excite phenomena like super-acceleration of the cosmic scale factor (Caldwell 2002). Theoretically, violation of NEC would have profound consequences not only for cosmology because all point-wise energy conditions would be broken. It cannot be achieved with a canonical Lagrangian *and* Einstein gravity. Phenomenologically, violation of NEC means $w_{\rm x} < -1$ for a *single* energy component with $\rho_{\rm x} > 0$. The sort of energy related to this state of a Friedmann-Robertson-Walker (FRW) spacetime is dubbed *phantom energy* and is described by *super-quintessence* models

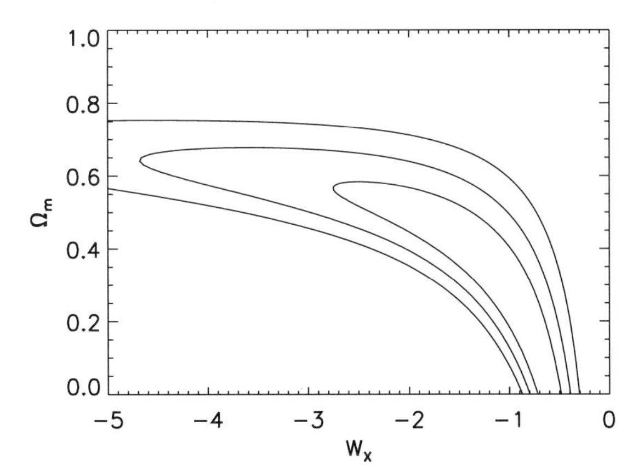


Figure 6: Likelihood contours ($1 - 3\sigma$) obtained with the Riess et al. (1998) sample of type-Ia SNe. The luminosities are corrected with the $\Delta m_{15}$ method. The equation of state parameter $w_{\rm x}$ is assumed to be redshift-independent.

(Caldwell 2002, see also Chiba, Okabe & Yamaguchi 2000). For $w_{\rm x} \geq -1$ NEC is not violated, and is described by *quintessence* models.

Assuming a spatially flat FRW geometry, $\Omega_{\rm m} + \Omega_{\rm x} = 1$, and $\Omega_{\rm m} \geq 0$ as indicated by the astronomical observations in Fig. 5 (left), the formal conditions for this two-component cosmic fluid translates into $w_{\rm x} \geq -1/3(1 - \Omega_{\rm m})$ for SEC, and $w_{\rm x} \geq -1/(1 - \Omega_{\rm m})$ for NEC (curved lines in Fig. 5 right). These energy conditions, characterizing the possible phases of a mixture of dark energy and cosmic matter, thus rely on the precise knowledge of $\Omega_{\rm m}$ and $w_{\rm x}$. Unfortunately, the effects of $w_{\rm x}$ are not very large. However, a variety of complementary observational approaches and their combination helps to reduce the measurement errors significantly.

The most direct (geometric) effect of $w_{\rm x}$ is to change cosmological distances. For example, for a spatially flat universe, comoving distances, $a_0 r = \int_0^z \frac{c dz'}{H(z')}$, are directly related to $w_{\rm x}$ via

$$\left[\frac{H(z)}{H_0}\right]^2 = \Omega_{\rm m}(1+z)^3 + (1 - \Omega_{\rm m}) \exp\left\{3 \int_0^z [1 + w_{\rm x}(z')]\, d\ln(1+z')\right\}. \quad (7)$$

A less negative $w_{\rm x}$ increases the Hubble parameter and thus reduces all cosmic distances. In general, $w_{\rm x}$ must evolve in time. To discuss Eq. (7) in terms of the resulting parameter degeneracy, let us assume $w_{\rm x}(z) = w_0 + w_1 \cdot z$ with the additional constraint that $w_0 = -1$ implies $w_1 = 0$. For this simple parameterization the same expansion rate at $z$ is obtained when $w_0$ and $w_1$ are related by $w_1 = -\frac{\ln(1+z)}{z - \ln(1+z)}(1 + w_0)$. The parameter degeneracy between $w_0$ and $w_1$ is a generic feature and can be seen in many proposed observational tests. Fortunately, its slope depends on $z$, so that the degeneracy can be broken with independent observations covering a large redshift range. Current observations have not the sensitivity to measure $w_0$ and $w_1$ separately so that basically all published measurements of the equation of state of the DE are on $w_0$ assuming $w_1 = 0$. The danger with this

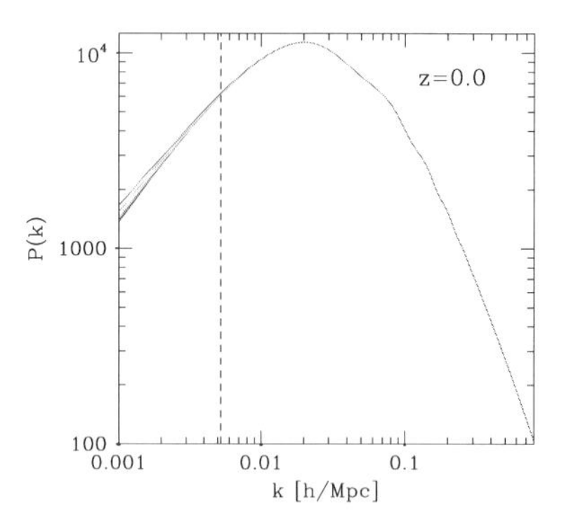


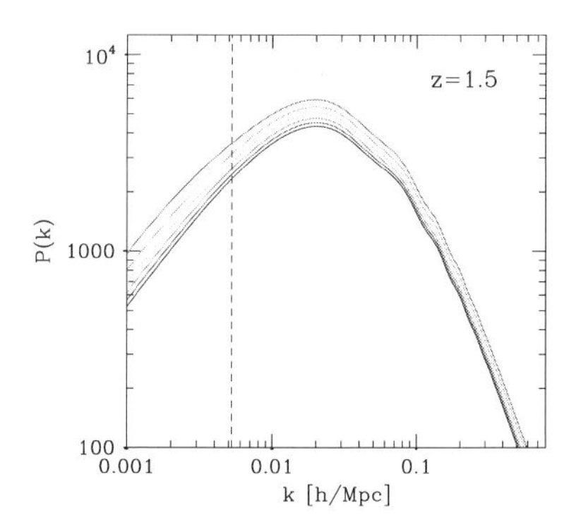


Figure 7: Matter power spectrum (left panel) and its evolution (right panel) for different redshift-independent equations of state $-1 \leq w_{\rm x} < 0$ of the DE. The lower curve in each panel is for $w_{\rm x} = -1$ and increases in amplitude with $w_{\rm x}$.

assumption is, however, that if the true $w_1$ would strongly deviate from zero then the estimated $w_0$ would be biased correspondingly (Maor et al. 2002). In addition, even when an explicit redshift dependency of $w_{\rm x}$ could be neglected, some parameter degeneracy between $\Omega_{\rm m}$ and $w_{\rm x}$ remains as suggested by Eq. (7) (see Fig. 6 obtained with the type-Ia SNe).

Structure growth via gravitational instability provides a further probe of $w_{\rm x}$. DE, not in form of a cosmological constant or vacuum energy density, is inhomogenously distributed - a smoothly distributed, time-varying component is unphysical because it would not react to local inhomogeneities of the other cosmic fluid and would thus violate the equivalence principle. An evolving scalar field with $w_{\rm x} < 0$ (e.g. quintessence) automatically satisfies these conditions (Caldwell, Dave & Steinhardt 1998a). The field is so light that it behaves relativistically on small scales and non-relativistically on large scales. The field may develop density perturbations on Gpc scales where sound speeds $c_{\rm s}^2 < 0$, but does not clump on scales smaller than galaxy clusters. Generally, perturbations come in either linear or nonlinear form depending on whether the density contrast, $\delta = (\rho/\bar{\rho}) - 1$, is smaller or larger than one.

In the linear regime, and when DE is modeled as a dynamical scalar field, the rate of growth of linear density perturbations in the CDM is damped by the Hubble parameter, $\delta''_{\rm cdm} + aH\delta'_{\rm cdm} = 4\pi G a^2 \delta\rho_{\rm cdm}$ ($a$ means scale factor and prime derivative with respect to conformal time). This evolution equation can be solved approximately by $\frac{d\ln\delta_{\rm cdm}}{d\ln a} \approx \left[1 + \frac{\rho_{\rm x}(a)}{\rho_{\rm cdm}(a)}\right]^{-0.6}$ (Caldwell, Dave & Steinhardt 1998b), provided that $\rho_{\rm x} < \rho_{\rm r}$ at radiation-matter equality. It is seen that $\rho_{\rm x}(a)$ and thus a more positive $w_{\rm x}$ delays structure growth. To reach the same fluctuations in the CDM field, structures must have formed at higher $z$ compared to the standard CDM model. For a redshift-independent $w_{\rm x}$, transfer and growth functions can be found in Ma et al. (1999). The effects of a constant $w_{\rm x}$ on $P(k)$ are shown in Fig. 7. The sensitivity of CMB anisotropies to $w_{\rm x}$ is limited to the integrated Sachs-Wolfe effect because $\Omega_{\rm x}$ dominates only at late $z$ (Eq. 7). Spergel et al. (2003) showed that the

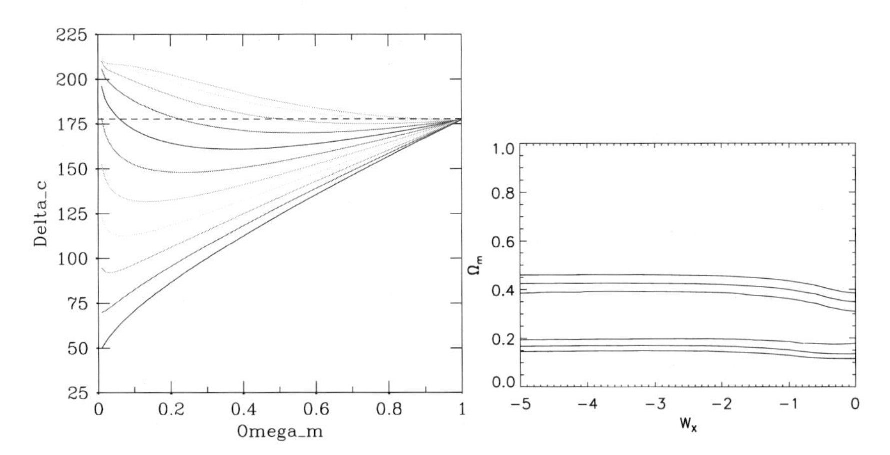


Figure 8: **Left:** Virial density in units of the critical matter density for a flat universe as a function of $\Omega_{\rm m}$ and $w_{\rm x}$. The $w_{\rm x}$ values range from $-1$ (lower curve) to zero (upper curve). **Right:** Likelihood contours (1-3$\sigma$) obtained from nearby cluster counts (REFLEX: Schuecker et al. 2003b) assuming a constant $w_{\rm x}$ and marginalized over $0.5 < \sigma_8 < 1$.

WMAP data could equally well fit with $\Omega_{\rm m} = 0.47$, $h = 0.57$, and $w_{\rm x} = -0.5$ once $w_{\rm x}$ is regarded as a free (constant) parameter.

In the nonlinear regime, the effects of DE are not very large. For the cosmological constant, Lahav et al. (1991) used the theory of peak statistics in Gaussian random fields and linear gravitational-instability theory in the linear regime and the spherical infall model to evolve the profiles to the present epoch. They found that the local dynamics around a cluster at $z = 0$ does not carry much information about $\Lambda$. However, DM haloes have core densities correlating with their formation epoch. Therefore, when $w_{\rm x}$ delays structure growth, then DM haloes are formed at higher $z$ with higher core densities and should thus appear for fixed mass and redshift more concentrated in $w_{\rm x} > -1$ models compared to $\Lambda$. This is reflected in the virial densities of collapsed objects in units of the critical density shown in Fig. 8 (left). The first semi-analytic computations of a spherical collapse in a fluid with DE with $-1 \leq w_{\rm x} < 0$ were performed by Wang & Steinhardt (1998). Schuecker et al. (2003b) enlarged the range to $-5 < w_{\rm x} < 0$, whereas Mota & van de Bruck (2004) discussed the spherical collapse for specific potentials of scalar fields. For recent simulations see Klypin et al. (2003) and Bartelmann et al. (2004).

These arguments have to be combined with the general discussion of Eq. (1) to understand the sensitivity of cluster counts on $w_{\rm x}$. Keeping the present-day cluster abundance and lower mass limit $M_{\rm min}$ in Eq. (1) fixed, the dominant effect of $w_{\rm x}$ comes from structure growth and volume (Haiman, Mohr & Holder 2001). For a larger $w_{\rm x}$, the DE field delays structure growth so that the number of distant clusters increases. However, a large $w_{\rm x}$ yields a small comoving count volume for the clusters which counteracts the growth effect. The compensation works mainly at small $z$ and leads to a comparatively small sensitivity of cluster counts at $z < 0.5$ on $w_{\rm x}$. For $z > 0.5$, the effect of a delayed structure growth starts to dominate and the number of high-$z$ clusters increases with $w_x$. However, the realistic case is when a redshift and cosmology-dependent lower mass limit is included. In this case, it could be shown that at high $z$, the $w_x$-dependence of the redshift distribution is mainly caused by the $w_x$-dependence of the lower mass limit in the sense that a larger $w_x$ decreases distances and therefore increases the number of high-$z$ clusters, whereas at small redshifts no strong dependency beyond the standard $\Omega_m$-$\sigma_8$ degeneracy remains. The inclusion of a $z$-dependent mass limit does only slightly damp the sensitivity on $\Omega_m$.

This high-$z$ behaviour of the number of clusters is very important for future planned cluster surveys (e.g. DUO Griffiths et al. 2004) where in the wide (northern) survey about 8 000 clusters will be detected over 10 000 square degrees on top of the SDSS cap up to $z = 1$, and where in the deep (southern) survey about 1 800 clusters will be detected over 176 square degrees up to $z = 2$ (if they exists at such high redshifts). REFLEX has most clusters below $z = 0.3$. For a constant $w_x$ the likelihood contours are shown in Fig. 8 (right) as a function of $\Omega_m$ (Schuecker et al. 2003b). The effects of yet unknown possible systematic errors are included by using a very large range of $\sigma_8$ priors ($0.5 < \sigma_8 < 1.0$). As expected, the $w_x$ dependence is very weak.

The past examples (Fig. 6 and Fig. 8 right) have shown that presently neither SNe nor galaxy clusters alone give an accurate estimate of the redshift-independent part of $w_x$. This is also true for CMB anisotropies. However, the resulting likelihood contours of SNe and galaxy clusters appear almost orthogonal to each other in the high-$w_x$ range. Their combination thus gives a quite strong constraint on both $w_x$ and $\Omega_m$ (Fig. 9 left). This is a typical example of cosmic complementarity which stems from the fact that SNe probe the homogeneous Universe whereas galaxy clusters test the inhomogeneous Universe as well. The final result of the combination of different SNe samples and REFLEX clusters yields the $1\sigma$ constraints $w_X = -0.95 \pm 0.32$ and $\Omega_m = 0.29 \pm 0.10$ (Schuecker et al. 2003b). Averaging all results obtained with REFLEX and various SN-samples yields $w_x = -1.00^{+0.18}_{-0.25}$ (Fig. 9 left). The figure shows that the measurements suggest a cosmic fluid that violates SEC and fulfills NEC. In fact, the measurements are quite consistent with the cosmological constant and leave not much room for any exotic types of DE. The violation of the SEC gives a further argument that we live in a Universe in a phase of accelerated cosmic expansion.

Ettori, Tozzi & Rosati (2003) used the baryonic gas mass fraction of clusters in the range $0.72 \leq z \leq 1.27$ and obtained $w_x \leq -0.49$. The combination with SN data yields $w_x < -0.89$, erroneously referring to the constraint $w_x \geq -1$. Henry (2004) used the X-ray temperature function and found $w_x = -0.42 \pm 0.21$, assuming $w_x \geq -1.0$. In a preliminary analysis, Sereno & Longo (2004) used angular diameter distance ratios of lensed galaxies in rich clusters, and shape parameters of surface brightness distributions and gas temperatures from X-ray data, and obtained $w_x = -0.83 \pm 0.14$, assuming $w_x \geq -1.0$. Rapetti, Allen & Weller (2004) combined cluster X-ray gas mass fractions with WMAP data and obtained the constraints $w_x = -1.05 \pm 0.11$. A formal average of the most accurate und unconstrained $w_x$ measurements using galaxy clusters (Schuecker et al. 2003b, Rapetti et al. 2004) gives

$$w_x = -1.00 \pm 0.05 \,. \tag{8}$$

Lima, Cunha & Alcaniz (2003) give a summary of the results of the $w_\mathrm{x}$-$\Omega_\mathrm{m}$ tests obtained with various methods, all assuming a redshift-independent $w_\mathrm{x}$. A clear trend is seen that $w_\mathrm{x} > -0.5$ is ruled out by basically all observations. The large degeneracy seen in Fig. 6 (left) towards $w_\mathrm{x} < -1$ translates into a less well-defined lower bound. Hannestad & Mörtsell (2002) found $w_\mathrm{x} > -2.7$ by the combination of CMB, SNe and large-scale structure data.

Melchiorri et al. (2003) combined seven CMB experiments including WMAP with the Hubble parameter measurements from the Hubble Space Telescope and luminosity measurements of type-Ia SNe, and found the 95% confidence range $-1.45 < w_\mathrm{x} < -0.74$. If they include also 2dF data on the large-scale distribution of galaxies they found $-1.38 < w_\mathrm{x} < -0.82$. More recent measurements support the tendency that $w_\mathrm{x}$ is close to the value expected for a cosmological constant as found by the combination of REFLEX and SN data. Spergel et al. (2003) used a variety of different combinations between WMAP and galaxy data and obtained the $1\sigma$ corridor $w_\mathrm{X} = -0.98 \pm 0.12$. Riess et al. (2004) combined data from distant type-Ia SNe with CMB and large-scale structure data, and found $w_\mathrm{x} = -1.02^{+0.13}_{-0.19}$. Their results are also inconsistent with a rapid evolution of the DE. Combining Ly-$\alpha$ forest and bias analysis of the SDSS with previous constraints from SDSS galaxy clustering, the latest SN and WMAP data, Seljak et al. (2004) obtained $w_\mathrm{x} = -0.98^{+0.10}_{-0.12}$ at $z = 0.3$ (they also obtained $\sigma_8 = 0.90 \pm 0.03$). A combination of the $w_\mathrm{x}$ measurements of REFLEX, Rapetti et al. (2004), Spergel et al. (2003), Riess et al. (2004), and Seljak et al. (2004) yields $w_\mathrm{x} = -0.998 \pm 0.038$. Independent from this more or less subjective summary, it is still save to conclude that all recent measurements are consistent with a cosmological constant, and that the most precise estimates suggest that $w_\mathrm{x}$ is very close to $-1$. This points towards a model where DE behaves very similar to a cosmological constant, i.e., that the time-dependency of the DE cannot be very large. In fact, Seljak et al. have also tested $w_\mathrm{x}$ at $z = 1$, and found $w_\mathrm{x}(z = 1) = -1.03^{+0.21}_{-0.28}$ and thus no significant change with $z$.

Cluster abundance measurements have not yet reached the depth to be very sensitive to the normalized cosmological constant $\Omega_\Lambda$ or $\Omega_\mathrm{x}$. The most reliable estimates todate come from the X-ray gas mass fraction. Vikhlinin et al. (2003) used the cluster baryon mass as a proxy for the total mass, thereby avoiding the large uncertainties on the M/T or M/L relations, yielding with 17 clusters with $z \approx 0.5$ the degeneracy relation $\Omega_\mathrm{m} + 0.23\Omega_\Lambda = 0.41 \pm 0.10$. For $\Omega_\mathrm{m} = 0.3$, this would give $\Omega_\Lambda = 0.48 \pm 0.12$. Allen et al. (2002) obtained with the X-ray gas mass fraction in combination with the other measurements described above the constraint $\Omega_\Lambda = 0.95^{+0.48}_{-0.72}$. Ettori et al. (2003) obtained $\Omega_\Lambda = 0.94 \pm 0.30$, and Rapetti et al. (2004) $\Omega_\Lambda = 0.70 \pm 0.03$. Combining lensing and X-ray data, Sereno & Longo (2004) obtained $\Omega_\Lambda = 1.1 \pm 0.2$. The formal average and $1\sigma$ standard deviation of these measurements is

$$\Omega_\Lambda = 0.83 \pm 0.24\,. \tag{9}$$

The last effect of DE and thus $w_\mathrm{x}$ discussed here is interesting by its own, but also offers a possibility for cross-checks of $w_\mathrm{x}$ measurements. The effect is related to a possible non-gravitational interaction between DE and ordinary matter (e.g. Amendola 2000). We showed above (e.g., Eq. 8) that the most obvious candidate for DE is presently the cosmological constant with all its catastrophic problems (Sect. 6).

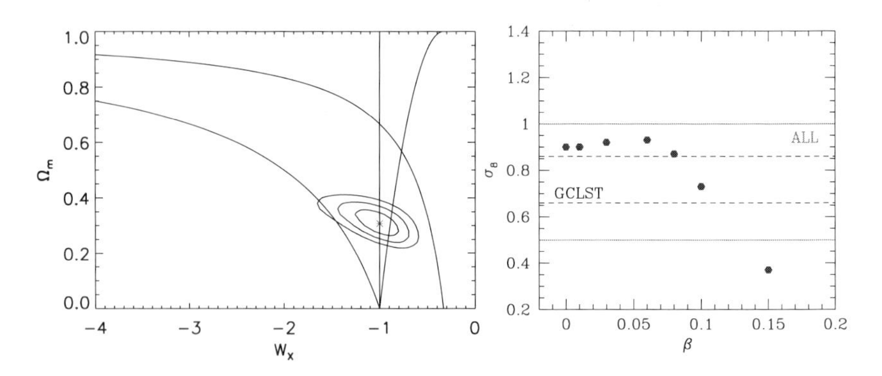


Figure 9: **Left:** Combination of $w_x$ measurements based various SN samples and the REFLEX sample assuming a redshift-independent $w_x$. The likelihood contours $(1-3\sigma)$ are centred around $w_x = -1$ which corresponds to the cosmological constant (vertical line). The two curved lines correspond to the SEC (upper line) and the NEC (lower line). The curved line in the right part of the diagram corresponds to a specific holographic DE model of Li (2004). **Right:** Normalization parameter of the matter power spectrum $\sigma_8$ compared to the coupling strength $\beta$ where $\beta = 0$ means no coupling between DE and DM. The inner region marked by the dashed horizontal lines (GCLST) marks observational constraints from the scatter of all $\sigma_8$ estimates obtained from galaxy clusters during the past 2 years. The broader range marked by the continuous horizontal lines (ALL) is a plausible interval which takes into account also $\sigma_8$ measurements from other observations.

However, a very small redshift-dependency of the DE density cannot be ruled out. Based on this possible residual effect, a further explanation would be a light scalar (quintessential) field $\phi$ where its potential can drive the observed accelerated expansion similar as in the de-Sitter phase of inflationary scenarios. In general, $\phi$ interacts beyond gravity to baryons and DM with a strength similar to gravity. However, some (unkown) symmetry could signficantly reduce the interaction (Carroll 1998) – otherwise it would have already been detected – so that some coupling could remain. The following discussion is restricted to a possible interaction between DE and DM.

The general covariance of the energy momentum tensor requires the sum of DM $(m)$ and DE $(\phi)$ to be locally conserved so that we can allow for a coupling of the two fluids, e.g., in the simple linear form,

$$\begin{aligned} T^{\mu}_{\nu(\phi);\mu} &= C(\beta)T_{(m)}\phi_{;\nu}\,, \\ T^{\mu}_{\nu(m);\mu} &= -C(\beta)T_{(m)}\phi_{;\nu}\,, \end{aligned} \tag{10}$$

with the dimensionless coupling constant $\beta$ in $C(\beta) = \sqrt{\frac{16\pi G}{3c^4}}\beta$, but more complicated choices are, however, possible. Observational constraints on the strength of a nonminimal coupling $\beta$ between $\phi$ and DM are $|\beta| < 1$ (Damour et al. 1990). For a given potential $V(\phi)$, the corresponding equation of motion of $\phi$ can be solved. Amendola (2000) discussed exponential potentials which yield a present accelerating phase. A generic result is a saddle-point phase between $z = 10^4$ and $z = 1$ where the normalized energy density related to the scalar field, $\Omega_\phi$, is significantly higher compared to noncoupling models. The saddle-point phase thus leads to a further suppression of structure growth and thus to smaller $\sigma_8$ (when the models are normalized with the CMB) compared to noninteracting quintessence models (Fig. 9 right). The present observations appear quite stringent. The X-ray cluster constraint $\sigma_8 = 0.76 \pm 0.10$ (Eq. 3) obtained in Sect. 4 suggests a clear detection of a nonminimal coupling between DE and DM:

$$\beta = 0.10 \pm 0.01 \,. \tag{11}$$

This would provide an argument that DE cannot be the cosmological constant because $\Lambda$ cannot couple non-gravitationally to any type of matter. In this case, the quite narrow experimental corridor found for $w_{\rm x}$ (Eq. 8) would be responsible for the nonminimal coupling. However, a possibly underestimated $\sigma_8$ by galaxy clusters, and thus no nonminimal couplings and a DE in form of a cosmological constant seem to provide a more plausible alternative (see Sect. 7).

# 6 The Cosmological Constant Problem

Recent measurements of the equation of state $w_{\rm x}$ of the DE do not leave much room for any exotic type of DE (Eq. 8 in Sect. 5). In this section we take the most plausible assumption that the observed accelerated cosmic expansion is driven by Einstein's cosmological constant more serious. In this case, we are, however, confronted with the long-standing cosmological constant problem (e.g., Weinberg 1989). To some extent also DE models based on scalar fields suffer on this problem because they have to find a physical mechanism (symmetry) which makes the value of $\Lambda$ negligible. To illustrate the problem, separate the effectively observed DE density as usual into a gravitational and non-gravitational part,

$$\rho_\Lambda^{\rm eff} = \rho_\Lambda^{\rm GRT} + \rho_\Lambda^{\rm VAC} = 10^{-26}\,{\rm kg\,m^{-3}}\,, \tag{12}$$

for $\Omega_\Lambda = 0.7$. The non-gravitational part represents the physical vacuum. A free scalar field offers a convinient way to get an estimate of a plausible vacuum energy density. Interpreting this field as a physical operator and thus constraining it to Heisenberg's uncertainty relations, quantize the field in the canonical manner. The quantized field behaves like an infinite number of free harmonic oscillators. The sum of their zero particle (vacuum) states, up to the Planck energy, corresponding to a cutoff in physical (not comoving) wavenumber, is

$$\rho_\Lambda^{\rm VAC} = \frac{\hbar}{c}\int_0^{E_{\rm p}/\hbar c} \frac{4\pi k^2 dk}{(2\pi)^3}\frac{1}{2}\sqrt{k^2+(mc/\hbar)^2} \approx 10^{+93}\,{\rm kg\,m^{-3}}\,, \tag{13}$$

for $m = 0$. The cosmological constant problem is the extra-ordinary fine-tuning which is necessary to combine the effectively measured DE density in Eq. (12) with the physical vacuum (13). This simple (though quite naive) estimate immediately shows that something fundamentally has gone wrong with the estimation of the physical vacuum in Eq. (13). An obvious answer is related to the fact that for the estimation of the physical vacuum, gravitational effects are completely ignored. One could think of a quantum gravity with strings. However, present versions of such theories seem to provide only arguments for a vanishing or a negative cosmological constant (Witten 2000, but see below).

A hint how inclusion of gravity could effectively work in Eq. (13), comes from black hole thermodynamics (Bekenstein 1973, Hawking 1976). Analyzing quantized particle fields in curved but not quantized spacetimes, it became clear that the information necessary to fully describe the physics inside a certain region and characterized by its entropy, increases with the surface of the region. This is in clear conflict to standard non-gravitational theories where entropy as an extensive variable always increases with volume. Non-gravitational theories would thus vastly overcount the amount of entropy and thus the number of modes and degrees of freedom when quantum effects of gravity become important. Later studies within a string theory context could verify a microscopic origin of the black hole entropy bound (Strominger & Vafa 1996). Bousso (2002) generalizes the prescription how entropy has to be determined even on cosmological scales, leading to the Covariant Entropy Bound. 't Hooft (1993) and Susskind (1995) elevated the entropy bound as the Holographic Principle to a new fundamental hypothesis of physics.

A simple intuitive physical mechanism for this holographic reduction of degrees of freedom is related to the idea that each quantum mode in Eq. (13) should carry a certain amount of gravitating energy. If the modes were packed dense enough, they would immediately collapse to form a black hole. The reduction of the degrees of freedom comes from the ignorance of these collapsed states. Later studies of Cohen, Kaplan & Nelson (1999), Thomas (2002), and Horvat (2004) made the exclusion of states inside their Schwarzschild radii more explict which further strengthen the entropy bound so that a new estimate of the physical vacuum is

$$\rho_\Lambda^{\rm HOL} = \frac{c^2}{8\pi G}\frac{1}{R_{\rm EH}^2} \approx 3\cdot 10^{-27}\,{\rm kg\,m^{-3}}\,, \tag{14}$$

where $R_{\rm EH}$ is the present event horizon of the Universe. This is, however, not a solution of the cosmological constant problem because gravity and the exclusion of microscopic black hole states were put in by hand and not in a self-consistent manner by a theory of quantum gravity. Nevertheless, the similarity of Eqs. (12) and (14) might be taken as a hint that gravitational hologpraphy could be relevant to find a more complete theory of physics.

A method to test for consistency of present observations with gravitational holography, is closely related to the fact that gravitational holography as tested with the Covariant Entropy Bound on cosmological scales is based on the validity of the Null Energy Condition (NEC). However, in contrast to the NEC as discussed in Sec. 5 for the total cosmic fluid, Kaloper & Linde (1999) could show that for the Covariance Entropy Bound each individual component of the cosmic substratum must obey

$$-1 \le w_i \le +1\,. \tag{15}$$

The problematic component is the equation of state of the dark energy. The observed values summarized in Sect. 5 suggest $w_{\rm x} = -1.00 \pm 0.05$ which is consistent with the bound (15). One can take this as the first consistency test of probably the most important assumption of the Holographic Principle on macroscopic scales. However, a direct measurement of *cosmological entropy* on light sheets as defined in Bousso (2002) is still missing.

Li (2004) recently combined holographic ideas with DE to 'solve' the cosmological constant problem. Applying the stronger entropy bound as suggested by Thomas (1998) and Cohen et al. (1999), and using the cosmic event horizon as a characteristic scale of the Universe, accelerating solutions of the cosmic scale factor at low $z$ could be found together with relations between the density of cosmic matter and $w_{\mathrm{x}}$ as shown in Fig. 9 (left). This model of holographic DE appears to be quite consistent with present observations and was in fact used in Eq. (14) to estimate the density of the physical vacuum.

't Hooft (1993) and Susskind (1995) give arguments suggesting that M-theory should satisfy the Holographic Principle. Horava (1999) in his 'conservative' approach to M-theory, defined by specific gauge symmetries and invariance under spacetime diffeomorphisms and parity, could show that the entropy bound and thus holography emerges quite naturally. Therefore, any astronomical test supporting gravitational holography more directly or some of its basic assumptions like the NEC as described above should give important hints towards the development of a more complete theory of physics.

There is a class of models based on higher dimensions which follow the Holographic Principle. Brane-worlds emerging from the model of Horava & Witten (1996a,b) are phenomenological realizations of M-theory ideas. Recent theoretical investigations concentrate on the Randall & Sundrum (1996a,b) models where gravity is used in an elegant manner to compactify the extra dimension. Some of these models also follow the Holographic Principle. Here, matter and radiation of the visible Universe are located on a $(1+3)$-dimensional brane. Expressed in a simplified manner, non-gravitational forces, described by open strings, are attached with their endpoints on branes. Gravity, described by closed strings, can propagate also into the $(1+4)$-dimensional bulk and thus 'dilutes' differently than Newton or Einstein gravity. Table-top experiments of classical gravity (and BBN) confine the size of an extra dimension to $< 0.16\,\mathrm{mm}$ (Hoyle et al. 2004). Einstein gravity formulated in a five dimensional spacetime and combined with a five-dimensional cosmic line element carrying the symmetries of the assumed brane-world, can yield FL-like solutions with the well-known phenomenology at low $z$ (Binetruy et al. 2000).

The analysis of perturbations in brane-world scenarios is not yet fully understood (Maartens 2004). Difficulties arise when perturbations created on the brane propagate into the bulk and react back onto the brane. Only on large scales are the computations under control because here the effects of the backreaction are small and can be neglected. It is thus not yet clear, whether the resulting effects on the power spectrum described below are mere reflections of such approximations or generic features of higher dimensions.

Brax et al. (2003) and Rhodes et al. (2003) discussed the effects of extra dimensions on CMB anisotropies and large scale structure formation. Models with extra dimensions can at low energies be described as scalar-tensor theories where the light scalar fields (moduli fields) couple to ordinary matter in a manner depending on the details of the higher dimensional theory. An illustration of the expected effects on

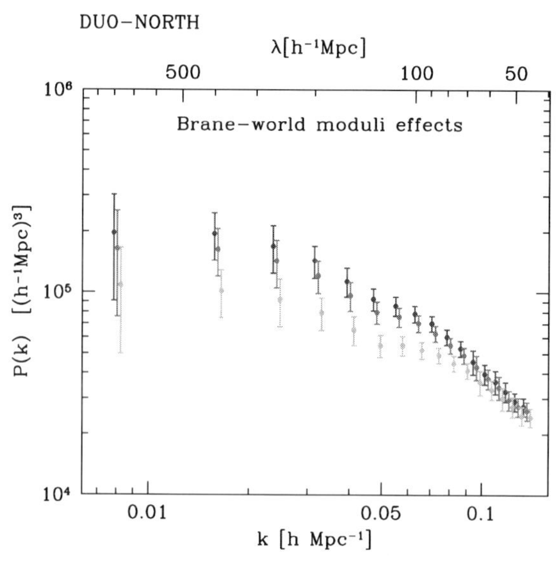


Figure 10: Predicted cluster power spectra based on matter power spectra of Rhodes et al. (2003). The effect of the extra-dimension decreases the $P(k)$ amplitudes at large scales. The error bars are typical for a DUO-like X-ray cluster survey. In order to show the differences more clearly, power spectra for each extra dimension are slidely shifted relative to each other along the comoving $k$ axis.

the cluster power spectrum is given in Fig. 10. The error bars are computed with cluster samples selected from the Hubble Volume Simulation under the conditions of the DUO wide survey (P. Schuecker, in prep.). It is seen that $P(k)$ gets flatter on scales around $300\,h^{-1}$ Mpc with increasing size of the extra dimension. A careful statistical analysis shows that more than 30 000 galaxy clusters are needed to clearly detect the presence of an extra dimension on scales below 0.16 mm.

# 7 Summary and conclusions

X-ray galaxy clusters give, in combination with other measurements, the observational constraints and their $1\sigma$ errors on the matter density $\Omega_{\rm m} = 0.31 \pm 0.03$, the normalized cosmological constant $\Omega_\Lambda = 0.83 \pm 0.23$, the normalization of the matter power spectrum $\sigma_8 = 0.76 \pm 0.10$, the neutrino energy density $\Omega_\nu = 0.006 \pm 0.003$, the equation of state of the DE $w_{\rm x} = -1.00 \pm 0.05$, and the linear interaction $\beta = 0.10 \pm 0.01$ between DE and DM. These estimates suggest a spatially flat universe with $\Omega_{\rm tot} = \Omega_{\rm m} + \Omega_\Lambda = 1.14 \pm 0.24$, as assumed in many cosmological tests based on galaxy clusters.

They do, however, not provide an overall consistent physical interpretation. The problem is related to the low $\sigma_8$ which leads to an overestimate of the neutrino mass compared to laboratory experiments and to an interaction between DE and DM. Such a high interaction is not consistent with a DE with $w_{\rm x} = -1.00 \pm 0.05$ because the latter indicates that DE behaves quite similar to a cosmological constant which cannot exchange energy beyond gravity.

A more convincing explanation is that $\sigma_8 = 0.76$ should be regarded as a lower limit so that DE would be the cosmological constant without any nonminimal couplings. Systematic underestimates of $\sigma_8$ by 10-20% are not unexpected from recent simulations (e.g., Randall et al. 2002, Rasia et al. 2004). Present data do not allow any definite conclusion, especially in the light of the partially obscured effects of non-gravitational processes in galaxy clusters and because of our ignorance of a possible time-dependency of $w_x$. However, the inclusion of further parameters obviously improves our abilities for consistency checks.

Energy conditions form the bases of many phenomena related to gravity and holography. M-theory should also come holographic, as well as brane-world gravity as a phenomenological realization of M-theory ideas. Tests of the resulting cosmologies will in the end confront alternative theories of quantum gravity. Observational tests on cosmological scales as illustrated by the effects of an extra-dimension on the cluster power spectrum probably need the 'ultimate' cluster survey, i.e. a census of possibly all $10^6$ rich galaxy clusters which might exist down to redshifts of $z = 2$ in the visible Universe.

*Acknowledgements: I would like to thank Hans Böhringer and the REFLEX team for our joint work on galaxy clusters and cosmology.*

## References

Allen, S. W., Schmidt, R. W., Fabian, A. C. 2002, MNRAS, 334, L11

Allen, S. W., Schmidt, R. W., Fabian, A. C., Ebeling, H. 2003, MNRAS, 344, 43

Allen, S. W., Schmidt, R. W., Bridle, S. L. 2003, MNRAS, 346, 593

Allen, S. W., Schmidt, R. W., Ebeling, H., Fabian, A. C., van Speybroeck, L. 2004, MNRAS, 353, 457

Amendola, L. 2000, PhRvD, 62, 043511

Amossov, G., Schuecker, P. 2004, A&A, 421, 425

Ashie, Y. et al. 2004, PhRvL, 93, 101801

Bahcall, N. A. 1999, in Formation of structure in the universe, eds. A. Dekel & J. P. Ostriker, Cambridge Univ. Press, Cambridge, p. 135

Bahcall, N. A., Fan, X. 1998, ApJ, 504, 1

Bahcall, N. A. et al. 2003, ApJ, 585, 182

Barceló, C., Visser, M. 2001, PhLB, 466, 127

Bartelmann, M., Dolag, K., Perrotta, F., Baccigalupi, C., Moscardini, L., Meneghetti, M., Tormen, G. 2004, astro-ph/0404489

Bekenstein, J. 1973, PhRvD, 9, 3292

Binetruy, P., Deffayet, C., Ellwanger, U., Langois, D. 2000, PhLB, 477, 285

Birkingshaw, M., Gull, S. F., Hardebeck, H. 1984, Nature, 309, 34

Böhringer, H., Voges, W., Fabian, A. C., Edge, A. C., Neumann, D. M. 1993, MNRAS, 264, L25

Böhringer, H. et al. 2000, ApJS, 129, 35

Böhringer, H. et al. 2001, A&A, 369, 826

Böhringer, H. et al. 2002, ApJ, 566, 93

Böhringer, H. et al. 2004, A&A, 425, 367

Bond, J. R. 1995, PhRvL, 74, 4369

Borgani, S., Guzzo, L. 2001, Nature, 409, 39

Borgani, S., Rosati, P., Tozzi, P., Norman, C. 1999, ApJ, 517, 40

Borgani, S. et al. (2001), ApJ, 561, 13

Borgani, S., Murane, G., Springel, V., Diaferio, A., Dolag, K., Moscardini, L., Tormen, G., Tornatore, L., Tozzi, P. 2004, MNRAS, 348, 1078

Bousso, R. 2002, RvMP, 74, 825

Brax, Ph., van de Bruck, C., Davis, A.-C., Rhodes, C. S. 2003, PhRvD, 67, 023512

Briel, U. G., Henry, J. P., Böhringer, H. 1992, A&A, 259, L31

Briel, U. G., Finoguenov, A., Henry, J. P. 2004, A&A, 426, 1

Burles, S., Nollett, K. M., Turner, M. S. 2001, ApJL, 552, L1

Caldwell, R. R. 2002, PhLB, 545, 17

Caldwell, R. R., Dave, R., Steinhardt, P. J. 1998a, PhRvL, 80, 1582

Caldwell, R. R., Dave, R., Steinhardt, P. J. 1998b, Ap&SS, 261, 303

Carlstrom, J. E., Holder, G. P., Reese, E. D. 2002, ARAA, 40, 643

Carroll, S. M. 1998, PhRvL, 81, 3067

Chaboyer, B., Krauss, M. 2002, ApJL, 567, L45

Chiba, T., Okabe, T., Yamaguchi, M. 2000, PhRvD, 62, 023511

Cohen, A. G., Kaplan, D. B., Nelson, A. E. 1999, PhRvL, 82, 4971

Collins, C. A. et al. 2000, MNRAS, 319, 939

Cruz, M., Martinez-Gonzalez, E., Vielva, P., Cayon, L. 2004, MNRAS (in press)

Damour, T., Gibbons, G. W., Gundlach, C. 1990, PhRvL, 64, 123

David, L. P., Jones, C., Forman, W. 1995, ApJ, 445, 578

Ebeling, H., Edge, A. C., Böhringer, H., Allen, S. W., Crawford, C. S., Fabian, A. C., Voges, W., Huchra, J. P. 1998, MNRAS, 301, 881

Edge, A. C. 2004, in Clusters of Galaxies, eds. J. S. Mulchaey, A. Dressler, and A. Oemler, Cambridge Univ. Press, Cambridge, p. 58

Efstathiou, G., Frank, C. S., White, S. D. M., Davis, M. 1988, MNRAS, 235, 715

Eke, V. R., Cole, S., Frenk, C. S. 1996, MNRAS, 282, 263

Epstein, H., Glaser, V., Jaffe, A. 1965, NCim, 36, 2296

Ettori, S., Fabian, A.C. 1999, MNRAS, 305, 834

Ettori, S., Tozzi, P., Rosati, P. 2004, A&A, 398, 879

Ettori, S. et al. 2004, MNRAS, 354, 111

Evrard, A. E. 1997, MNRAS, 292, 289

Evrard, A. E., Metzler, C. A., Navarro, J. F. 1996, ApJ, 469, 494

Fabian, A. C. et al. 2000, MNRAS, 318, L65

Fabian, A. C., Sander, J. S., Allen, S. W., Crawford, C. S., Iwasawa, K., Johnstone, R. M., Schmidt, R. W., Taylor, G. B. 2003, MNRAS, 344, L43

Feretti, L., Gioia, I. M., Giovannini, G., I. Gioia 2002, Merging processes in galaxy clusters, Eds., Astrophysics and Space Science Library, Vol. 272, Kluwer Academic Publisher, Dordrecht

Flanagan, É.É., Wald, R. M. 1996, PhRvD, 54, 6233

Forman, W. et al. 2003, ApJ (submitted), astro-ph/0312576

Friedman, J. L., Schleich, K., Witt, D. M. 1993, PhRvL, 71, 1486

Fukuda, Y. et al. 1998, PhRvL, 81, 1562

Fukugita, M., Liu, G.-C., Sugiyama, N. 2000, PhRvL, 84, 1082

Fukugita, M., Peebles, J. P. E. 2004, astro-ph/0406095

Gladders, M. D., Yee, D., Howard, K. C. 2004, ApJS (in press), astro-ph/0411075

Goto, T. et al. 2002, AJ, 123, 1807

Griffiths, R. E., Petre, R., Hasinger, G. et al. 2004, in Proc. SPIE conference (submitted)

Haiman, Z., Mohr, J. J., Holder, G. P. 2001, ApJ, 553, 545

Hawking, S. W., Penrose, R. 1970, in Proc. of the Royal Society of London. Series A. Mathematical and Physical Sciences. Vol. 314, Issue 1519, p. 529

Hawking, S. W., Ellis, G. F. R. 1973, The large scale structure of space-time, Cambridge Monographs on Mathematical Physics, Cambridge Univ. Press, London

Hawking, S. W. 1976, PhRvD, 13, 191

Henry, J. P. 2004, ApJ, 609, 603

Horava, P. 1999, PhRvD, 59, 046004

Horava, P., Witten, E. 1996a, NuPhB, 460, 506

Horava, P., Witten, E. 1996b, NuPhB, 475, 94

Horvat, R. 2004, PhRvD, 70, 087301

Hoyle, C. D., Kapner, D. J., Heckel, B. R., Adelberger, E. G., Gundlach, J. H., Schmidt, U., Swanson, H. E. 2004, PhRvD, 70, 042004

Ikebe, Y., Reiprich, T. H., Böhringer, H., Tanaka, Y., Kitayama, T. 2002, A&A, 383, 773

Jenkins, A., Frenk, C. S., White, S. D. M. et al. 2001, MNRAS, 321, 372

Kaiser, N. 1984, ApJL, 284, L9

Kaiser, N. 1986, MNRAS, 222, 323

Kaiser, N. 1987, MNRAS, 227, 1

Kaiser, N., Squires, G. 1993, ApJ, 404, 441

Kaloper, N., Linde, A. 1999, PhRvD, 60, 103509

Komatsu, E. et al. 2003, ApJS, 148, 119

Kim, R. S. J. et al. 2002, AJ, 123, 20

Klypin, A., Maccio, A. V., Mainini, R., Bonometto, S. A. 2003, ApJ, 599, 31

Lacey, C. G., Cole, S. M. 1993, MNRAS, 262, 627

Lacey, C. G., Cole, S. M. 1994, MNRAS, 271, 676

Lahav, O., Rees, M. J., Lilje, P. B., Primack, J. R. 1991, MNRAS, 251, 128

Li, M. 2004, Phys. Lett. B (submitted), astro-ph/0403127

Lima, J. A. S., Cunha, J. V., Alcaniz, J. S. 2003, PhRvD, 68, 023510

Ma, C.-P., Caldwell, R. R., Bode, P., Wang, L. 1999, ApJ, 521, L1

Maartens, R. 2004, LRR, 7, 7

Maor, I., Brustein, R., McMahon, J., Steinhardt, P. J. 2002, PhRvD, 65, 123003

Matarrese, S., Coles, P., Lucchin, F., Moscardini, L. 1997, MNRAS, 286, 115

Mather, J. C. et al. 1990, ApJ, 354, L37

Mayo, A. E., Bekenstein, J. D. 1996, PhRvD, 54, 5059

Melchiorri, A., Mersini, L., Ödman, C. J., Trodden, M. 2003, PhRvD, 68, 043509

Mo, H. J., White, S. D. M. 1996, MNRAS, 282, 347

Morris, M. S., Thorne, K. S. Yurtsever, U. 1988, PhRvL, 61, 1446

Mota, D. F., van de Bruck, C. 2004, A&A, 421, 71

Mullis, C. R., Rosati, P., Lamer, G., Böhringer, H., Schwope, A., Schuecker, P., Fassbender, R. 2005, ApJL, 623, 85

Peebles, P. J. E. 1980, The Large-Scale Structure of the Universe, Princeton Univ. Press, Princeton

Peebles, P. J. E. 1993, Principles of Physical Cosmology, Univ. Press, Princeton, Princeton

Peebles, P. J. E., Ratra, B. 2004, RvMP, 75, 559

Pierpaoli, E., Borgani, S., Scott, D., White, M. 2003, MNRAS, 242, 163

Ponman, T. J., Cannon, D. B., Navarro, J. F. 1999, Nature, 397, 135

Pope, A. C. et al. 2004, ApJ, 607, 655

Postman, M., Lubin, L. M., Gunn, J. E., Oke, J. B., Hoessel, J. G., Schnieder, D. P., Christensen, J. A. 1996, AJ, 111, 615

Randall, L., Sundrum, R. 1996a, PhRvL, 83, 3370

Randall, L., Sundrum, R. 1996b, PhRvL, 83, 4690

Randall, S. W., Sarazin, C. L., Ricker, P. M. 2002, ApJ, 577, 579

Rapetti, D., Allen, S. W., Weller, J. 2004, MNRAS (submitted), astro-ph/0409574

Rasia, E., Mazzotta, P., Borgani, S., Moscardini, L., Dolag, K., Tormen, G., Diaferio, A., Murante, G. 2004, ApJL (submitted), astro-ph/0409650

Ratra, B., Peebles, P. J. E. 1988, PhRvD, 37, 3406

Reiprich, T. H., Böhringer, H. 2002, ApJ, 567, 716

Rhodes, C. S., van de Bruck, C., Brax, Ph., Davis, A.-C. 2003, PhRvD, 68, 3511

Richstone, D., Loeb, A., Turner, E. 1992, ApJ, 363, 477

Riess, A. G., Filippenko, A. V., Challis, P. et al. 1998, AJ, 116, 1009

Riess, A. G. et al. 2004, ApJ, 607, 665

Rosati, P., Borgani, S., Norman, C. 2002, ARAA, 40, 539

Szalay, A. S. et al. 2003, ApJ, 591, 1

Schuecker, P., Böhringer, H. 1998, A&A, 339, 315

Schuecker, P., Böhringer, H., Arzner, K., Reiprich, T. H. 2001a, A&A, 370, 715

Schuecker, P., Böhringer, H., Reiprich, T. H., Feretti, L. 2001b, A&A, 378, 408

Schuecker, P. et al. 2001c, A&A, 368, 86

Schuecker, P., Guzzo, L., Collins C. A., Böhringer, H. 2002, MNRAS, 335, 807

Schuecker, P., Böhringer, H., Collins, C. A., Guzzo, L. 2003a, A&A, 398, 867

Schuecker, P., Caldwell, R. R., Böhringer, H., Collins, C. A., Guzzo, L., Weinberg, N. N. 2003b, A&A, 402, 53

Schuecker, P., Böhringer, H., Voges, W. 2004, A&A, 420, 425

Schuecker, P., Finoguenov, A., Miniati, F., Böhringer, H., Briel, U. G. 2004, A&A, 426, 387

Seljak, U. et al. 2004, PhRvD (submitted) astro-ph 0407372

Sereno, M., Longo, G. 2004, MNRAS, 354, 1255

Sheth, R. K., Tormen, G. 2002, MNRAS, 329, 61

Spergel, D. et al. 2003, ApJS, 148, 175

Springel, V. et al. 2005, Nature (in press), astro-ph/0504097

Steigman, G., 2002 as cited in Peebles, P. J. E., Ratra, B. 2003, RvMP, 75, 559

Strominger, A., Vafa, C. 1996, PhL B, 379, 99

Susskind, L. 1995, JMP, 36, 6377

Suwa, T., Habe, A., Yoshikawa, K., Okamoto, T. 2003, ApJ, 588, 7

Tegmark, M., Zaldarriaga, M. 2002, PhRvD, 66, 103508

Thomas, S. 2002, PhRvL, 89, 081301

'tHooft, G. 1993, in Salamfestschrift: a collection of talks, World Scientific Series in 20th Century Physics, Vol. 4, eds. A. Ali, J. Ellis and S. Randjbar-Daemi, World Scientific, 1993, e-print gr-qc/9310026

Vauclair, S. C. et al. 2003, 412, 37

Viana, P. T. P., Liddle, A. R. 1996, MNRAS, 281, 323

Vikhlinin, A., Markevitch, M., Murray, S. S. 2001, ApJ, 551, 160

Vikhlinin, A. et al. 2003, ApJ, 590, 15

Visser, M. 1997, PhRvD, 56, 7578

Vogeley, M. S., Szalay, A. S. 1996, ApJ, 465, 34

Voit, G. M. 2004 RMP (in press), astro-ph/0410173

Wald, R. M. 1984, General Relativity, The University of Chicago Press, Chicago and London

Wang, L., Steinhardt, P. J. 1998, ApJ, 508, 483

Weinberg, S. 1989, RvMP, 61, 1

Wetterich, C. 1988, NuPhB, 302, 668

White, S. D. M., Frenk, C. S. 1991, ApJ, 379, 52

White, S. D. M., Efstathiou, G., Frank, C. S. 1993, MNRAS, 262, 1023

White, S. D. M., Navarro, J. F., Evrard, A. E., Frank, C. S. 1993, Nature, 366, 429

White, S. D. M., Fabian, A. C. 1995, MNRAS, 273, 72

Witten, E. 2000, hep-ph/0002297

Wu, X.-P., Chiueh, T., Fang, L.-Z., Xue, Y.-J. 1998, MNRAS, 301, 861

Zhang, Y.-Y., Finoguenov, A., Böhringer, H., Ikebe, Y., Matsushita, K., Schuecker, P. 2004, A&A, 413, 49

Zlatev, I., Wang, L., Steinhardt, P. J. 1999, PhRvL, 82, 896

# The Evolution of Field Spiral Galaxies over the Past 8 Gyr

Asmus Böhm and Bodo L. Ziegler

University Observatory Göttingen
Geismarlandstr. 11, 37083 Göttingen, Germany
boehm@uni-sw.gwdg.de

**Abstract**

*We have performed a large observing campaign of intermediate–redshift disk galaxies including multi–object spectroscopy with the FORS instruments of the Very Large Telescope and imaging with the Advanced Camera for Surveys onboard the Hubble Space Telescope. Our data set comprises 113 late–type galaxies in the redshift range $0.1 < z < 1.0$ and thereby probes galaxy evolution over more than half the age of the universe. Spatially resolved rotation curves have been extracted and fitted with synthetic velocity fields that account for geometric distortions and blurring effects. With these models, the intrinsic maximum rotation velocity $V_{\max}$ was derived for 73 spirals within the field–of–view of the ACS images. Combined with the structural parameters from two-dimensional surface brightness profile fitting, the scaling relations (e.g., the Tully–Fisher Relation) at intermediate redshift were constructed. The evolution of these relations offers powerful tests of the predictions of simulations within the Cold Dark Matter hierarchical scenario.*

*By comparing our sample to the Tully–Fisher Relation of local spiral galaxies, we find evidence for a differential luminosity evolution: the massive distant galaxies are of comparable luminosity as their present-day counterparts, while the distant low–mass spirals are brighter than locally by up to $>2^m$ in rest–frame $B$. Numerous tests applied to the data confirm that this trend is unlikely to arise from an observational bias or systematic errors. Discrepancies between several previous studies could be explained as a combination of selection effects and small number statistics on the basis of such a mass–dependent luminosity evolution. On the other hand, this evolution would be at variance with the predictions from numerical simulations. For a given $V_{\max}$, the disks of the distant galaxies are slightly smaller than those of their local counterparts, as expected for a hierarchical structure growth. Hence, the discrepancy between the observations and theoretical predictions is limited to the properties of the stellar populations. A possible explanation could be the suppression of star formation in low–mass disks which is not yet properly implemented in models of galaxy evolution.*

*Reviews in Modern Astronomy 18.* Edited by S. Röser
 ISBN 3-527-40608-5

# 1 Introduction

Within the last few years, our knowlegde of the basic parameters which determine the past, present and future of the universe has improved significantly. Thanks to the combined results from studies of the Cosmic Microwave Backgrund, the Large Scale Structure, Big Bang Nucleosynthesis and distant supernovae, we now have strong evidence for a flat metric of spacetime (Spergel et al. 2003 and references therein). According to the observations, 73% of the mean density of the universe originate from Dark Energy, 23% are contributed by Cold Dark Matter and only 4% by "ordinary" baryonic matter. In such a cosmology, structure growth proceeds hierarchically, with small structures forming first in the early cosmic stages, followed by the successive build-up of larger structures via merger and accretion events.

Although the constituents of the Dark Energy and Dark Matter remain unknown, the $\Lambda$CDM or "concordance" cosmology has been a very successful tool for the interpretation of structures on Mpc scales and beyond (e.g., Peacock 2003). On scales of individual galaxies, however, several discrepancies between observational results and theoretical predictions have been found, a prominent of which is the "angular momentum problem". This term depicts the loss of angular momentum of the baryons to the surrounding DM halo, resulting in galactic disks within numerical simulations which are smaller than observed (e.g., Navarro & White 1994), however more recent studies made progress in this respect (e.g., Governato et al. 2004). Aiming at a quantitative test of the hierarchical scenario at the scale of individual galaxies, we performed an observational study which covers a significant fraction of the Hubble time.

For this purpose, we utilised scaling relations like the Tully–Fisher relation (TFR, Tully & Fisher 1977) between the luminosity $L$ and the maximum rotation velocity $V_{\rm max}$ of spiral galaxies. Basically, this correlation can be understood as a combination of the virial theorem and the rotational stabilisation of late–type galaxies. By comparing local and distant spirals of a given $V_{\rm max}$, the luminosity evolution within the look–back time can be determined. Since the maximum rotation velocity is a measure for the total (virial) mass of a disk galaxy ($V_{\rm max}^2 \propto M_{\rm vir}$, e.g. van den Bosch 2002), the TF analysis relates the evolution of stellar population properties to the depth of the gravitational potential well.

Numerical simulations within CDM-dominated cosmologies have been successfully used to reproduce the observed slope of the local TFR, whereas the zero points were offset due to dark halos with too high concentrations (e.g., Steinmetz & Navarro 1999). The TFR slope was predicted to remain constant with cosmic look–back time in such $N$-body simulations; nevertheless the modelling of realistic stellar populations at sufficient resolution remains a challenge — typically, the masses of individual particles are of the order of $10^5 M_\odot \ldots 10^6 M_\odot$. Other theoretical approaches focussed more on the chemo–spectrophotometric aspects of disk galaxy evolution. For example, Boissier & Prantzos (2001) calibrated their models to reproduce the observed colors of local spirals. Compared to these, the authors predicted higher luminosities for massive disks and lower luminosites for low–mass disks at redshifts $z > 0.4$. A similiar evolution was found by Ferreras & Silk (2001). By modelling the mass–dependent chemical enrichment history of disk galaxies with the local TFR as a constraint, the authors found a TFR slope that increases with look–back time (i.e., for a parameterisation $L \propto V_{\mathrm{max}}^{\alpha}$, $\alpha$ increases with redshift).

In the last decade, many observational studies of the *local* TFR have produced very large samples with $N_{\mathrm{obj}} \approx 1000$ (e.g. Haynes et al. 1999), not only to derive the slope and scatter with high accuracy, but also to map the peculiar velocity field out to $cz \approx 15000\,\mathrm{km\,s^{-1}}$ (e.g. Mathewson & Ford 1996). Other groups used spirals, partly with cepheid–calibrated distances, to measure the Hubble constant. For example, Sakai et al. (2000) derived a value of $H_0 = (71 \pm 4)\,\mathrm{km\,s^{-1}\,Mpc^{-1}}$ with this method.

At higher redshifts, robust measurements of rotation velocities become increasingly difficult, which is mainly for two reasons. Firstly, because the objects are very faint. Given a redshift of $z = 0.5$, the surface brightness at galactocentric radii of $\sim 3\,r_{\mathrm{d}}$ — where the regime of constant rotation velocity is reached — is typically $\mu_B \approx 27\,\mathrm{mag\,arcsec^{-2}}$. Spatially resolved spectroscopy at this level has become feasible just with the generation of 10m-class telescopes. The second difficulty coming into play arises from the small apparent sizes of the galaxies, this issue will be described in detail in Sect. 4.

A number of samples with 10-20 objects in the regime $0.25 < \langle z \rangle < 0.5$ have been observed to estimate a possible evolution in luminosity by comparison to the local TFR. The results of these studies were quite discrepant: e.g. Vogt et al. (1996, 1997) found only a modest increase in luminosity of $\Delta M_B \approx -0.5^m$, whereas Simard & Pritchet (1998) and Rix et al. (1997) derived a much stronger brightening with $\Delta M_B \approx -2.0^m$. A more recent study of 19 field spirals by Milvang-Jensen et al. (2003) yielded a value of $\Delta M_B \approx -0.5^m$ and showed evidence for an increase of this brightening with redshift.

It seems likely that some of these results are affected by the selection criteria. For example, Rix et al. selected blue colors with $(B-R)_{\mathrm{obs}} < 1.2^m$, Simard & Pritchet strong [O II] emission with equivalent widths >20 Å, while Vogt et al. partly chose large disks with $r_{\mathrm{d}} > 3\,\mathrm{kpc}$. The two former criteria prefer late–type spirals, whereas the latter criterion leads to the overrepresentation of large, early–type spirals. Additionally, due to the small samples, all these studies had to assume that the local TFR slope holds valid at intermediate redshifts — we will adress this topic again in Sect. 6.

Based on a larger data set from the DEEP Groth Strip Survey (Koo 2001) with $N \approx 100$ spirals in the range $0.2 < z < 1.3$, Vogt (2001) found a constant TFR slope and only a very small rest–frame $B$-band brightening of $\lesssim 0.2^m$. On the other hand, in a more recent study based on the same survey, an evolution of the luminosity–metallicity relation both in slope and zero point was observed (Kobulnicky et al. 2003). The authors argued that low–luminosity galaxies probably have undergone a decrease in luminosity combined with an increase in metallicity during the last $\sim 8\,\mathrm{Gyr}$.

Throughout this article, we will assume the concordance cosmology with $\Omega_{\mathrm{m}} = 0.3$, $\Omega_\Lambda = 0.7$ and $H_0 = 70\,\mathrm{km\,s^{-1}\,Mpc^{-1}}$.

# 2 Sample Selection & Observations

The sample described here has been selected within the FORS Deep Field (FDF, see Heidt et al. 2003), an $UBgRIJK$ photometric survey covering a sky area of $\sim 6 \times 6\,\mathrm{arcmin}^2$ near the southern Galactic pole. The imaging was performed with the Very Large Telescope (optical bands) and the New Technology Telescope (Near Infrared bands). Based on a catalogue with spectral types and photometric redshift estimates (Bender et al. 2001), we chose objects for follow–up spectroscopy which satisfied the following criteria: *1)* late–type Spectral Energy Distribution, i.e., galaxies with emission lines, *2)* total apparent $R$-band magnitude $R \leq 23^m$, *3)* photometric redshift $z_{\mathrm{phot}} \leq 1.2$ to ensure that at least the [O II]3727 doublet falls within the wavelength range of the spectra, *4)* disk inclination angle $i \geq 40°$ and *5)* deviation between slit direction and apparent major axis of $\delta \leq 15°$. The two latter constraints were chosen to limit the geometric distortions of the observed rotation curves. For some objects, however, these limits had to be exceeded during the construction of the spectroscopic setups.

After a pilot observation in 1999, the spectroscopy was performed in 2000 and 2001 using the FORS1 & 2 instruments of the VLT in multi–object spectroscopy mode with a total integration time of 2.5 hrs per setup. Using the grism `600R`, a spectral resolution of $R \approx 1200$ was achieved with a spectral scale of 1.07 Å/pix and a spatial scale of 0.2 arcsec/pix. The seeing ranged between 0.4 and 1.0 arcsec with a median of 0.74 arcsec. In total, 129 late–type galaxies were observed.

For an accurate derivation of the galaxies' structural parameters, like disk inclination, scale length etc., we also took Hubble Space Telescope images of the FDF with the Advanced Camera for Surveys using the F814W filter. To cover the complete FDF area, a $2 \times 2$ mosaic was observed.

# 3 Spectrophotometric Analysis

The spectra of 113 galaxies were reliable for redshift determination. Out of these, 73 objects eventually yielded maximum rotation velocities (see next section) and were covered by the HST/ACS imaging; these objects will be referred to as the FDFTF sample in the following. They span the redshift range $0.09 < z < 0.97$ with a median of $\langle z \rangle = 0.45$ corresponding to look–back times $1.2\,\mathrm{Gyr} < t_1 < 7.6\,\mathrm{Gyr}$ with $\langle t_1 \rangle = 4.7\,\mathrm{Gyr}$. This data set covers all spectrophotometric types from very early–type spirals (Sa or $T = 1$) to very late–type galaxies (Sdm/Im or $8 \leq T \leq 10$).

An analysis of the galaxies' surface brightness profile profiles was conducted with the GALFIT package (Peng et al. 2002). To fit the disk component, an exponential profile was used, while a potential bulge was approximated with a Sérsic profile. In the case of 13 FDFTF galaxies, the fit residual images and large fit errors indicated an irregular component that could not be approximated properly with a Sérsic law. The bulge–to–total ratios of these galaxies were assumed to be undefined. The $B/T$ ratios of the other 60 FDFTF galaxies ($0 \leq B/T \leq 0.53$ with $\langle B/T \rangle = 0.04$) confirm that the vast majority of these galaxies are disk-dominated.

Total apparent magnitudes were determined using the `mag_auto` algorithm of the Source Extractor package (Bertin & Arnouts 1996). For the computation of absolute $B$-band magnitudes $M_B$, we used the filter which, depending on the redshift of a given object, best matched the rest–frame $B$-band. For galaxies at $z \leq 0.25$, $0.25 < z \leq 0.55$, $0.55 < z \leq 0.85$ and $z > 0.85$, we thus utilised the $B$, $g$, $R$ and $I$ magnitudes, respectively. Thanks to this strategy, the $k$-correction uncertainties $\sigma_k$ — usually a substantial source of error to the luminosities of distant galaxies — are smaller than $0.1^m$ for all types and redshifts in our sample. For the correction of intrinsic dust absorption, we followed the approach of Tully & Fouqué (1985) assuming a face–on ($i = 0°$) extinction of $A_B = 0.27^m$. The absolute magnitudes of the FDFTF galaxies computed this way span the range $-18.0^m \geq M_B \geq -22.7^m$.

The spectra of 12 objects in our sample cover a wavelength range that simultaneously shows emission in [O II]3727, H$\beta$, [O III]4959 and [O III]5007 at sufficient signal–to–noise to determine the equivalent widths. These lines can be used to estimate the gas-phase metallicity. We adopted the analytical expressions given by McGaugh (1991) to compute the abundances O/H from the $R_{23}$ and $O_{32}$ parameters. Since all the galaxies have $M_B < -18^m$, we assumed that they fall on the metal-rich branch of the $R_{23}$–O/H relation. The galaxies have abundances $8.37 < \log(\mathrm{O/H}) < 8.94$. We will use these estimates to investigate the luminosity–metallicity relation in Sect. 5.

# 4 Derivation of V$_{\mathrm{max}}$

We extracted spatially resolved rotation curves from the two–dimensional spectra by fitting Gaussians to the emission lines stepwise along the spatial axis. Line fits at any projected radius which, compared to the instrumental broadening (FWHM$_{\mathrm{ins}} \approx 4.5$ Å), had very small (FWHM$_{\mathrm{fit}} < 2$ Å) or very large (FWHM$_{\mathrm{fit}} > 12$ Å) line widths were assumed to be noise and therefore neglected.

The analysis of spatially resolved rotation curves from optical spectroscopy of *local* spiral galaxies is relatively straightforward. But in the case of distant galaxies with very small apparent sizes, the effect of the slit width on the observed rotation velocities $V_{\mathrm{rot}}(r)$ must be considered. At redshift $z = 0.5$, a scale length of 3 kpc — typical for an $L^*$ spiral — corresponds to $\sim$0.5 arcsec only, which is half the slit width used in our observations. Any value of $V_{\mathrm{rot}}(r)$ is therefore an integration perpendicular to the spatial axis (slit direction), a phenomenon which is the optical equivalent to "beam smearing" in radio observations. The seeing has an additional blurring effect on the observed rotation curves. If not taken into account, these two phenomena would lead to an underestimation of the *intrinsic* rotation velocities and, in particular, the *intrinsic* $V_{\mathrm{max}}$.

We overcame this problem by generating synthetic rotation curves. For the intrinsic rotational law, we used a simple shape with a linear rise of $V_{\mathrm{rot}}(r)$ at small radii, turning over into a region of constant rotation velocity where the Dark Matter halo dominates the mass distribution. Alternatively, we also tested the so–called "Universal Rotation Curve" shape (Persic et al. 1996), a parameterisation which introduces a velocity gradient in the outer regions of the disk which is positive for

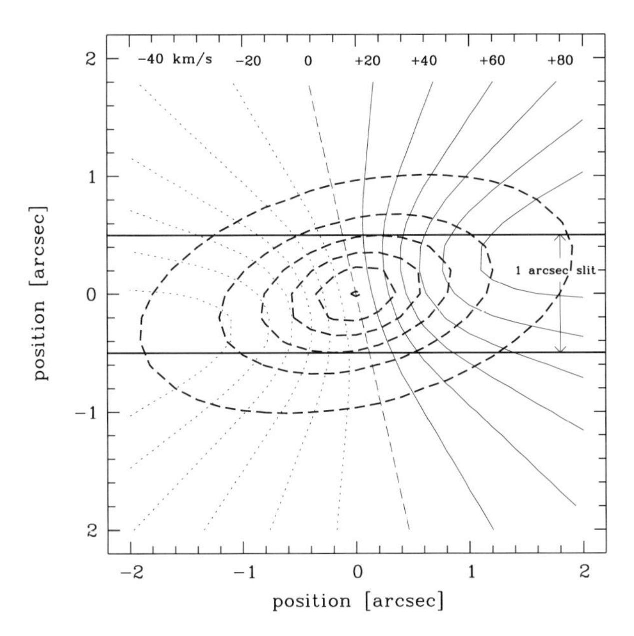


Figure 1: Example of a simulated rotation velocity field for an object from our data set with a disk inclination $i = 64°$ and a misalignment angle between apparent major axis and slit direction of $\delta = +13°$. The dashed ellipses denote the isophotes of the disk, with the outermost corresponding to an $I$-band surface brightness of $\mu_I \approx 25\,\mathrm{mag\,arcsec}^{-2}$. The curved dotted and solid lines correspond to line–of–sight rotation velocities ranging from $-120$ km/s to $+120$ km/s. The two solid horizontal lines visualise the position of the slit used for spectroscopy.

sub-$L^*$ objects and negative for objects much more luminous than $L^*$. However, the results given here are not sensitive to the form of the intrinsic rotational law — see Böhm et al. (2004) for a detailed discussion of this topic — and we therefore only use $V_{\mathrm{max}}$ values determined with an intrinsic "rise–turnover–flat" shape here.

Given the observed inclination, position angle and scale length of an object, the intrinsic rotation velocity field was constructed, Fig. 1 shows an example. In the next step of the simulation, the velocity field was weighted with the surface brightness profile. The effect of this was that, just like for the observed data, brighter regions contributed stronger to the rotation velocities in direction of dispersion than fainter regions (the "beam smearing" effect). Following the weighting, the velocity field was convolved with the Point Spread Function to simulate the blurring due to seeing. Finally, a "stripe" was extracted from the velocity field, with a position and width that corresponded to the slit used during the observations, and integrated perpendicular to the spatial axis. The results of the whole procedure was a synthetic rotation curve which introduced the same geometric and blurring effects as the corresponding observed rotation curve. By fitting the *simulated* rotation curve to the *observed* rotation curve, we derived the intrinsic value of $V_{\mathrm{max}}$. Four examples of observed rotation curves along with the best-fitting synthetic rotation curves are shown in Fig. 2.

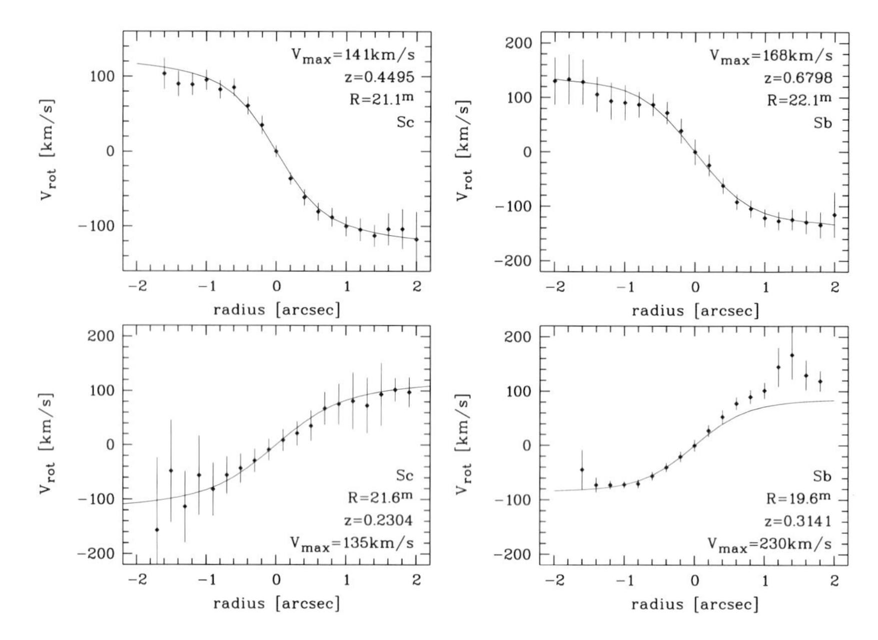


Figure 2: Examples of rotation curves from our data set. The solid lines are the synthetic rotation curves fitted to the observed rotation velocity as a function of radius (solid symbols) used to derive the intrinsic maximum rotation velocity. For each object the spectrophotometric type, total apparent $R$ magnitude, redshift and $V_{\max}$ are given. The two upper curves were classified as high quality data, the two lower ones as low quality data due to the large measurement errors (lower left) or an asymmetric shape (lower right).

36 galaxies had to be rejected from the further analysis because the $S/N$ was too low to probe the regime of constant rotation velocity at large radii, or because the rotation curves were perturbated. Four objects were reliable for the $V_{\max}$ determination, but their positions were located outside the field–of–view of the HST/ACS mosaic imaging. 34 objects had curves with a high degree of symmetry and clearly reached into the "flat" regime, we consider these as high quality data. 39 curves had a relatively small spatial extent or mild asymmetries, these will be referred to as low quality data in the following. In total, our kinematic data set thus comprises 73 late–type galaxies at a mean look–back time of $\sim 5\,\mathrm{Gyr}$. The objects span the range $25\,\mathrm{km/s} \leq V_{\max} \leq 450\,\mathrm{km/s}$ with a median of 129 km/s (high quality data only: $62\,\mathrm{km/s} \leq V_{\max} \leq 410\,\mathrm{km/s}$ and $\langle V_{\max} \rangle = 154\,\mathrm{km/s}$).

## 5 Scaling Relations at Intermediate Redshift

In Fig. 3, the maximum rotation velocities and absolute magnitudes of the distant FDFTF galaxies are compared to the local $B$-band Tully–Fisher relation by Pierce

& Tully (1992):

$$M_B = -7.48 \log V_{\rm max} - 3.52 \tag{1}$$

with a scatter of $\sigma_B = 0.41^m$. Note that, at variance with the original relation given by these authors, we have calibrated the zero point to a face-on extinction of $0.27^m$ to achieve consistency with the computation of the distant galaxies' absolute magnitudes. We emphasize that the further analysis is not sensitive to the choice of the local reference sample: e.g., for the large data set of Haynes et al. (1999, comprising 1097 objects), we find a very similar relation of

$$M_B = -7.85 \log V_{\rm max} - 2.78, \tag{2}$$

using a bisector fit (two geometrically combined least–square fits with the dependent and indepedent variable interchanged). We will utilise the Pierce & Tully sample here for the sake of comparability to intermediate–redshift TF studies in the literature which mostly have used this sample as a local reference.

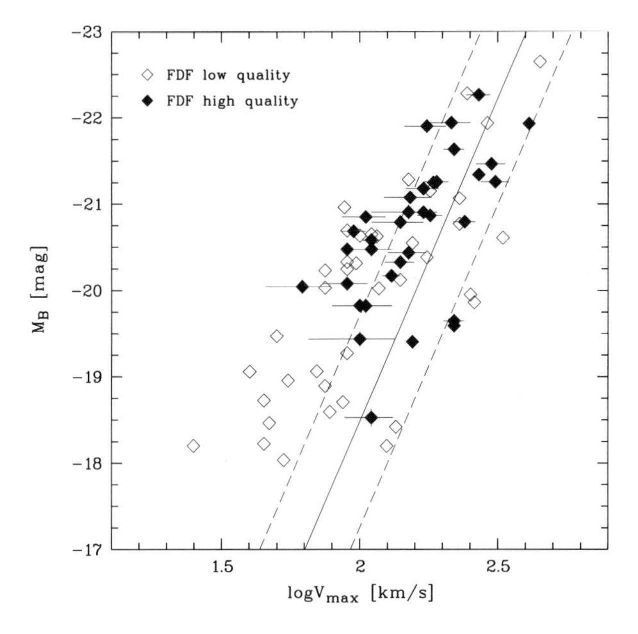


Figure 3: FORS Deep Field sample of spirals in the range $0.1 \leq z \leq 1.0$ in comparison to the local Tully–Fisher relation by Pierce & Tully (1992); the solid line denotes the fit to the local data, the dashed lines give the $3\,\sigma$ limits. The distant sample is subdivided according to rotation curve quality: high quality curves (solid symbols) extend well into the region of constant rotation velocity at large radii and therefore give robust values of $V_{\rm max}$. Error bars are shown for the high quality data only.

On the average, the distant galaxies are overluminous with respect to their local counterparts, we find a median offset of $\langle \Delta M_B \rangle = -0.98^m$ for the total FDFTF sample and $\langle \Delta M_B \rangle = -0.81^m$ for the high quality data only. But we find also evidence for a differential evolution. Fig. 3 indicates a relatively good agreement between the intermediate–redshift galaxies and the local TFR in the regime of fast rotators, i.e. high masses, while the distant low–mass galaxies systematically deviate from the relation of present-day spirals. For low quality data, this may partly be due to underestimated maximum rotation velocities, since the corresponding curves have a relatively small spatial extent and do not robustly probe the region of constant rotation velocity at large radii. In the case of high quality rotation curves, this is however unlikely, since these extent well into the "flat region".

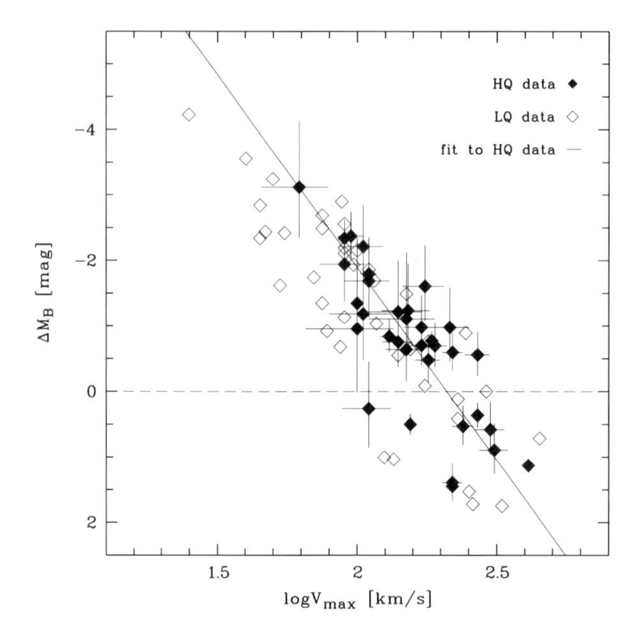


Figure 4: Offsets of the distant FORS Deep Field galaxies from the local TFR by Pierce & Tully (1992). The distant sample is subdivided according to rotation curve quality: high quality curves (solid symbols) extend well into the region of constant rotation velocity at large radii and therefore give robust values of $V_{\max}$. The dashed horizontal line corresponds a zero luminosity evolution. While high–mass galaxies are in agreement with the local TFR or even slightly underluminous given their $V_{\max}$, the objects are increasingly overluminous towards small values of $V_{\max}$ (error bars are shown for the high quality data only).

A 100 iteration bootstrap bisector fit (average of 100 bisector fits with randomly removed objects in each iteration) to the 34 FDFTF objects with high quality rotation curves yields

$$M_B = -(4.05 \pm 0.58) \log V_{\max} - (11.8 \pm 1.28), \tag{3}$$

i.e. the TFR slope we find at intermediate redshift is significantly shallower than in the local universe. Since the derivation of the galaxies' structural parameters and of the $V_{\max}$ values has been based entirely on HST/ACS imaging, Eq. 3 is a confirmation of the results presented in Böhm et al. (2004) which were limited to ground–based imaging.

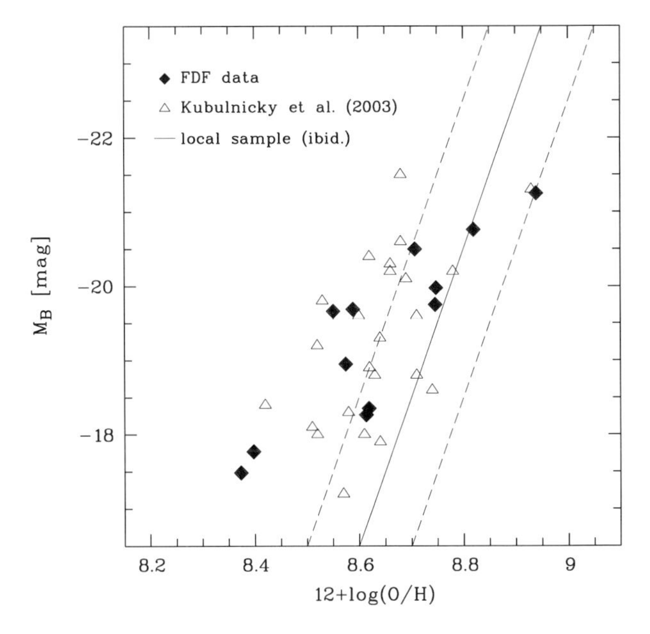


Figure 5: The gas-phase metallicities of 12 spirals from the FDF data set (filled symbols) in comparison to the local luminosity-metallicity relation constructed by Kobulnicky et al. (2003, solid line; dashed lines give the estimated $1\sigma$ scatter). Also shown are distant galaxies presented in Kobulnicky et al. (open symbols) which cover a similar redshift range as the FDF galaxies. Both distant samples show a "tilt" with respect to the local relation which likelywise indicates a combined evolution in luminosity and metallicity of low–luminosity galaxies.

We show the individual offsets $\Delta M_B$ of the FDF galaxies from the local TFR as a function of their maximum rotation velocity in Fig. 4. Even when restricting the sample to the high quality rotation curves, we find significant overluminosties of up to more than $2^m$ in the rest-frame $B$ for low–mass spirals. $L^*$ galaxies, corresponding to $\log V_{\rm max} \approx 2.3$ according to Eq. 1, scatter around a negligible evolution, while the fastest rotators are systematically underluminous.

In Fig. 5, we show the sub-sample of 12 FDF galaxies for which we could determine the oxygen abundances O/H in comparison to the local luminosity–metallicity relation as given in Kobulnicky et al. (2003, the displayed scatter is a rough estimate). In addition, a sub–sample of distant late–type galaxies from the DEEP survey (ibid.) is shown, which has been restricted to the same redshift intervall ($0.22 < z < 0.46$) that is covered by the FDF galaxies. Both sub–samples thus represent a look–back time of $\sim$4 Gyr. For the sake of comparability, the absolute magnitudes $M_B$ of the FDF galaxies given in this figure are *not* corrected for intrinsic absorption, as is the case for the Kobulnicky et al. data. Both distant samples indicate a "tilt" with respect to the local $L$–$Z$ relation. At given $\log{\rm (O/H)}$, high–metallicity galaxies at intermediate redshift agree relatively well with the local $L$–$Z$ relation, whereas low–metallicity objects are overluminous. Alternatively, the distributions may be in-

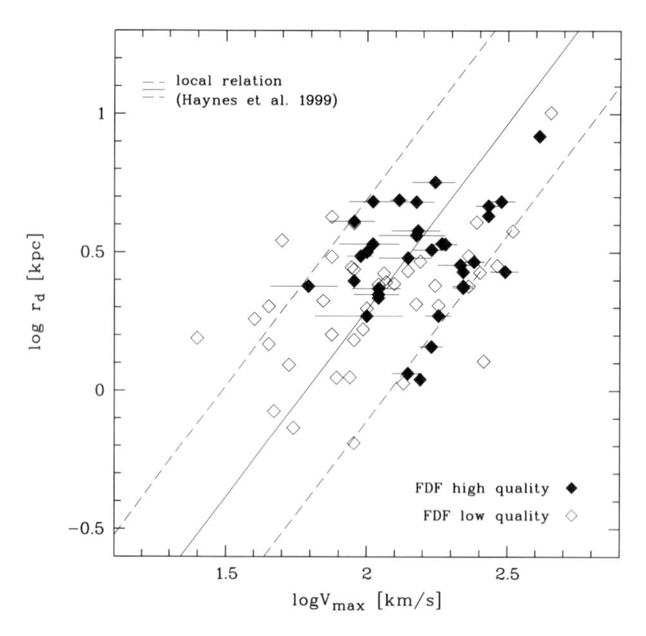


Figure 6: Comparison between the intermediate–redshift FDF galaxies and the local velocity–size relation of the Haynes et al. sample (1999, solid line; dashed lines correspond to the 3 $\sigma$ scatter). Error bars are shown for high quality data only.

terpreted such that the distant low–luminosity galaxies have smaller chemical yields than locally, while high–luminosity galaxies do not differ strongly in O/H between intermediate and low redshift. If the offsets of the FDF spirals we observe in the TF diagram are due to younger stellar populations than locally, it is probable that Fig. 5 shows a combined evolution in luminosity *and* metallicity. This indeed has been the conclusion of Kobulnicky et al. after comparison of their data to single–zone models.

The third scaling relation which we want to focus on here is presented in Fig. 6, where the FDF galaxies are shown with respect to the local velocity–size relation correlating $V_{\rm max}$ and $r_{\rm d}$. To derive the latter, we used the sample of Haynes et al. (1999). A bisector fit this data set yields

$$\log r_{\rm d} = 1.35 \log V_{\rm max} - 2.41. \tag{4}$$

Since the scale lengths of the FDF galaxies were determined in the $I$-band (HST-filter F814W), the data intrinsically probe shorter wavelengths towards higher redshifts. For direct comparability to the local sample ($I$-band as well), we had to account for this rest–frame shift. Otherwise, the measured FDF disk sizes could be overestimated, in particular for the more distant galaxies. Adopting the relations between scale lengths of local spirals at different wavelengths presented in de Jong (1996), we have transformed the observer's frame $I$-band scale lengths to the rest–frame $I$-band values. Note, however, that these factors are relatively small: for the FDF galaxies at $z \approx 1$ , the correction corresponds to only $\sim 10\%$. The characteristic disk sizes of the FDFTF sample cover the range $0.7\,\mathrm{kpc} \leq r_\mathrm{d} \leq 10.1\,\mathrm{kpc}$ with a median $\langle r_\mathrm{d} \rangle = 2.7\,\mathrm{kpc}$.

# 6 Discussion

The $V_\mathrm{max}$-dependent TF offsets we observe at redshift $\langle z \rangle \approx 0.5$ may be indicative for a significant decrease of the luminosity of low–mass galaxies — possibly combined with an increase in metallicity — over the past $\sim 5$ Gyr and a negligible evolution of high–mass galaxies. This evolution would be at variance with theoretical predictions: e.g., Steinmetz & Navarro (1999) find mass–independent TF offsets towards higher redshifts with an $N$-body Smoothed Particle Hydrodynamics code. Boissier & Prantzos (2001), who used a "backwards approach" model calibrated to the observed chemo–spectrophotometric properties of local spirals, predict overluminosties of high–mass spirals and underluminosities of low–mass spirals towards larger look–back times.

It has to be ruled out that our result might be induced by an observational bias or a systematic error. E.g., it is known that present–day spirals show a correlation between their TF residuals and broad–band colors (e.g. Kannappan et al. 2002), with blue galaxies preferentially populating the regime of overluminosities. We have therefore tested whether Fig. 4 may simply reflect an evolution of the color–residual relation with redshift, finding no evidence for such a trend (Böhm et al. 2004). Another issue that has to be adressed is the potential impact of sample incompleteness. Any magnitude–limited data set contains only a fraction of the objects that are located within the observed volume. Towards lower luminosities (or slower rotation velocities), this fraction becomes smaller. Furthermore, the magnitude limit corresponds to higher luminosities at higher redshifts. An incompleteness bias could therefore result in a flattening of the distant TFR with increasing redshift. However, dividing our sample into objects with $z \leq 0.449$ (37 galaxies) and $z > 0.449$ (36 galaxies), we find no evidence for such a redshift dependence, the respective slopes of the two redshifts bins are $-3.49$ and $-3.77$. For a more sophisticated test of sample incompleteness which is based on the work of Giovanelli et al. (1997), we refer to our results presented in Ziegler et al. (2002).

In the following, we will address another three examples of tests we performed. These are related to the influence of the intrinsic rotation curve shape, the impact of the intrinsic absorption correction and the issue of galaxy-galaxy interactions.

To derive the intrinsic maximum rotation velocity, we have assumed an intrinsic rotation curve shape with a linear rise of the rotation velocity at small radii which turns over into a region of constant rotation velocity at a radius that depends on the rest–frame wavelength of the used emission line. This shape is observed for kinematically unperturbated, massive ($\sim L^*$) local spirals (e.g., Sofue & Rubin 2001). For galaxies of very high or very low masses, on the other hand, it is observed that even in the outer parts, most rotation curves have a velocity gradient. While the rotation velocity keeps rising beyond the "turnover" radius in very low–mass spirals, the velocity gradient in very high–mass spirals is negative. Persic et al. (1996) have used $>$1000 curves of local spirals to derive a parameterisation that uses the luminosity of an object as an indicator for the rotation curve shape. To ensure that the observed TF offsets cannot be attributed to a false assumption on the intrinsic rotational law, we have alternatively used this so–called "Universal Rotation Curve" shape as input for the computation of our synthetic velocity fields. If we use the $V_{\max}$ values derived this way to recompute the offsets from the local TFR, the luminosity evolution we find is smaller by only $\sim 0.15^m$ at $V_{\max} \approx 100$ km/s. Since this is a very modest change of the offsets we found on the basis of the simple "rise–turnover–flat"–shape (which have a median of $\langle \Delta M_B \rangle \approx -1.74^m$ at $V_{\max} \approx 100$ km/s for the HQ data), we conclude that our results do not differ significantly between these two assumptions on the intrinsic rotation curve shape.

Similarly, we have tested whether a different approach to correct for the intrinsic absorption would have an effect on our results. All values given here were derived following Tully & Fouqué (1985), i.e., the amount of intrinsic absorption is assumed to depend only on the inclination of the disk. More recently, Tully et al. (1998) have found evidence that the dust reddening is – at least locally — stronger in high–mass spirals than in low–mass spirals. Using their results, we have recomputed the absolute magnitudes of the FDFTF galaxies. As a new local reference that is consistently corrected for intrinsic absorption following Tully et al., we adopted the sample of Verheijen (2001) which is slightly steeper (slope $-8.1$) than the Pierce & Tully (1992) sample. This is simply due to the fact that fast rotating, high–mass spirals are assumed to have a larger amount of intrinsic absorption than in the Tully & Fouqué approach, and vice versa in the low–mass regime. The offsets of the FDF high quality data from the Verheijen TFR are however very similar to the initial values ($\langle \Delta M_B \rangle \approx -1.77^m$ vs. $\langle \Delta M_B \rangle \approx -1.74^m$ at $V_{\max} \approx 100$ km/s). With respect to the two conventions of intrinsic absorption correction discussed here, the TF offsets therefore are robust.

A third aspect we want to focus on here concerns the interplay between the stellar population properties and the environment. From studies in the local universe, it is known that galaxies residing in close pairs can be subject to tidal interactions which can increase the star formation rates. In such cases, the fraction of high–mass stars would be enlarged and, in turn, the mass–to–light ratio would be decreased, resulting in overluminosities in the TF diagram. This triggering of star formation would be particularly efficient in low–mass galaxies (e.g., Lambas et al. 2003). Though we have selected our objects from a sky region that should be representative for low–density environments, it is not clear a priori whether the correlation between the TF offsets and $V_{\max}$ can at least in part be attributed to tidally induced star formation. Based on $>10^5$ galaxies from the 2dF survey, Lambas et al. have found that the star formation rates can be significantly increased in objects that have close companions with a separation $\Delta V_{\rm sys} \leq 250$ km/s in systematic velocity and a projected distance of $D_{\rm proj} \leq 100$ kpc. Using all 267 available spectroscopic redshifts of FDF galaxies at $z \leq 1$ from our own study and a low–resolution survey presented in Noll et al. (2004), and adopting the Lambas et al. constraints cited above, we have found 12 FDFTF objects with confirmed neighbors.

In Fig. 7, we show the TF offsets of these galaxies in comparison to the rest of the sample. Though the small sub–sample of pair candidates does not allow robust statistics, the galaxies with close companions appear to be similarly distributed as

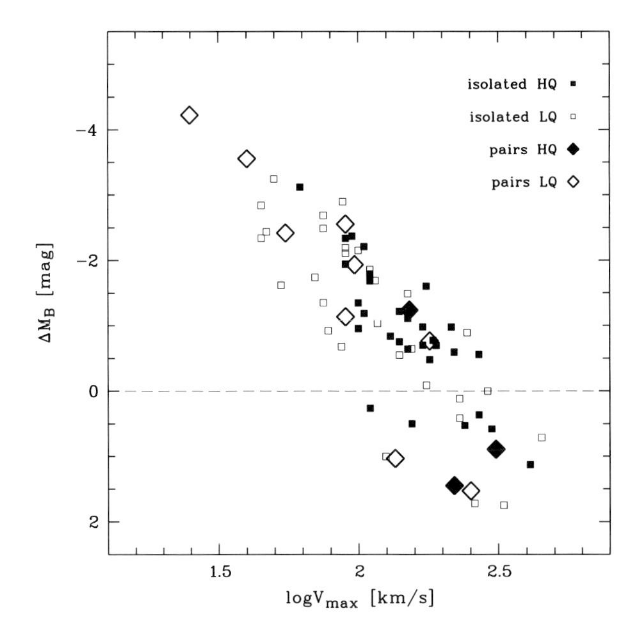


Figure 7: Offsets of the distant FORS Deep Field galaxies from the local TFR by Pierce & Tully (1992). Large symbols depict FDF objects which show spectroscopically confirmed neighbors within $\Delta V_{\mathrm{sys}} \leq 250$ km/s in systematic velocity and a projected separation $D_{\mathrm{proj}} \leq 100\,\mathrm{kpc}$. See text for details.

the rest of the sample. Moreover, we find a hint that the rotation curve quality is reduced with respect to the probably isolated galaxies — only 3 out of 12 (25%) pair candidates were classified to have high quality rotation curves, whereas for the rest of the sample, this fraction is 31 out of 61 (51%). In particular, the pair candidates with HQ curves are not systematically biased towards large overluminosities. Since we have spectroscopic redshifts only for a fraction of the galaxies within the probed volume, it is possible that we missed some close pairs. However, the aim was to test whether those pair candidates which *are* identified systematically differ from the remaining FDFTF objects. We thus conclude that our analysis which is based on the high quality rotation curves is very unlikely to be affected by tidally induced star formation.

To summarise all the tests performed, we find no evidence for any systematic error or bias that may be the source of the observed shallow slope of the intermediate–redshift TFR. We now will show that our findings can be used to interpret the rather discrepant results of previous TF studies of distant galaxies introduced in Sect. 1. For this, Fig. 8 shows our sample sub–divided according to the Spectral Energy Distribution into early–type spirals (Sa/Sb), intermediate–types (Sc) and very late–types (Sdm/Im). All three sub–samples show a correlation between the TF offsets and $V_{\mathrm{max}}$, but cover different mass regimes: galaxies with late–type spectra have smaller average $V_{\mathrm{max}}$ values than early–type spirals. The respective classes have median values of $\langle V_{\mathrm{max}} \rangle_{\mathrm{Sdm/Im}} = 91\,\mathrm{km/s}$, $\langle V_{\mathrm{max}} \rangle_{\mathrm{Sc}} = 140\,\mathrm{km/s}$ and $\langle V_{\mathrm{max}} \rangle_{\mathrm{Sa/Sb}} =$

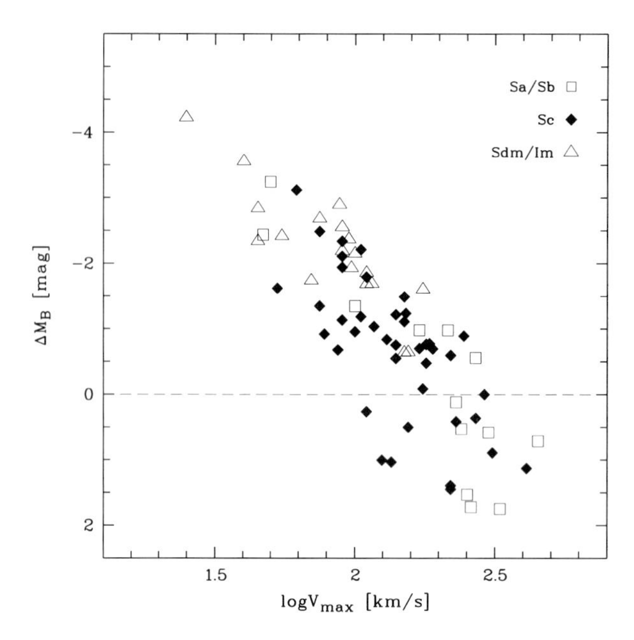


Figure 8: Offsets of the distant FORS Deep Field galaxies from the local TFR by Pierce & Tully (1992). The distant sample is subdivided according to the SED type. All three sub–samples indicate a $\Delta M_B$–$V_{\rm max}$ correlation, but cover different mass regimes. See text for details.

240 km/s. For samples which are too small to robustly test a correlation between $\Delta M_B$ and $V_{\rm max}$ but can only be used to determine *average* TF offsets, this would results in a correlation between SED type and $\langle \Delta M_B \rangle$.

Studies comprising only 10-20 galaxies which were selected on blue colors or strong emission lines and therefore preferentially contained late–type spirals (e.g., Rix et al. 1997, Simard & Pritchet 1998), found evidence for large luminosity offsets at intermediate redshift with respect to the local TFR. According to the above, this can be attributed to a combination of a selection effect and small number statistics. Similary, a study by Vogt et al. (1996) which was selected on large disks and therefore mainly contained early-type spirals (which on average are more massive than late–types), yielded a modest value for the mean TF offset.

A more recent study on ~100 intermediate–redshift spirals by Vogt (2001) did not find evidence for a slope evolution of the TFR, at variance with our results. Moreover, this study yields only a modest average TF offset of $-0.2^m$, compared to the $-0.8^m$ we find for the HQ data (note that these values correspond to the same cosmology and similar redshifts). It is difficult to speculate whether this could be attributed to different criteria of rotation curve quality or differences in the $V_{\rm max}$ derivation procedures. On the other hand, a sample of 64 galaxies drawn from the same survey (the DEEP project, see Koo 2001) finds, in comparison to local galaxies, a tilt of the intermediate-redshift luminosity–metallicity relation which is also indicated by our data (Fig. 5).

Is is a complicated issue to determine the key processes which can give rise to the mass–dependent TF offsets we observe. Several effects likelywise act in combination when local and distant spirals are compared. E.g., the stellar populations of the intermediate–redshift galaxies are probably younger than those of their local counterparts, the gas mass fractions and chemical yields also evolve with time etc. Since even a relatively small fraction of young, high–mass stars can have a significant effect on the luminosity in the blue bands, a straightforward interpretation would be that the TF offsets of the FDF galaxies point towards a correlation between mass and *age*.

As a first deeper insight into this matter, Ferreras et al. (2004) have used single–zone models of chemical enrichment on a sub–sample of the FDFTF galaxies at $z > 0.5$. These models were determined by only four free parameters: formation redshift, gas infall timescale, star formation efficiency and gas outflow fraction. Model star formation histories were generated which, combined with the latest Bruzual & Charlot models (2003), were used to compute simulated $UBgRIJK$ broad-band colors. Probing a large volume in parameter space, these synthetic colors were fitted to the observed broad-band colors, thus deriving the four model parameters for each of the $z > 0.5$ FDF galaxies. The best–fitting models indicated that high–mass galaxies on the average have higher star formation efficiencies, with a "break" at $V_{\rm max} \approx 140$ km/s which — for reasonable $M/L$ ratios — interestingly is good agreement with the results found by Kauffmann et al. (2003) for *local* galaxies from the Sloan Digital Sky Survey.

Moreover, the best–fitting formation redshifts of the Ferreras et al. models were found to be higher for the more massive FDFTF galaxies than for low–mass galaxies. The models hence yielded evidence that high–mass spirals started to convert their gas into stars at earlier cosmic epochs and on shorter timescales than low–mass ones. When evolved to zero redshift, the mean model stellar ages turned out to be older in high–mass galaxies than in low–mass galaxies. These results hint towards an *anti–hierarchical evolution* of the stellar (baryonic) component, a phenomenon that recently has been referred to as "down-sizing" (e.g., Kodama et al. 2004).

It is however improbable that this implies a contradiction to the hierarchical growth of the Cold Dark Matter halos. In Fig. 9, we show the offsets $\Delta \log r_{\rm d}$ of the FDFTF galaxies with high quality rotation curves from the local velocity–size relation (which we presented in Fig. 6) as a function of *redshift*. Given their maximum rotation velocity, distant galaxies with $\Delta \log r_{\rm d} < 0$ have disks that are smaller than in the local universe, while the disks of galaxies with $\Delta \log r_{\rm d} > 0$ are larger. Though the scatter in $\Delta \log r_{\rm d}$ is substantial, we observe a slight trend towards smaller disk sizes at higher redshifts. This is in compliance with the results from other observational studies (e.g., Giallongo et al. 2000, Ferguson et al. 2004). Moreover, as is depicted in Fig. 9, the fit to the data is in relatively good agreement with the prediction of disk growth in the hierarchical scenario (e.g. Mo et al. 1998).

On the one hand, we find that the observed luminosity evolution of the FDF galaxies deviates from the results of simulations which were used to predict the TFR of distant spirals. On the other hand, the intermediate–redshift disks are observed to be smaller than locally, which is in compliance with the CDM hierarchical scenario. The results from single–zone models might indicate that the star formation is sup-

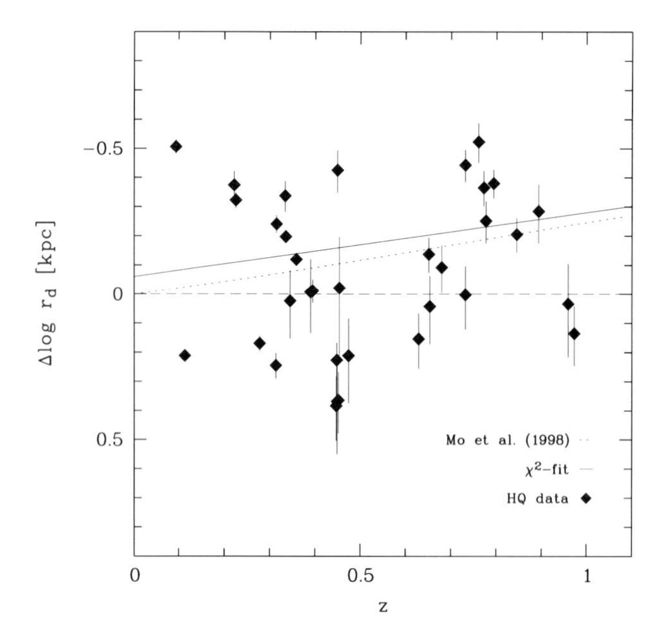


Figure 9: Offsets $\Delta \log r_{\rm d}$ of the distant FORS Deep Field galaxies (high quality rotation curves only) from the local velocity–size relation of the sample by Haynes et al. (1999) shown in Fig. 6, plotted as a function of redshift. Objects with $\Delta \log r_{\rm d} > 0$ have larger disks than local spirals at a given $V_{\rm max}$, whereas values $\Delta \log r_{\rm d} < 0$ correspond to disks which are smaller than in the local universe. As indicated by the fit to the data (solid line), we find a slight trend towards smaller disks at higher redshifts, in agreement with theoretical predictions within the hierarchical scenario (Mo et al. 1998, dotted line).

pressed in low–mass galaxies due to, e.g., SN feedback (cf. Dalcanton et al. 2004). Potentially, the discrepancies between observations and simulations arise from the fact that the mechanisms suppressing the star formation are not yet implemented realistically enough in models of galaxy evolution.

# 7 Conclusions

Using the FORS instruments of the VLT in multi–object spectroscopy mode and HST/ACS imaging, we have observed a sample 113 disk galaxies in the FORS Deep Field. The galaxies reside at redshifts $0.1 < z < 1.0$ and thereby probe field galaxy evolution over more than half the age of the universe. All spectrophotometric types from Sa to Sdm/Irr are comprised. Spatially resolved rotation curves have been extracted and fitted with synthetic velocity fields that account for geometric distortions as well as blurring effects arising from seeing and optical beam smearing. The intrinsic maximum rotation velocities $V_{\rm max}$ were derived for 73 galaxies within the field–of–view of the ACS images. Two–dimensional surface brightness profile fits were performed to measure the structural parameters like disk inclinations, position angles etc.

The massive distant galaxies fall onto the local Tully–Fisher Relation, while the low–mass distant galaxies are brighter than locally by up to $>2^m$ in rest–frame $B$. This trend might be combined with an evolution in metallicity. We find no evidence for a bias or systematic errors that could induce the observed shallow slope of the Tully–Fisher Relation at intermediate redshifts. Discrepancies between several previous studies could be explained as a combination of selection effects and small number statistics on the basis of such a mass–dependent luminosity evolution. On the other hand, this evolution would be at variance with the predictions from numerical simulations. For a given $V_{\rm max}$, the disks of the distant galaxies are slightly smaller than those of their local counterparts, as expected for a hierarchical structure growth. Our results therefore are discrepant with theoretical predictions only in terms of the stellar populations properties. A possible explanation could be the suppression of star formation in low–mass disks which is not yet properly implemented in models of galaxy evolution.

**Acknowledgements**

This study is based on observations with the European Southern Observatory Very Large Telescope (observing run IDs 65.O-0049, 66.A-0547 and 68.A-0013), and observations with the NASA/ESA Hubble Space Telescope, PID 9502. We are grateful for the continuous support of our project by the PI of the FDF consortium, Prof. I. Appenzeller (LSW Heidelberg), and by Prof. K. J. Fricke (USW Göttingen). We also thank Drs. J. Heidt (LSW Heidelberg), D. Mehlert (LSW Heidelberg) and S. Noll (USW München) for performing the spectroscopic observations and ESO for the efficient support during the observations. Furthermore we want to thank J. Fliri and A. Riffeser (both USW München) for the cosmic ray rejection on the ACS images and Dr. B. Milvang-Jensen and M. Panella (both MPE Garching) for fruitful discussions. Our work was funded by the Volkswagen Foundation (I/76 520) and the Deutsches Zentrum für Luft- und Raumfahrt (50 OR 0301).

## References

Bender, R., Appenzeller, I., Böhm, A., et al. 2001, The FORS Deep Field: Photometric redshifts and object classification, in Deep Fields, ed. S. Cristiani, A. Renzini, & R. E. Williams, ESO astrophysics symposia (Springer), 96

Bertin, E., & Arnouts, S. 1996, A&AS, 117, 393

Boissier, S., & Prantzos, N. 2001, MNRAS, 325, 321

Böhm, A., Ziegler, B. L., Saglia, R. P., et al. 2004, A&A, 420, 97

Bruzual, A. G., & Charlot, S. 2003, MNRAS, 344, 100

de Jong, R. S. 1996, A&A, 313, 377

Dalcanton, J. J., Yoachim, P., & Bernstein, R. A. 2004, ApJ, 608, 189

Ferguson, H. C. F., Dickinson, M., Giavalisco, M., et al. 2004, ApJ, 600, L107

Ferreras, I., & Silk, J. 2001, ApJ, 557, 165

Ferreras, I., & Silk, J., Böhm, A., & Ziegler, B. L. 2004, MNRAS, 355, 64

Giallongo, E., Menci, N., Poli, F., et al. 2000 ApJ, 530, L73

Giovanelli, R., Haynes, M. P., Herter, T., et al. 1997, AJ, 113, 53

Governato, F., Mayer, L., Wadsley, J., et al. 2004, ApJ, 607, 688

Haynes, M. P., Giovanelli, R., Chamaraux, P., et al. 1999, AJ, 117, 2039

Heidt, J., Appenzeller, I., Gabasch, A., et al. 2003, A&A, 398, 49

Kannappan, S. J., Fabricant, D. G., & Franx, M. 2002, AJ, 123, 2358

Kauffmann, G., Heckman, T. M., White, S. D. M., et al. 2003, MNRAS, 341, 54

Kobulnicky, H. A., Willmer, C. N. A., Phillips, A. C., et al. 2003, ApJ, 599, 1006

Kodama, T., Yamada, T., Akiyama, M., et al. 2004, MNRAS, 350, 1005

Koo, D. C. 2001, DEEP: Pre–DEIMOS Surveys to I∼ 24 of Galaxy Evolution and Kinematics, in Deep Fields, ed. S. Cristiani, A. Renzini, & R. E. Williams, ESO astrophysics symposia (Springer), 107

Lambas, D. G., Tissera, P. B., Sol Alonso, M., & Coldwell, G. 2003, MNRAS, 346, 1189

Mathewson, D. S., & Ford, V. L. 1996, ApJS, 107, 97

McGaugh, S. 1991, ApJ, 380, 140

Milvang-Jensen, B., Aragón-Salamanca, A., Hau, G. K. T., Jørgensen, I., & Hjorth, J. 2003, MNRAS, 339, L1

Mo, H. J., Mao, S., & White, S. D. M. 1998, MNRAS, 295, 319

Navarro, J. F., & White, S. D. M. 1994, MNRAS, 267, 401

Noll, S., Mehlert, D., Appenzeller, I., et al. 2004, A&A, 418, 885

Peacock, J. A. 2003, RSPTA, 3671, 2479

Peng, C. Y., Ho, L. C., Impey, C. D., & Rix, H.-W. 2002, AJ, 124, 266

Persic, M., & Salucci, P., & Stel, F. 1996, MNARS, 281, 27

Pierce, M. J., & Tully, R. B. 1992, ApJ, 387, 47

Rix, H.-W., Guhathakurta, P., Colless, M., & Ing, K. 1997, MNRAS, 285, 779

Sakai, S., Mould, J. R., Hughes, S. M. G., et al. 2000, ApJ, 529, 698

Simard, L., & Pritchet, C. J. 1998, ApJ, 505, 96

Sofue, Y. & Rubin, V. 2001, ARA&A, 39, 137

Spergel, D. N., Verde, L., Peiris, H. V., et al. 2003, ApJS, 148, 175

Steinmetz, M., & Navarro, J. F. 1999, ApJ, 513, 555

Tully, R. B., & Fisher, J. R. 1977, A&A, 54, 661

Tully, R. B., & Fouqué, P. 1985, ApJS, 58, 67

Tully, R. B., Pierce, M. J., Huang, J.-S., et al. 1998, AJ, 115, 2264

van den Bosch, F. 2002, MNRAS, 332, 456

Verheijen, M. A. W. 2001, ApJ, 563, 694

Vogt, N. P., Forbes, D. A., Phillips, A. C., et al. 1996, ApJ, 465, L15

Vogt, N. P., Phillips, A. C., Faber, S. M., et al. 1997, ApJ, 479, L121

Vogt, N. P. 2001, Distant Disk Galaxies: Kinematics and Evolution to z∼1, in Deep Fields, ed. S. Cristiani, A. Renzini, & R. E. Williams, ESO astrophysics symposia (Springer), 112

Ziegler, B. L., Böhm, A., Fricke, K. J., et al. 2002, ApJ, 564, L69

# Galaxy Collisions, Gas Stripping and Star Formation in the Evolution of Galaxies

Jan Palouš

Astronomical Institute, Academy of Sciences of the Czech Republic
Boční II 1401, 141 31 Praha 4, Czech Republic
palous@ig.cas.cz

## Abstract

*I review gravitational and hydrodynamical processes during formation of clusters and evolution of galaxies. Early, at the advent of N-body computer simulations, the importance of tidal fields in galaxy encounters has been recognized. Orbits are crowded due to tides along spiral arms, where the star formation is enhanced. Low relative velocity encounters lead to galaxy mergers. The central dominating galaxies in future clusters form before the clusters in a merging process in galaxy groups. Galaxy clusters are composed in a hierarchical scenario due to relaxation processes between galaxies and galaxy groups. As soon as the overall cluster gravitational potential is built, high speed galaxy versus galaxy encounters start to play a role. These harassment events gradually thicken and shorten spiral galaxy disks leading to the formation of S0 galaxies and ellipticals. Another aspect of the high speed motion in the hot and diluted Intracluster Medium (ICM) is the ram pressure exerted on the Interstellar Matter (ISM) leading to stripping of the ISM from parent spirals. The combinations of tides and ram pressure stripping efficiently removes the gas from spirals, quenching the star formation in galactic disks, while triggering it in the tidal arms and at the leading edge of gaseous disk. Gas stripping from disks transports the metals to the ICM. In some cases, the gas extracted from the galactic disks becomes self-gravitating forming tidal dwarf galaxies.*

*Star formation (SF) is another important ingredient in the evolution of galaxies. Young stars provide the energy, mass feedback and the metals to the ISM. SF also drives the supersonic turbulence and triggers SF at other places in the galaxy. Examples of supershells in the ISM, resultant from the evolution of stellar cluster, are given. Structures, supershells, filaments and sheets are also produced when the ISM is compressed during galaxy collisions and when high velocity clouds (HVC) fall and ram through gaseous galaxy disks. In some cases, the compressed structures become self-gravitating and fragment to produce clumps. When a low density medium is compressed, the clump masses and sizes correspond to those of giant molecular clouds (GMC). When already existing GMC's are compressed they are driven to collapse forming massive superstar clusters. The fragmentation process in the dense environment of GMC's produces fragments of stellar masses with the mass function having a slope similar to the stellar Initial Mass Function.*

*Reviews in Modern Astronomy 18.* Edited by S. Röser
 ISBN 3-527-40608-5

# 1 Introduction

The internal evolution of isolated galaxies is the subject of many studies. The evolution of stellar disks depends on the ratio between the radial component of the velocity dispersion $\sigma_R$ multiplied with the local epicyclic frequency $\kappa$, which is an analog of pressure in the stellar disk, and the disk surface density $\Sigma$ multiplied with the constant of gravity $G$, which gives the local gravitational force in the flat disk. To evaluate the local stability of a rotating, self-gravitating disk, the above quantities are combined in the Toomre (1964, 1977) $Q$ parameter

$$Q = \frac{\sigma_R \kappa}{3.36 G \Sigma} \tag{1}$$

For $Q < 1$ the disk is locally unstable and forms large scale deviations from the axial symmetry - spiral arms. A purely stellar disk heats up, when individual stars scatter on spiral arms, and self-stabilizes.

If the fraction of the disk kinetic energy in random motions is small enough, the bar forms in its central part (Ostriker & Peebles 1973). The bar exchanges the angular momentum with stars, which results in radial redistribution of mass leading to the formation of a central mass concentration. The growing central mass concentration partially dissolves the bar itself (Shen & Sellwood 2004).

A dissipative gaseous interstellar medium (ISM) is an additional component of the disk moving in a common gravitational field with stars. Energy dissipation, supersonic shocks and collisions of interstellar clouds reduce random motions of the ISM agitating spiral-like instabilities, which gradually heat the stellar disk. In an isolated spiral galaxy, the heating by a bar and spiral arms increases the stellar velocity dispersion by less than a factor of 2 (Bournaud et al. 2004).

The ISM complements the influence of the central mass concentration with viscous forces that shift systematically the gas flow-lines relative to the stellar orbits. The evolution of a bar and spiral structure in an isolated galaxy, their growth and dissolution, is driven by a combination of the increase in stellar velocity dispersion due to scattering of individual stars on large scale deviations from the axial symmetry, by the growth of the central mass concentration and by the torque between the gaseous and stellar components (Bournaud & Combes 2004).

Star formation, the mass recycling between stars and ISM, and the energy feedback are also recognized as fundamental processes. They are included in closed box models the evolution of isolated galaxies (Jungwiert et al. 2001, 2005). Several issues are addressed, remaining as open question if, when, and how much the box has to be open to gas infall, environmental effects, and galaxy major and minor mergers:

- Persistence of star formation. With the present rate, star formation would consume all the gas within less than a few Gyr. Fresh gas supplies are needed to keep the star formation active.

- The metallicity distribution and chemical evolution of the disk. The G-dwarf problem requires gas infall.

- The existence of a thin gaseous disk, where recurrently the instabilities operate also needs accreted fresh gas.

- Thin disks are destroyed in minor merger and harassment events. They have to be recreated.

- Renewal of bars driving the mass to the galactic center also requires gas accretion.

In this review, the tides, major, intermediate and minor mergers events, harrasment events, gas stripping, star formation and feedback are described. We discuss the relevance of these processes in the formation, and evolution of galaxies, and in the evolution of galaxy groups and clusters.

# 2 Tides

The importance of tides has been shown in early numerical N-body simulations by Toomre & Toomre (1972). Their restricted N-body simulations of an encounter between two spiral galaxies led to the basic features. As the galaxies pass by each other, tidal forces provide the stars and ISM of the disks with sufficient energy to escape the inner potential well. Two streams of disk material appear on both near and far side of the disks relative to the position of the peri-center. In the near side, stars and gas form a tidal bridge between the disks while the far side material forms long tidal tails. The length and the shape of the tails is a sensitive function of the relative velocity and of the geometry of the encounter: the most effective are the low relative velocity prograde encounters, where the spin and angular momentum vectors are aligned. In this case, the momentum and energy exchange in resonant orbits is most effective, leading to long tidal arms and a close interconnection between the central parts of the original disks, to merge in the future.

The evolution of tidal debris is described by Mihos (2004). Most of the remnant material remains bound. It follows elliptical orbits and only a small fraction of the orbits are unbound. The fraction of gravitationally bound material is proportional to the extend of the dark matter halo: large halos result in less unbound material. The bound tidal debris return after some time to the galaxy disk in the form of gas infall and high velocity clouds, which may trigger the star formation. A fraction of the disk gas on unbound orbits becomes part of the intracluster medium (ICM). The metallicity of the ICM, which is about 1/4 $Z_\odot$, gives an evidence of the connection between galaxy disks and ICM. The slow encounters are the most effective in driving the gas of the original disks inward to the center of the future galaxy produced by the merger, which triggers a nuclear starburst.

# 3 Major mergers

Computer simulations of low velocity encounters of spiral galaxies with similar masses - major merger events - produce remnants that are in surprisingly good agreement with the observed shapes, density profiles and velocity distribution of the observed giant elliptical galaxies. The mass ratio of the progenitor disks determines the global properties of the remnant (Burkert & Naab 2003a, b). Giant elliptical galaxies can be subdivided into disky and boxy types (Naab & Burkert 2003): the less massive disky giant ellipticals show disk-to-bulge ratios of S0 galaxies and rotation in their outer parts. They can be understood as resulting from major mergers with mass ratios from 1/1 to 1/4. The more massive boxy giant ellipticals form by $1:1$ mergers with special initial orientations only, which weakens the disk merger scenario as the possible formation mechanism. Other processes have to be added, such as the star formation and the energy dissipation in the gaseous components of progenitor disks.

The present day giant elliptical galaxies may have been formed by major mergers of gas-rich galaxies with a subsequent starburst, or by mergers of less gas-rich galaxies, which is more common in galaxy clusters. The second scenario assumes that a fraction of stars forms before ellipticals in smaller gas-rich galaxies. In the $\Lambda$CDM cosmological simulations the number of mergers varies with z. The giant haloes of future elliptical exist already at $z = 6$ with the relatively small variation of mass at $z < 6$. Giant ellipticals exist at redshifts $z = 3$ with about $50\%$ of stellar mass, however, a typical cD galaxy has suffered significant merging events even at redshifts $z < 1$ (Gao et al. 2004). Bimodal distribution of metallicity observed in elliptical galaxies is a product of gas-rich mergers, when the globular clusters form in the tidal arms (Li et al. 2004).

The Antennae, the colliding pair of galaxies NGC 4038/39, provide an nearby example of an early phase of a low velocity encounter between galaxies. The velocity field of Antennae galaxies, showing the details of the interaction, has been measured by Amram et al. (1992). Later, when the collision partners will not be distinguished any more, it should result in a field elliptical galaxy. Another example of a slightly more advanced ongoing merger event, is the Hickson compact group 31 (Amram et al. 2004). HCG 31 is a group in early phase of merger growing through slow and continuous acquisition of galaxies from the associated environment. One of the best examples of remnants after a recent merger event is the giant elliptical galaxy NGC 5018. Its inner part shows a uniform $\sim$ 3 Gyr old stellar population presumably produced in a merger induced starburst (Buson et al., 2004).

# 4 Intermediate and minor mergers

Collisions of intermediate (1/5 - 1/10) mass ratio partners show the intermediate merger events, which are explored by Bournaud et al. (2004). They result in a hybrid system with spiral-like morphology and elliptical-like galaxy kinematics similar to some of the objects identified by Chitre & Jog (2002).

Minor mergers, i.e. merger of a galaxy with a satellite 1/10 or less massive than the galaxy itself, are discussed in the context of formation of thick disk and of disk globular clusters (Bekki & Chiba 2002; Bertschik & Burkert 2003). The disk globular clusters are formed in the high-pressure dense central region of the gas-rich dwarf galaxy, which is compressed in a tidal interaction with the thin disk of the spiral galaxy, and later they are dispersed in the disk region, when the original dwarf merges with the galaxy. The origin of thick galactic disks may be the result of the same minor merger event producing the disk globular clusters.

# 5 Evolution of galaxies in clusters

An overabundance of spiral galaxies and an under abundance of S0 galaxies in high-redshift clusters, when it is compared to low $z$ clusters, is a consequences of galaxy merger events and of other environmental effects. In $\Lambda$DCM cosmological models with $\Omega_0 = 0.4$ and $\Omega_\Lambda = 0.6$, the galaxy interactions and tidal effects play a rôle mainly at intermediate redshifts $0.5 < z < 5$ (Gnedin 2003). Galaxies entering clusters with low relative velocities merge their halos with the cluster and their subsequent dynamical evolution is due to perturbations along their orbit. The infalling galaxy groups experience merger events of their members. After virialization, when the galaxy velocity in the cluster increases to a few $10^3$ km s$^{-1}$ the mergers are suppressed and high-speed galaxy encounters and interactions with the intracluster medium (ICM) start to play a role.

The ICM consists of hot ($\sim 10^7$K) diluted ($10^{-3} - 10^{-4}$ cm$^{-3}$) gas, which is detected in X-ray observations. The optical/infrared observations (Renzini, 1997) monitor the stellar component of galaxies, and also their chemical composition. The bulk of the cluster light is produced in bright giant elliptical galaxies and in galactic bulges by intermediate and low mass stars. The content of Fe in clusters scales with their total light and the abundances of various elements does not seems to change from cluster to cluster (Renzini, 2004). If the metals are mainly produced in SN Type Ia and SN Type II, the constancy of metallic ratios implies the same universal ratio of the two types of SN. It may be interpreted as a sign that majority of cluster metals have been produced in the dominating giant galaxies, and that the stars formation holds a universal initial mass function (IMF). It appears that the total mass of metals in the ICM supersedes the total mass of metals in stars (Renzini 2004). The galaxies lose more metals to the ICM than they are able to retain. Their total abundance does not correspond to the actual SN rates, the observations could be reproduced if the SN Type Ia in ellipticals have 5 - 10 time larger rates in the past compared to present and if the slope of the IMF is not too shallow: Salpeter (1955) IMF with the power law slope -2.35 matches the requirements.

How the metals were transported to the ICM? Potential mechanisms are the ICM ram pressure stripping of the galactic ISM, which may become more effective in combination with gravitational tides, or the star formation feedback connected to early stellar winds driven by the starburst forming majority of the stellar galaxy itself. An evidence for early stellar winds serve the Lyman-break galaxies (Pettini et al. 2003). Another possibility are late local winds due to massive starbursts (Heckmann 2003) or flows driven by a declining rate of SN type Ia (Ciotti et al. 1991). Both, gas extraction by stripping and gas ejection by winds will be described below.

# 6 Galaxy harassment

Moore et al. (1996) discuss the heating effects influencing sizes and profiles of dark matter haloes of galaxies in clusters: tidal heating of haloes on eccentric orbits and impulsive heating from rapid encounters between haloes. N-body simulations show that both processes restrict the halo sizes. Frequent high speed galaxy encounters in clusters, galaxy harassment, drive the morphological transformation of galaxies in clusters. Moore et al. (1995) show the dramatic evolution, which happened in clusters during the recent $\sim$ 5 Gyr. Both, young clusters at $z \sim 0.4$ and old nearby clusters have central dominant elliptical galaxies, which have been formed before clusters. The young clusters are populated by many spiral galaxies, but the old, large virialized systems, have inside of a deeper potential valley all other galaxy types, like S0, dwarf Ellipticals, dwarf S0, dwarf Spheroidals and Ultra Compact Dwarfs (Moore 2003). Numerical simulations show that if the cluster velocity dispersion is more than a few times the internal velocity dispersion of galaxies, they do not merge in encounters. The observed morphological change is due to impulsive interactions during high speed encounters restricting the dark matter haloes and pumping the kinetic energy into disks of spirals changing them to S0, dE, dS0, dSph, UCD galaxies. Hydrodynamical processes, like gas stripping, also matters, and help with the change, particularly in relation to star formation and ICM.

# 7 Gas stripping

Gunn & Gott (1972) in a description of events during and after the collapse, or formation of a galaxy cluster, predict the formation of hot and diluted intracluster medium (ICM) produced in shock randomization of infalling gaseous debris. The interstellar medium (ISM) in a galaxy moving through the ICM feels the ram pressure $P_r$, which is proportional to

$$P_r \sim \rho_{ICM} v^2, \tag{2}$$

where $\rho_{ICM}$ is the density of the ICM and $v$ is the velocity of the galaxy relative to the ICM. The ISM may be stripped away the parent galaxy if at given distance from the galactic center the ram pressure exceeds the gravitational restoring force $F_r$

$$F_r = 2\pi G \Sigma_s \Sigma_{ISM}, \tag{3}$$

where $\Sigma_s$ is the surface density of stars, $\Sigma_{ISM}$ is the surface density of the ISM and $G$ is the constant of gravity.

The ram pressure gas stripping influences the presence of the ISM, which may affect the star formation rates in galaxies. It is still an open question, if the observation of Butcher & Oemler (1978, 1984), showing that distant clusters contain a far higher fraction of blue star forming galaxies than their near-by counter-parts, can be explained with the ICM ram pressure stripping of the ISM from galaxies.

A mechanism that suppresses the star formation in local galaxy clusters is apparently connected to the morphological transformation of galaxies: S0 galaxies are under-represented in distant galaxy clusters, luminous spiral galaxies are in deficit near the centers of local galaxy clusters (Dressler et al. 1997; Couch et al. 1998). This suggests that the cluster environment removes the ISM from galaxies, suppressing star formation and transforming spirals to S0's.

Numerical simulations of ram pressure gas stripping using 3-dimensional SPH/N-body code have been preformed by Abadi et al. (1999). They confirm the predictions of Gunn & Gott (1972) that the radius to which the gas is removed from the parent

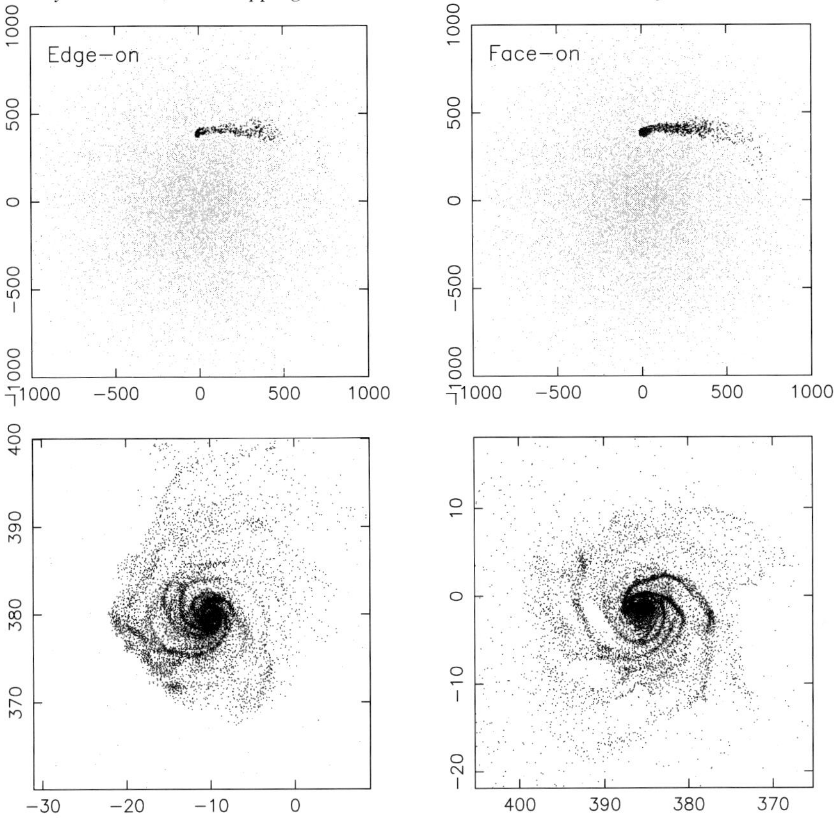

Figure 1: Simulations of ram pressure gas stripping (Jachym, 2005). Edge-on orientation of the galaxy plane relative to the orbit (left panel) is compared to the pole-on orientation of the galaxy plane (right panel).

galaxy depends on the relation of the ram pressure to the restoring force. But in any case a substantial part of the cold gas remains sufficiently bound to the stellar disk. The star formation rate is reduced by a factor of 2 only, which brings them to a conclusion that the simple ram pressure stripping does not adequately explain the sharp decline of star formation seen in Butcher-Oemler effect.

Volmer et al. (2001) show in simulations using sticky particles that the gas stripping is rather sensitive to the galaxy orbit, particularly to the minimum distance to the cluster center, and also to the orientation of the galactic disk relative to the orbit inside the cluster. In some cases the removed ISM is re-accreted and it falls back to the galactic disk, possibly triggering star formation in the central part of the disk within the remaining gas. Simulations of ram pressure gas stripping along an orbit in a cluster by Jachym (2005) give similar result: the orbital parameters and the orientation of the galaxy are important. The gas is more effectively removed when the pole of the galactic plane is near the direction of the galaxy orbital motion in a cluster (Fig. 1) compared to the situation when the galaxy moves edge-on along its orbit.

# 8 Gas stripping and tides

Recently reviews on the efficiency of the stripping mechanism were given by van Gorkom (2004) and Combes (2004). The theoretical considerations stress the effects of viscosity and thermal conduction including Kelvin-Helmholtz instability (Nulsen, 1982). The turbulence, shells and supershells formed by star formation increase the effect of the ram pressure stripping due to erosion at the edges of ISM holes (Bureau & Carignan, 2002). Ram pressure also contributes to the overpressure inside of expanding structures increasing their sizes.

A proto-type of overwhelming gas stripping in the Virgo cluster is the galaxy NGC 4522 with a truncated HI disk, enhanced star formation in the central region, extra-planar gas, while undisturbed stellar disk. The gas distribution suggests an ICM-ISM interaction, while the undisturbed disk rules out a gravitational interaction (Kenney & Koopmann, 1999).

The transformation of galaxies in dense environment results from the combined action of gas stripping and gravitational tides. C 153 galaxy in the cluster Abel 2125 shows an ongoing gas stripping: A tail of ionized gas is seen in [O II] emission, which extends at least 70 kpc toward the cluster core along C 153 orbit, coincide with a soft X-ray feature seen in the Chandra observations (Keel et al. 2004). At the same time the HST optical picture shows clumpy morphology, including luminous star-forming complexes and chaotic dust features. The perturbed stellar disk with enhanced star formation activity suggests a possibility that a burst of star formation has been initiated during the close passage of C 153 to the cluster center.

More examples of the stripped dwarf irregular galaxies (van Zee et al. 2004) and of galaxies with truncated star formation disks in Virgo cluster (Koopmann & Kenney 2004a, b) show both the gas stripping and gravitational interactions including the induced star formation, however, the dominant environmental effect on galaxies in clusters is the ram pressure gas stripping.

# 9 Tidal Dwarf Galaxies

The long tidal tails observed in many cases of interacting galaxies show massive clumps of $10^9 M_\odot$ (e.g. IC 1182, NGC 3561, NGC 4676, etc.). These massive blobs has been named Tidal Dwarf Galaxies (TDG), since they have galactic masses and the chemical composition corresponds to the recycled matter pushed out of the disks of the interacting partners. To become a long living independent galaxy they should be gravitationally bound systems (Duc et al. 2000). The most prominent interacting system, the Antennae galaxies, has been studied by Mirabel et al. (1992), who describe a TDG candidate at the tip of their long tidal arm. Hibbard et al. (2001) also analyzed this concentration of gas and star forming regions, however, from observations it is very difficult to assess if it is gravitationally bound.

To decide if a TDG candidate will be a new galaxy formed out of an interaction, several questions have to be addressed (Bournaud et al. 2004):

- Are the blobs real concentrations in three dimensions? They may be just projection effects due to tidal arm geometry.

- Are they kinematically decoupled from the tidal arms? Are they long living?
- Do they contain dark matter?

Bournaud et al. (2004) give the answer at least to the first question: The simulations of galaxy collisions provide shapes of tidal arms, which may be virtually observed from all the sides. They conclude that some observed $10^9 M_\odot$ mass concentrations are real TDG candidates. Some of them are self-gravitating, but to decide on their future is still difficult with the current resolution of simulations. The star formation, energy and mass feedback have to be included in the future numerical experiments. The third problems on the content of the dark matter also remains open due to the uncertainty in the internal kinematics of TDG candidates.

The simulations provide one more important conclusion (Duc et al. 2004): the existence of $10^9 M_\odot$ mass concentrations at the tips of tidal arms is rather sensitive to the extent and the density profile of the halo. It has to be extended enough ($\sim$150 kpc from the center of a collision partner) so that the collision happens within it. Then the flow lines of the perturbed gas from different galactocentric distances in the original disk concentrate at the tips of the tidal arms, giving a kinematical origin to the TDG. When the halo is too concentrated so that the collision happens at the distance, where the rotation curve already decreases, the perturbed gas populates all the tidal arm and there is no place where $10^9 M_\odot$ may gather.

The kinematical gathering of stars and gas distinguishes TDG from super-star clusters (SSC), which are seen in the interacting galaxies. SSC are not only less massive ($10^5 - 10^7 M_\odot$) but they arise from gravitational instabilities in the stellar or gaseous components along the tidal arms.

# 10 SMC, LMC and the Milky Way system

The Small (SMC), and Large Magellanic clouds (LMC) and the Milky Way form the nearest interacting system of galaxies, where the gravitational and hydrodynamical processes can be studied. We see the result of a combination of gravitational tidal forces with gaseous ram pressure. The star formation and the mass and energy feedback is also involved as demonstrated with many expanding ISM shells and supershells.

The high resolution HI surveys of the LMC (Kim et al. 1998, 1999), of the SMC (Staveley-Smith et al. 1997; Stanimirovich et al. 1999), of the Magellanic bridge region (Muller et al. 2003) and of all the system (Bruns et al. 2005) have been preformed with the Australia Telescope Compact Array and with the Parkes radiotelescope. The following large-scale features are distinguished on the (l, b) integrated HI intensity maps: LMC, SMC, Magellanic Bridge joining the two clouds, Magellanic Stream starting at the SMC and following an almost polar plane passing less than about 10° of the galactic south pole and stretching more than 100°, the Leading Arms – the HI gas preceding the motion of the clouds, and the Interface Region – the HI between Magellanic Bridge and Magellanic Stream (Fig. 2). All these features are also distinguished on the average radial velocity maps. The radial velocity changes smoothly from $RV_{LSR} = -400\,\mathrm{km\,s^{-1}}$ at the end of the Magellanic Bridge to $+100$ km s$^{-1}$ at the SMC, to $+240$ km s$^{-1}$ at the LMC, while the Leading Arm does not show a clear gradient. Deprojection correcting for the solar motion reduces the velocity difference between the LMC and SMC from $\Delta RV_{LSR} = 123$ km s$^{-1}$ to $\Delta RV_{GSR} = 67$ km s$^{-1}$. A further correction taking into account motions of the LMC relative to the MW reduces the difference to $\Delta RV_{LMCSR} = 10$ km s$^{-1}$. The HI average radial velocities show that the encounter between LMC and SMC happens at a small velocity not much larger that 10 km s$^{-1}$, which makes the interaction rather long, giving the time to the gas to flow away from its parent cloud and form the observed features.

The LMC HI disk seems to be compressed at the side opposite to the SMC, and the LMC shows rotation almost perpendicular to SMC - LMC direction. The LMC disk has a diameter of about 7.3 kpc with a rotation curve rising rapidly to 55 km s$^{-1}$ in the inner 1.5 kpc, more smoothly to a 63 km s$^{-1}$ peak at 2.4 kpc and declining thereafter (Kim et al. 1998). The SMC also shows rotation: SMC disk includes the bar-like feature about 4 kpc in extent, with a velocity gradient of about 100 km s$^{-1}$ (Stanimirović et al. 2004). The rotation curve rises to about 60 km s$^{-1}$ up to the turnover radius of 3 kpc. The velocity dispersion is high along the high column density axis of the Magellanic Bridge. LMC, Magellanic Bridge and SMC show similar velocity fields, which can result from the rotation of all the three partners with the same orientation along the axis LMC - SMC. The velocity field is rather broken at the southern part (in galactic coordinates) of the SMC and of the Magellanic Bridge. There, the Magellanic Stream and Interface Region start. At the other end, in front of the LMC, the Leading Arm can be split into three parts. Individual features in the Leading Arm show head-tail structure with the orientation along the direction of possible space motion of the LMC as it is given by Kroupa & Bastian (1997) and van der Marel (2001).

The HI observation should be complemented with the studies of distribution of planetary nebulae (Dopita et al. 1985), of carbon stars (Kunkel et al. 2000), and of Cepheids (Groenewegen 2000), and with the near-infrared star counts from the Two Micron All Sky Survey (2MASS) and the Deep Near-Infrared Southern Sky Survey (DENIS). van der Marel (2001) shows that LMC disk is not circular at larger radii, it is elongated in the direction of the Galactic center, suggesting the influence of the tidal forces of the Milky Way. The data should be compared to models of the interaction in attempts not to be in contradiction.

The N-body simulations modeling the gravitational interaction of the SMC with the LMC and Milky Way (Gardiner & Noguchi 1996) show that the last two close encounters between the interaction partners, 1.5 and 0.2 Gyr ago, are able to explain many of the observed structures. Magellanic Stream and Leading Arm have been created as a consequence of the former close encounter, 1.5 Gyr ago. The Magellanic Bridge and the Interface Region have been formed later during the last close encounter between LMC and SMC 0.2 Gyr ago. The discovery of the Leading Arm stressed the importance of gravitational tides and questioned the role of ram pressure stripping (Putman et al. 1998). However, the separation between gaseous and stellar features, absence of stars in tidal Magellanic Bridge and Magellanic Stream, filamentary structures in the SMC and head-tail structures in the Leading Arm and in the Magellanic Stream show that also hydrodynamical forces like ram pressure gas

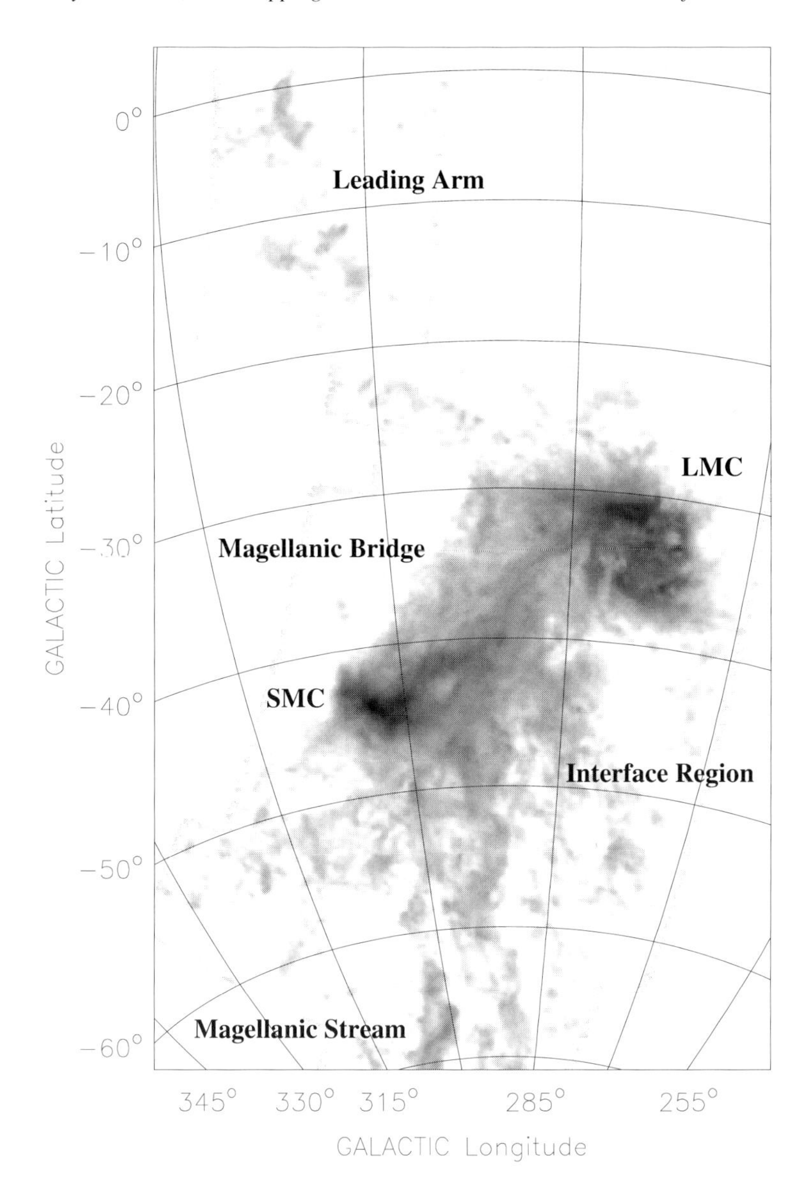


Figure 2: LMC & SMC: an interacting system of dwarf galaxies (Bruns et al. 2005).

stripping have to act there. Most of the gas in Magellanic Bridge and Leading Arm is coming from the SMC. The preencouter SMC, more that 1.5 Gyr ago, must have a gas disk of about 10 kpc in diameter (Stanimirović 2004), which shrunk forming the Magellanic Stream and Leading Arm 1.5 Gyr ago and Magellanic Bridge and Interface Region 0.2 Gyr ago. However, still a substantial amount of angular momentum of the original disk is left corresponding to other simulations of dwarf galaxies (Mayer et al. 2001) demonstrating that the tidal stripping removes the angular momentum rather slowly, at the timescale of 10 Gyr. Consequently, the recent two encounters have not been able to remove a substantial part of the original angular momentum making possible to see the rotation of the SMC at present times. The present rotation curve of the SMC shows that the dynamical mass within 4 kpc is about $2.4\ 10^9 M_\odot$ three-quarters of which is stellar. The dark matter is not needed for an explanation of the rotation speeds in the SMC.

Careful analysis of the HI distribution in the Magellanic Stream (Putman et al. 2003) suggests that the Magellanic Bridge is older that assumed above and the LMC and the SMC are bound together for at least two orbits. The dual filaments emanating from the SMC and from the Magellanic Bridge are of tidal origin and shaped by a small amount of ram pressure.

Star formation and energy feedback from young stars are included in the so far most sophisticated N-body model of the LMC - SMC - Milky Way encounter by Yoshizawa & Noguchi (2003). This model agrees well with several observed features including the Magellanic Stream, which apparently has tidal origin, it reproduces the presence of young stars in the south-east wing of the SMC and it also reproduces the acceleration in the star formation activity, which is due to recent close encounters between the clouds. Some open and unsolved questions remain: bimodal or many peak distribution in the main gaseous body of the SMC and Magellanic Bridge remain to be interpreted, it probably originates in the numerous expanding shells. The shells and supershells, if they trigger star formation, also remain an open question.

The influence of other MW satellites is unclear. The tidal stream of HI clouds connected to disruption of the Sagittarius dwarf galaxy on the polar orbit aloud the Milky Way provides a possible explanation for the anomalous velocity distribution of HI clouds near the south galactic pole (Putman et al. 2003). The possibility of formation of the local group dwarf members including LMC and SMC out of the Milky Way encounter with the M31 galaxy is discussed (Sawa & Fujimoto, 2004). Another picture describes the interaction of the Fornax-LeoI-LeoII-Sculptor-Sextans stream with the Magellanic Stream causing the gas stripping from the Fornax (Dinescu et al. 2004). Kroupa et al. (2005) propose the origin of the whole Local Group with the local dwarf galaxies in a common great circle.

# 11 Star formation, energy and mass feedback

Star formation is a complex process of the gravitational collapse and fragmentation, where the thermal and magnetic support competes with supersonic shock waves and energy dissipation. The density increases by 20 orders of magnitude from that of a molecular cloud core. The interplay between gravity, magnetic forces, hydrodynamical processes, radiative transfer and chemistry happens in a turbulent interstellar medium. Supersonic flows form sheets and filaments involving mass concentrations, which in some cases are bound by self-gravity. Some of them collapse forming single or binary stars, others disperse.

There are different sources powering the interstellar structures. On a small scale the pre-main sequence stellar winds and stellar radiation, on somewhat larger scale the main sequence stellar winds, and on even larger scale the supernovae. Young stars pump energy back to the interstellar medium, which influences the conditions for further star formation. The energy released by young OB associations compresses the ambient medium into shells and super-shells, which may collapse and trigger new star formation. The energy feedback triggering shell collapse and further star formation is a self-regulating mechanism of the galaxy evolution.

## 11.1 The observation of shells

Shells and supershells and holes in the HI distribution have been discovered in the Milky Way by Heiles (1979, 1984), in M31 by Brinks and Bajaja (1986), in M33 by Deul and Hartog (1990), in LMC by Kim et al. (1999), in SMC by Stanimirovič (1999), in HoII by Puche et at. (1992), in Ho I by Ott et al. (2001) and in IC 2574 by Walter and Brinks (1999). Most probably they are created by an energy release from massive stars, however, an alternative explanations, infall of high velocity clouds (Tenorio-Tagle and Bodenheimer, 1988), or gamma ray bursts (Efremov et al., 1998; Loeb and Perma, 1998) has been invoked in some cases. The majority of the observed shells is due to star formation (Ehlerová and Palouš, 1996). In a new search by Ehlerová and Palouš (2005), more than 600 shells have been identified in the Leiden-Dwingeloo HI survey of the Milky Way. In Fig. 3 we show the re-identification of a shell previously discovered by Heiles (1979) and a newly discovered shell. The distribution of them in the radial galactocentric direction and in the direction perpendicular to the galactic disk is similar to stellar distribution supporting the idea of a connection between massive stars and shells.

## 11.2 The collapse of shells

We discuss the supersonic expansion of shells and sheets from regions of localized deposition of energy and address the question if and when they fragment and collapse due to gravitational instability. The energy input from an OB association creates a blast-wave which propagates into the ambient medium (Ostriker & McKee 1988; Bisnovatyi-Kogan & Silich, 1995). The schematic representation of the situation is shown in Fig. 4. After the initial fast expansion the mass accumulated in the shell cools and collapses to a thin structure, which is approximated as infinitesimally thin layer surrounding the hot medium inside. Neglecting the external pressure and assuming the constant energy input $L$, the self-similar solution for radius $R$, expansion velocity $V$ and column density $\Sigma_{sh}$ is (Castor et al. 1975; Ehlerová & Palouš 2002):

$$R(t) = 53.1 \times \left(\frac{L}{10^{51}\ \mathrm{erg\ Myr^{-1}}}\right)^{\frac{1}{5}} \times \left(\frac{\mu}{1.3}\frac{n}{\mathrm{cm^{-3}}}\right)^{-\frac{1}{5}} \times \left(\frac{t}{\mathrm{Myr}}\right)^{\frac{3}{5}} \mathrm{pc} \quad (4)$$

$$V(t) = 31.2 \times \left(\frac{L}{10^{51}\ \mathrm{erg\ Myr^{-1}}}\right)^{\frac{1}{5}} \times \left(\frac{\mu}{1.3}\frac{n}{\mathrm{cm^{-3}}}\right)^{-\frac{1}{5}} \times \left(\frac{t}{\mathrm{Myr}}\right)^{-\frac{2}{5}} \mathrm{kms^{-1}} \quad (5)$$

$$\Sigma(t)_{\mathrm{sh}} = 0.564 \times \left(\frac{L}{10^{51}\ \mathrm{erg\ Myr^{-1}}}\right)^{\frac{1}{5}} \times \left(\frac{\mu}{1.3}\frac{n}{\mathrm{cm^{-3}}}\right)^{\frac{4}{5}} \times \left(\frac{t}{\mathrm{Myr}}\right)^{\frac{3}{5}} \mathrm{M_{\odot}pc^{-2}}, \quad (6)$$

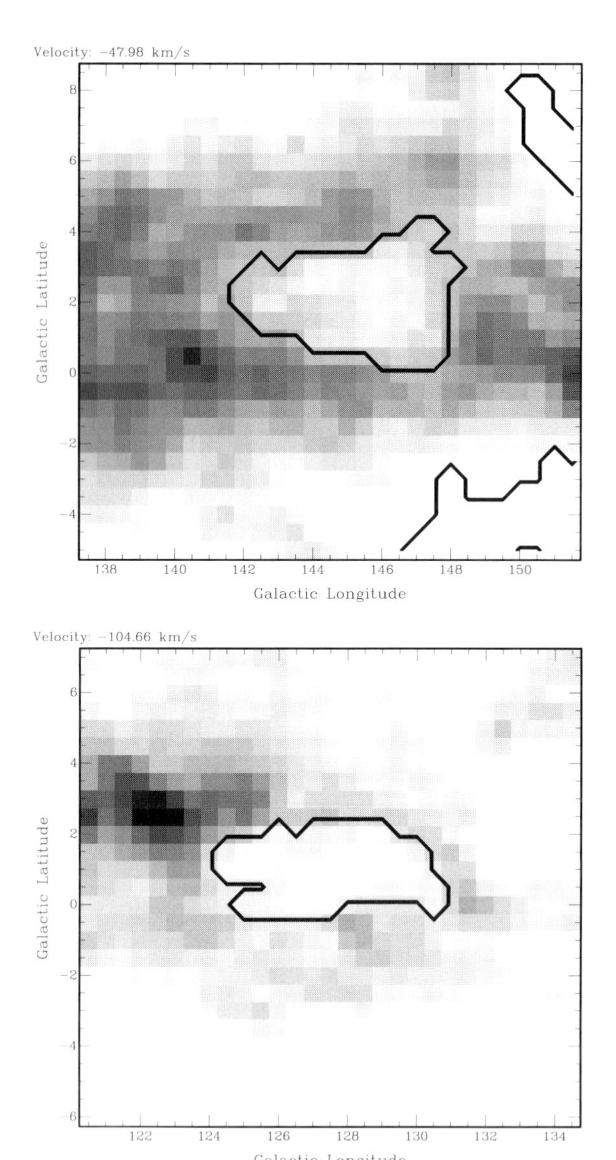


Figure 3: Re-identification of the shell GS 128+01-105 discovered by Heiles (1979) - upper frame, and a newly discovered shell - lower frame.

where $n$ is the density, $\mu$ is the mean atomic weight of the ambient medium and $t$ is the time since the beginning of an expansion.

The linear analysis of hydrodynamical equations including perturbations on the surface of the shells has been performed by Elmegreen (1994) and Wünsch & Palouš

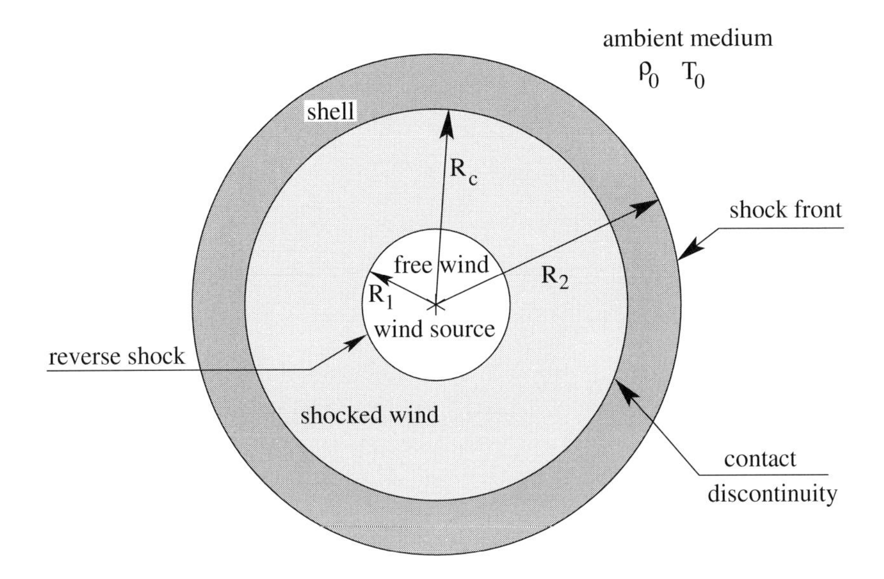


Figure 4: Schematic representation of a supershell expanding around an OB association.

(2001). The fastest growing mode is:

$$\omega = -\frac{3V}{R} + \sqrt{\frac{V^2}{R^2} + \left(\frac{\pi G \Sigma_{sh}}{c_{sh}}\right)^2}, \tag{7}$$

where $c_{sh}$ is the sound speed inside of the expanding shell.

In Fig. 5. we give the time evolution of the fastest mode. At early times, for $t < t_b$, the shell is stable. $t_b$ is the time, when the fastest mode starts to be unstable:

$$t_\mathrm{b} = 28.8 \times \left(\frac{c_\mathrm{sh}}{\mathrm{km\ s}^{-1}}\right)^{\frac{5}{8}} \times \left(\frac{L}{10^{51}\ \mathrm{erg\ Myr}^{-1}}\right)^{-\frac{1}{8}} \times \left(\frac{\mu}{1.3}\frac{n}{\mathrm{cm}^{-3}}\right)^{-\frac{1}{2}} \mathrm{Myr}. \tag{8}$$

Later, for $t > t_b$, when the expansion slows down and reduces the stretching, which acts against gravity, and when the shell column density increases, the shell starts to be gravitationally unstable. For ambient densities similar to values in the solar vicinity, $n \sim 10^{-1} - 10^2$, $t_b$ is a few $10^7$ yr, which means that the gravitational instability is rather slow compared the turbulent collision times and the galactic differential rotation. $t_b$ is much smaller in high density medium of GMC and dense cores, where it is $\sim 10^4$ yr only. Thus the shell gravitational instability is particularly important inside the GMCs.

The dispersion relation of the shell gravitational instability is:

$$\omega(\eta, t) = -\frac{3V}{R} + \sqrt{\frac{V^2}{R^2} - \frac{c_{sh}^2 \eta^2}{R^2} + \frac{2\pi G \Sigma_{sh} \eta}{R}}, \tag{9}$$

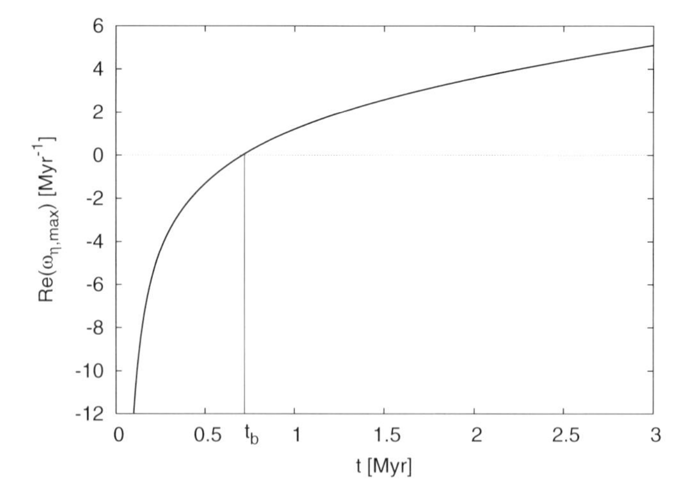


Figure 5: The fastest growing mode.

where $\eta$ is the dimensionless wavenumber and $\lambda$ is the wavelength of the perturbation: $\eta = 2\pi R/\lambda$. It is shown in Fig. 6: it gives the wavelength interval of unstable perturbations.

The resulting number of fragments is inversely proportional to the fragment growth time $t_{growth} = \frac{2\pi}{\omega(\eta,t)}$. Rapidly growing fragments are more frequent in the final mass spectrum than the slowly growing fragments.

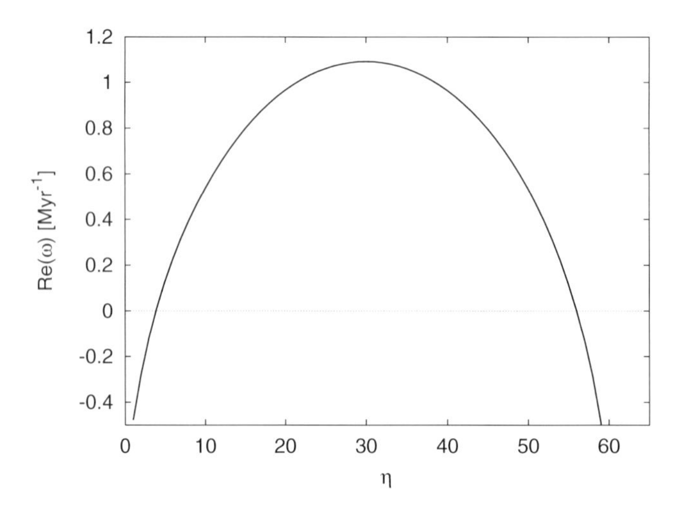


Figure 6: The dispersion relation

Thus the number of fragments in a given volume of radius $R$ is

$$N = \omega \frac{R^3}{(\lambda/4)^3}. \tag{10}$$

A fragment with the wavelength $\lambda$ has the mass

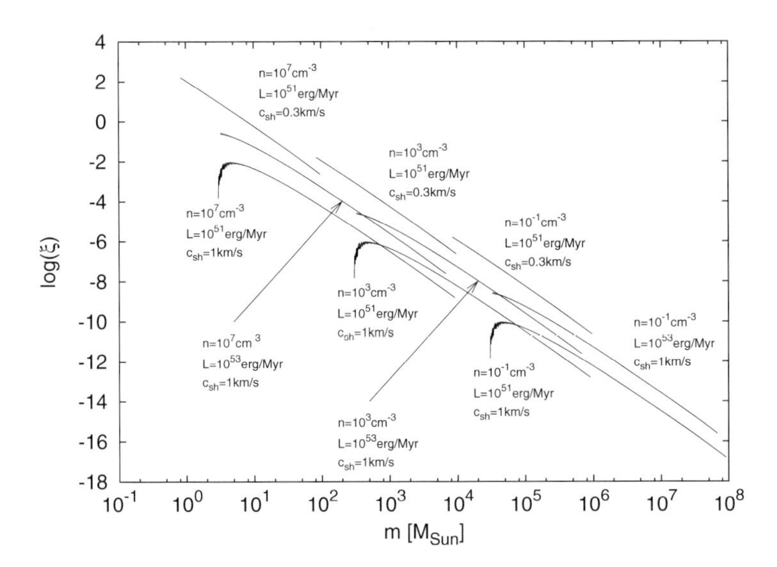


Figure 7: The mass spectrum of fragments of an expanding shell at time $5t_{\text{b}}$.

$$m = \frac{4}{3}\pi(\lambda/4)^3\rho. \tag{11}$$

We derive the mass spectrum $\xi(m) = \frac{dN}{dm}$:

$$\xi(m) = -\frac{4}{3}\,\pi\,R^3\,\rho\,\omega\,m^{-2}. \tag{12}$$

If the dispersion relation $\omega(\eta)$ were a constant, the slope of the mass spectrum $\xi(m) \propto m^{-\alpha}$ would be exactly approximated with a power law slope $\alpha$= 2. But in our case $\omega$ is not only the function of $\eta$ but also of the time $t$.

We assume that $t_{growth}$ for a given $\eta$ is inversely proportional to the average value of $\omega$ for this $\eta$ since the time $t_b(\eta)$ when a given mode starts to be unstable. The time average $\bar{\omega}$ for a given $\eta$ is calculated using the equation:

$$\bar{\omega}(\eta) = \frac{\int_{t_b(\eta)}^{t} \omega(\eta, t')dt'}{t - t_b(\eta)}. \tag{13}$$

The resulting mass spectra for different values of $n$, $c_{sh}$ and $L$, as they have been derived using the thin-shell approximation (4) - (6), are shown in Fig. 7. The high mass parts are well approximated by the power law with a slope $\alpha = 2.2 - 2.4$.

In low density medium, the collapse of expanding shells forms fragments with masses comparable to GMC, however, the collapse time is rather long, a few $10^7$ yr. In high density medium of GMC cores, the collapse time is rather short, $\sim 10^4$ yr, the masses of individual fragments are close to stellar and the initial mass function of fragments has a power law slope close to the Salpeter (1955) value -2.35. We conclude that the fragmentation of expanding shells qualifies as a possible process triggering the star formation in environments, where the density is high enough, or where it has been increased due to mass accumulation.

# 12 Formation of super-star clusters

Young and massive ($10^5 - 10^7 M_\odot$) super-star clusters are observed along tidal arms and bridges of the colliding galaxies. A model of a star forming factory is proposed by Tenorio-Tagle et al. (2003). This model invokes pressure-bounded, self-gravitating, isothermal cloud (Ebert 1955; Bonner 1956), which becomes gravitationally unstable when sufficiently compressed. The gravitational instability allows the cloud to enter isothermal collapse. As the collapse proceeds a first generation of stars is formed in the center of the cloud. The mass and energy feedback of the first generation of stars has an important impact on the collapsing cloud. Stellar winds and supernovae compress the infalling material forming a dense shell (see Fig. 8). The shell is able to trap the ionizing radiation and winds of the first generation of young stars. At the same time the shell fragments forming stars with a high efficiency. Thus the cloud, which is compressed in the case of the galaxy encounter, forms a new super-star cluster.

# 13 Evolution of the star formation rate

The investigation of colors in Hubble deep fields (Madau et al. 1996; 1998; Rowan-Robinson 2003) provides extinction as a function of $z$: it has been higher at $z = 0.5 - 1.5$ than locally, and lower at $z > 2$. Related models of the star formation history show the steep decline of the star formation rate since $z = 1$, the present star formation has in average about an order of magnitude smaller rate compared to the level at $z = 1 - 1.5$. Even deeper in the past the correction for the dust extinction remains highly uncertain and a conclusion on the evolution of the star formation rate for $z = 2 - 6$ is difficult.

This opens the question what drives the star formation. Bars in the central parts of galaxies does not seem to change substantially between $z = 1$ and present. They do trigger the star formation, however, there has to be more partners in the game. The fuel - the gas - is depleted in galaxies not only due to star formation, but also since the ISM is removed from the galaxy disks with tides, with the ICM ram pressure stripping and with the star formation feedback. Number of galaxy interactions also decreases. We conclude that the decline of the star formation is due to lower rate of triggering from galaxy interactions in a combination with the starvation, since the amount of gas in the star forming disks is reduced by environmental effects and by the star formation itself.

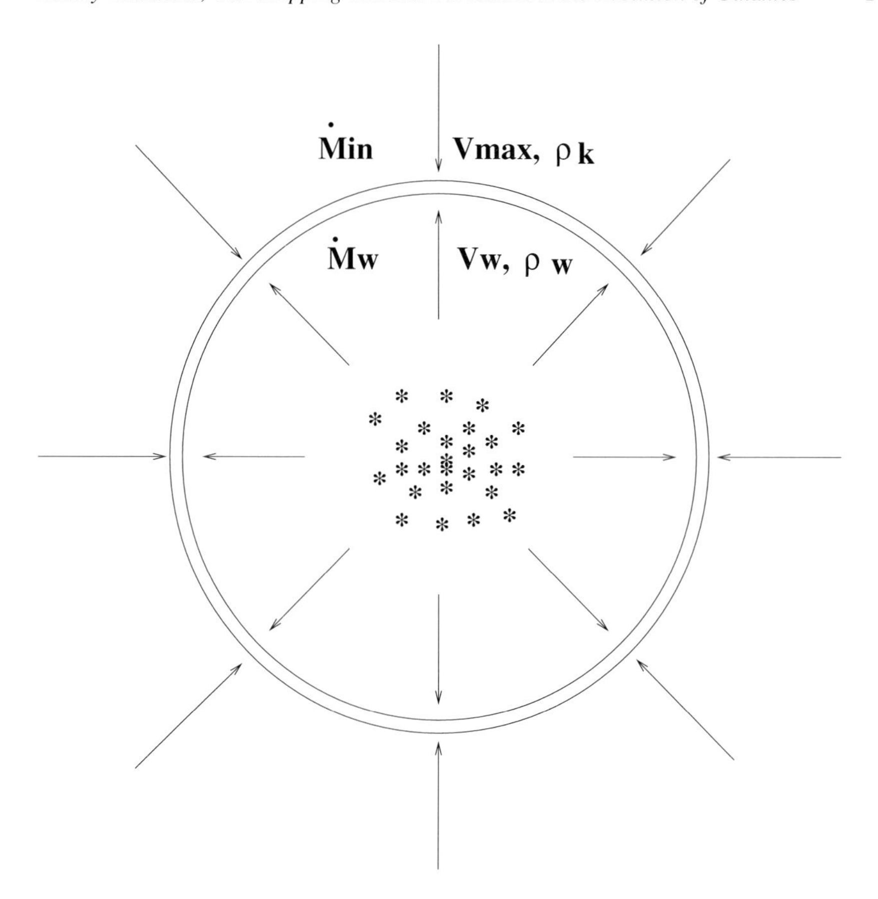


Figure 8: The shell inside of a collapsing cloud

## Acknowledgements

I would like to express my thanks to Guillermo Tenorio-Tagle for comments to the early version of the text. I also acknowledge the support of the Astronomical Institute, Academy of Sciences of the Czech Republic, through the project AV0Z10030501.

## References

Abadi, M. G., Moore, B., & Bower, R. G. 1999, MNRAS, 308, 947

Amram, P., Mercelin, M., Boulsteix, J., & Le Coarer, E. 1992, A&A, 266, 106

Amram, P., Mendes de Oliveira, C., Plana, H., Balkowski, C., Hernandez, C., Carignan, C., Cypriano, E. S., Sodré, L. Jr., Gach, J. L., & Boulsteix, J. 2004, ApJ, 612, L5

Bekki, K., & Chiba, M. 2002, 566, 245

Bertschik, M., & Burkert, A. 2003, RevMexAA, 17, 144

Bisnovatyi-Kogan, G. S. & Silich, S. A. 1995, RvMP 67, 661

Bonner, W. B. 1956, MNRAS, 116, 351

Bournaud, F., & Combes, F. 2004, A&A, submitted

Bournaud, F., Combes, F., & Jog, J. 2004, A&A, 418, L27

Bournaud, F., Duc, P.-A., Amram, P., Combes, F., & Gach, J.-L. 2004, A&A, 425, 813

Brinks, E., & Bajaja, E. 1986, A&A, 169, 14

Bruns, C., Kerp, J., Staveley-Smith, L., Mebold, U., Putman, M. E., Haynes, R. F., Kalberla, P. M. W., Muller, E., & Filipovic 2005, A&A, 432, 45

Bureau, M., & Carignan, C. 2002, AJ, 123, 1316

Burkert, A. & Naab, T. 2003a, in Coevolution of Black Holes and Galaxies, ed. L. C. Ho, Carnegie Obs.

Burkert, A. & Naab, T. 2003b, in Galaxies and Chaos, ed. G. Contopoulos & N. Voglis, Springer

Buson, L. M., Bertola, F., Bressan, A., Burstein, D., & Cappellari, M. 2004, A&A, 423, 965

Butcher, H., & Oemler, A. 1978, ApJ, 219, 18

Butcher, H., & Oemler, A. 1984, ApJ, 285, 426

Castor, J., McCray, R., & Weaver, R. 1975, ApJ, 200, L107

Chitre, A., & Jog, C. J. 2002, A&A, 388, 407

Ciotti, L., Pellegrini, S., Renzini, A., & D'Ercole, A. 1991, ApJ, 376, 380

Combes, F. 2004, IAU Symp. 217, 440

Couch, W. J., Barger, A. J., Smail, I., Ellis, R. S., & Sharples, R. M. 1998, ApJ, 497, 188

Deul, E. R., & den Hartog, R. H. 1990, A&A, 229, 362

Dinescu, D. I., Keeney, B. A., Majewski, S. R, & Terrence, G. M. 2004, AJ, 128, 687

Dopita, M. A., Ford, H. C., Lawrence, C. J., & Webster, B. L. 1985, ApJ, 296, 390

Dressler, A., Oemler, A., Couch, W. J., et al. 1997, ApJ, 490, 577

Duc, P.-A., Bournaud, F.,& Masset, F. 2004, A&A, 427, 803

Duc, P.-A., Brinks, E., Springel, V., et al. 2000, AJ, 10, 1238

Ebert, R. 1955, Z. Astrophys., 36, 222

Efremov, Yu. N., Elmegreen, B. G., & Hodge, P. W. 1998, ApJ, 501, L163

Ehlerová, S., & Palouš, J. 1996, A&A, 313, 478

Ehlerová, S., & Palouš, J. 2002, MNRAS, 330, 1022

Ehlerová, S., & Palouš, J. 2005, A&A, in press

Elmegreen, B. G. 1994, ApJ, 427, 384

Gao, L., Loeb, A., Peebles, P. J. E., White, S. D. M., & Jenkins, A. 2004, ApJ, 614, 17

Gardiner, L. T., & Noguchi, M. 1996, MNRAS, 278, 191

Gnedin, O. Y. 2003, ApJ, 582, 141

Groenewegen, M. A. T. 2000, A&A, 363, 901

Gunn, J. E., & Gott III, J. R. 1972, ApJ, 176, 1

Heckman, T. M. 2003, RMxAC, 17, 47

Heiles, C. 1979, ApJ, 229, 533

Heiles, C. 1983, ApJS, 55, 585

Hibbard, J. E., van der Hulst, J. M., Barnes, J. M., & Rich, R. M. 2001, AJ, 122, 2969

Jachym, P. 2005, in preparation

Jungwiert, B., Combes, F., & Palouš, J. 2001, A&A, 376, 85

Jungwiert, B., Combes, F., & Palouš, J. 2005, in preparation

Keel, W., Owen, F., Ledlow, M., & Wang, D. 2004, AAS, 203, 4705

Kenney, J. D. P., & Koopmann, R. A. 1999, AJ, 117, 181

Kim, S., Staveley-Smith, L., Dopita, M. A., Freeman, K. C., Sault, R. J., Kesteven, M. J., & McConnell, D. 1998, ApJ, 503, 674

Kim, S., Dopita, M. A., Staveley-Smith, L., & Bessell, M. S. 1999, AJ, 118, 2797

Koopmann, R. A., & Kenney, J. D. P. 2004, ApJ, 613, 851

Koopmann, R. A., & Kenney, J. D. P. 2004, ApJ, 613, 866

Kroupa, P., & Bastian, U. 1997, New A, 2, 77

Kroupa, P., Theis, Ch., & Boily, C. M. 2005, A&A, 431, 517

Kunkel, W. E., Demers, S., & Irwin, M. J. 2000, AJ, 119, 2789

Li, Y., Mac Low, M.-M., & Klessen, R. S. 2004, ApJ, L29

Loeb, A., & Perma, R. 1998, ApJ, 503, L35

Madau, P., Ferguson, H. C., Dickinson, M. E., Giavalisco, M., Steidel, Ch. C., & Fruchter, A. 1996, MNRAS, 283, 1388

Madau, P., Pozzetti, L., & Dickinson, M. 1998, ApJ, 498, 106

Mayer, L., Governato, F., Colpi, M., Moore, B., Quinn, T., Wadsley, J., Stadel, J., & Lake, G. 2001, ApJ, 559, 754

Mihos, J.Ch. 2004, IAU Symp. 217, eds P.-A. Duc, J. Braine and E. Brinks, p. 390

Mirabel, I. F., Dottori, H., & Lutz, D. 1992, A&A, 256, L19

Moore, B. 2003, in Clusters of Galaxies: Probes of Cosmological Structure and Galaxy Evolution, ed. J. S. Mulchaey, A. Dressler, and A. Oemler, CUP

Moore, B., Katz, N., & Lake, G. 1996, ApJ, 457, 455

Moore, B., Katz, N., Lake, G., Dressler, A., & Oemler, A. Jr. 1995, Nature, 379, 613

Muller, E. Staveley-Smith, L., Zealey, W., & Stanimirović, S. 2003, MNRAS, 339, 105

Naab, T., & Burkert, A. 2003, ApJ, 597, 893

Nulsen, P. 1982, MNRAS, 198, 1007

Ostriker, J. P., & McKee, C. F. 1988, RvMP 60, 1

Ostriker, J., & Peebles, J. A. 1973, ApJ, 186, 467

Ott, J., Walter, F., Brinks, E., van Dyk, S. D., Klein, U. 2001, AJ, 122, 3070

Pettini, M., Rix, S. A., Steidel, Ch. C., Shapley, A. E., & Adelberger, K. L. 2003, IAU Symp. 212, 671

Puche, D., Westpfhal, D., Brinks, E., & Roy, J. 1992, AJ, 103, 1841

Putman, M. E., Staveley-Smith, L., Freeman, K. C., Gibson, B. K., & Barnes, D. G. 2003, ApJ, 586, 170

Putman, M. E., Gibson, B. K., Staveley-Smith, L. + 23 co-authors 1998, Nature, 394, 752

Renzini, A. 1997, ApJ, 488, 35

Renzini, A. 2004, in Clusters of Galaxies: Probes of Cosmological Structure and Galaxy Evolution, Cambridge Univ. Press, eds J. S. Mulchaey, A. Dressler, & A. Oemler, p. 261

Rowan-Robinson, M. 2003, MNRAS, 345, 819

Salpeter, E. E., 1955, ApJ, 121, 161

Sawa, T., & Fujimoto, M. 2004, astro-ph/0404547

Shen, J., & Sellwood, J. 2004, ApJ, 604, 614

Stanimirović, S., Staveley-Smith, L., Dickey, J. M., Sault, R. J., & Snowden, S. L. 1999, MNRAS 302, 417

Stanimirović, S., Staveley-Smith, L., & Jones, P. A. 2004, ApJ, 604, 176

Staveley-Smith, L., Sault, R. J., Hatzidimitriou, D., Kesteven, M. J., & McConnell, D. 1997, MNRAS, 289, 225

Tenorio-Tagle, G., & Bodenheimer, P. 1988, ARA&A, 26, 145

Tenorio-Tagle, G., Palouš, J., Silich, S., Medina-Tanco, G. A., & Muñoz-Tuñon 2003, A&A, 411, 397

Toomre, A. 1964, ApJ, 139, 1217

Toomre, A. 1977, ARA&A, 15, 437

Toomre, A. & Toomre, J. 1972, ApJ, 178, 623

van der Marel, R. P. 2001, AJ, 122, 1827

van Zee, L., Skillman, E. D., & Haynes, M. P. 2004, AJ 128, 121

van Gorkom, J. H. 2004, in Clusters of Galaxies: Probes of Cosmological Structure and Galaxy Evolution, Cambridge University Press, Carnegie Observatories Astrophysics Series, eds. J. S. Mulchaey, A. Dressler, and A. Oemler, p. 306

Vollmer, B., Cayatte, V., Balkowski, C., & Duschl, W. 2001, ApJ, 561, 708

Walter, F., & Brinks, E. 1999, AJ, 118, 273

Wünsch, R., & Palouš, J. 2001, A&A, 374, 746

Yoshizawa, A., & Noguchi, M. 2003, MNRAS 339, 1135

# Star formation in merging galaxy clusters

Chiara Ferrari

Intitut für Astrophysik, Universität Innsbruck
Technikerstraße 25, A-6020 Innsbruck
Chiara.Ferrari@uibk.ac.at

**Abstract**

*In the framework of the hierarchical model of structure formation, galaxy clusters form through the accretion and merging of substructures of smaller mass. Merging clusters are a privileged laboratory to test the physics of evolutionary effects on their galaxy members. Shock waves in the ICM driven by the merging event may trigger star formation (Evrard 1991). Other physical mechanisms could trigger star formation within clusters: the infall of galaxies in the high pressure ICM (Dressler & Gunn 1983), the encounters and interactions between galaxies (Lavery & Henry 1988), the rapid variation of the gravitational tidal field (Bekki 1999). On the other side, gas stripping in galaxies due to ram pressure exerted by the ICM could weaken the star-burst phenomenon during cluster collision (Fujita et al. 1999). So far it is not yet clear which one of these competing effects is the dominant one. In this picture, I present the analysis of the dynamical state and star formation properties of the merging cluster Abell 3921 based on the comparison of new optical (Ferrari et al. 2005) and X-ray (Belsole et al. 2005) observations with numerical simulation results.*

# 1 Introduction

In the currently favoured cosmological model ($\Lambda$CDM, $\Omega_m$=0.3 and $\Omega_\Lambda$=0.7), small structures are the first to form, and then they merge giving rise to more and more massive systems in a hierarchical way. Cosmological simulations show that galaxy clusters, which are the most massive gravitationally bound systems of the present Universe, form and evolve through the merging of sub-clusters and groups of galaxies along filamentary structures (e.g. Borgani et al. 2004).

Major cluster-cluster collisions are the most energetic events since the Big Bang, with total releases of gravitational binding energies of the order of $10^{64}$ ergs (Sarazin 2003). Therefore, the merging event can strongly affect the physical properties of the different cluster components. While in the recent past the effects of mergers on the intra-cluster medium (ICM) and on the internal dynamics of clusters have been analysed in some detail both from the numerical (e.g. Schindler and Böhringer 1993, Schindler and Müller 1993, Ricker & Sarazin 2001) and from the observational point

*Reviews in Modern Astronomy 18.* Edited by S. Röser
 ISBN 3-527-40608-5

of view (e.g. Davis et al. 1995, Lemonon et al. 1997, Röttiger et al. 1998, Durret et al. 1998, Arnaud et al. 2000, Donnelly et al. 2001, Ferrari et al. 2003), at present the increasing spectral and spatial resolution of both X-ray and optical observations, combined with more and more detailed numerical simulations, open the possibility to analyse the effect of mergers not only on the global cluster properties, but also on the evolution of their galaxies.

The interest in understanding if and how the cluster environment plays a role in the evolution of galaxies begun with the discovery of the so-called "Butcher & Oemler effect". In 1978, Butcher & Oemler reported a strong evolution from bluer to redder colours in cluster galaxies, detecting an excess of blue galaxies at z=0.5 with respect to lower redshift systems. At the beginning of the 80's, Dressler & Gunn pointed out for the first time that the blue colour of the population detected by Butcher & Oemler was the result of star-formation (SF) activity (Dressler & Gunn 1983). Since then, different physical mechanisms have been proposed to be responsible for the triggering and the possible cessation of SF in cluster galaxies. They can be due to the interaction with other galaxies (e.g. mergers and collisions), or to effects specific to the environment, involving either the ICM (e.g. ram-pressure[1] stripping and compression) or the cluster gravitational potential (e.g. tidal effects).

## 2 Merging clusters: privileged laboratories to test the physics of evolutionary effects

Due to the strong energies involved during cluster-cluster collisions, the merging event can amplify and make the physical mechanisms responsible for the evolution of SF properties in galaxies more efficient. Merging clusters are therefore privileged laboratories to test the physics of evolutionary effects.

Due to the high relative velocities of the colliding sub-clusters ($\approx$3000 km/s), both the ICM density and the relative velocity between the galaxies and the ICM are increased during the merging event. As a consequence, the ram-pressure exerted by the ICM on the galaxy inter-stellar medium (ISM) can significantly increase. Both numerical simulations and observations suggest that the ram-pressure can have a double effect on SF. On one side it can sweep the ISM of galaxies away, thus decreasing their SFR due to a lack of gas. On the other hand, the pressure exerted by the ICM can enhance SF through the compression of the ISM: when gas rich galaxies fall into the central region of a cluster for the first time, they experience a rapid increase of ICM external pressure, quickly exceeding the ISM internal pressure, which triggers the star-burst by compressing the molecular clouds in galaxies (the so-called "first infall" model by Dressler & Gunn 1983). Of course, we could have the same kind of effect whenever a gas rich galaxy moves in a dense ICM with a high relative velocity between them, e.g. during cluster-cluster collisions. Up to now, numerical and observational results have shown that either the ram-pressure stripping (e.g. Fujita et al. 1999, Bartholomew et al. 2001, Gomez et al. 2001), or the ISM com-

[1] $\mathrm{P_{ram}} \propto \rho_{\mathrm{ICM}} \mathrm{v_{rel}}^2$, where $\rho_{\mathrm{ICM}}$ is the density of the inter-cluster medium, and $\mathrm{v_{rel}}$ is the relative velocity between the ICM and the galaxy.

pression (e.g. Dressler & Gunn 1983, Evrard 1991, Poggianti et al. 2004) could be the dominant mechanism acting on SF in galaxy clusters. The two mechanisms can also be both efficient, but with different time scales, since, during the fast infall of a galaxy in a dense ICM, we could have an initial increase and a following drop of SFR (Quilis et al. 2000).

During cluster-cluster collisions, the SF properties of galaxies can also be strongly affected both by tidal effects due to the merging event (Bekki 1999), and by interactions between galaxies (e.g. Lavery & Henry 1988, Bekki, Shioya and Couch 2001). In the first case, the time-dependent tidal gravitational field of the merger gives strong non-axisymmetric perturbations to the disk galaxies in the colliding clusters. Subsequently this tidal field induces an efficient transfer of gas to the central region of the galaxy, and finally triggers the star-burst episode. For what concerns the second point, even if the dense environment of clusters seems to assist galaxy interactions, the effects of collisions between galaxies are weaker in clusters than in the field, since the random velocities of the cluster members are generally greater than the internal velocity dispersions of galaxies (Struck 1999 and references therein). However, if groups of galaxies with a smaller velocity dispersion are falling into major clusters, the effects of interactions may be enhanced. Observational results confirm that a fraction of active galaxies (i.e. with recent or on-going SF) shows significant signs of interactions with other galaxies (e.g. Blake et al. 2005, Zabludoff et al. 1996).

Other mechanisms that have been proposed to affect galaxy properties are: a) "harassment" (Moore et al. 1996), a mechanism that can severely damage the disk of spiral galaxies due to fast (several thousands of km/s) and close encounters with bright galaxies, which can cause impulsive gravitational shocks and a global tidal heating, and b) "strangulation" (or "starvation"), for which a halo of hot gas is stripped from galaxies in dense environments, leading to a gradual winding down of SF as the remaining cold, disk gas is consumed (Larson, Tinsley and Caldwell 1980, Bekki, Couch and Shioya 2002).

In practice, more than one of the above mechanisms is probably responsible for the SF properties of active galaxies in clusters, and in particular in merging clusters, and a general consensus has not yet been reached about the net role played by cluster collisions in the evolution of the SFR.

## 2.1 Observations and simulations of merging clusters

In this picture, a combined observational and numerical approach is the only way to shed light upon the role played by cluster-cluster collisions on SF. In collaborations with different researchers of several European (Intitut für Astrophysik, Innsbruck, Austria; Observatoire de la Côte d'Azur, Nice, France; INAF, Bologna, Italy; SAp/CEA, Saclay, France), and Australian (Sidney University) institutes, I have begun a multi-wavelength and numerical analysis of a sample of merging galaxy clusters. We aim at investigating which one of the different physical mechanisms summarised in the previous section plays a role in affecting the SFR of cluster galaxies, and at testing if the merging event makes these mechanisms more efficient.

For this we need to determine: a) the dynamical state of the observed galaxy clusters; b) how the star-formation history (SFH) of cluster members varies inside the interacting systems, verifying if there exists a correlation between the spatial distribution of active galaxies and the regions of the cluster mainly affected by the interaction (i.e. characterised by compression and heating of the ICM, and by strong perturbations in the galaxy dynamics); c) evaluate if and how the fraction of active galaxies evolves during the different phases of the merging event. While following all the steps of sub-cluster collisions is in principle feasible from the numerical point of view, it is not possible from the observational side, due to the extremely long time scales of mergers, i.e. several Gyr. We have therefore decided to observe a sample of galaxy clusters going from pre-merging to nearly virialised systems, and to study these clusters by combining optical (imaging: WFI@2.2m ESO, CFH12k@CFHT; spectroscopy: EFOSC2@3.6m ESO, VIMOS@VLT ESO, 2dF@AAT), X-ray (Chandra, XMM) and radio (VLA, ATCA) data.

### 2.1.1 Observations

In order to fulfil the main objectives of our programme, the observational analysis requires several steps.

- First of all we need to reconstruct very precisely the merging scenario of each observed cluster, determining the mass ratio between the interacting sub-clusters, their impact parameter, the angle between the collision axis and the plane of the sky, and the time elapsed since the beginning of the interaction. The comparison between optical and X-ray observations is essential for characterising the dynamical state of galaxy clusters. The ICM and galaxy density maps allow us to estimate roughly which phase of the merging event we are observing due to the strong difference in the relaxation time scales of the gas and of the collisionless component in clusters (i.e. galaxies and dark matter). A detailed reconstruction of the merging scenario is then possible through the dynamical and kinematic analysis of member galaxies combined with the study of the density and temperature maps of the ICM (e.g. Arnaud et al. 2000, Maurogordato et al. 2000).

- By combining our optical and radio observations we then identify the active galaxies of the cluster, and we study their velocity and spatial distributions. In order to individuate star-forming and post-star-forming objects, we adopt the classification scheme of Dressler et al. (1999) based on the presence and strength of [OII] emission and Balmer absorption (in particular $H_\delta$) lines in the galaxy spectrum. The objects characterised by the presence of strong Balmer absorption lines and absent [OII] emission are classified as post-starburst (PSB) galaxies. They are often referred to "E+A" or "k+a/a+k" galaxies, since their spectral features are typical of ellipticals with a large population of recently formed A stars. However, the absence of [OII] emission suggests that SF is no longer going on. The conclusion is that these galaxies have previously (<1-1.5 Gyr ago) undergone a burst of SF, which has recently been truncated rather suddenly (Poggianti et al. 1999). The presence of [OII] and, in general, of emission lines indicates that an episode of SF is going on in the observed galaxy. Following Poggianti et al. (1999), emission line galaxies are normally divided in three categories. Those presenting moderate Balmer absorption and weak to moderate [OII] emission are classified as "e(c)" objects. They have spectra similar to those of typical present-day spirals, but an alternative interpretation is that they could be long star-bursts observed in the late phase of their star-bursting episode. Spectra with very strong [OII] emission are classified as "e(b)" types, and they correspond to star-burst galaxies, since it has been shown that so strong emission lines cannot be reached with a normal spiral-like SFH (Poggianti et al. 1999). Finally, another class of objects has been detected: those with strong Balmer absorption and measurable [OII] emission. Different explanations on the nature of these galaxies have been proposed: they could be PSB galaxies with some residual SF, but, since [OII] emission suffers from strong dust-extinction, they are more likely dusty star-bursts, i.e. e(b)-type galaxies whose strong [OII] emission is obscured by dust. The radio continuum luminosity is on the contrary a tracer of on-going SF that is unbiased by dust (Miller & Owen 2001). Therefore, our radio observations can reveal the presence of star-forming objects, and they can help to disentangle between real post-star forming galaxies and dusty star-forming galaxies. We are also planning to obtain other SFR indicators that are less affected or completely unaffected by dust absorption, i.e. the $H_\alpha$ emission line and IR fluxes.

- Finally, we verify if there exists a link between the presence and the properties of the detected active galaxies and the dynamical state of the cluster. If a correlation is found, we need to compare the observational results to numerical simulations of merging clusters in order to understand which are the physical mechanisms acting on the SFR of cluster galaxies.

### 2.1.2 N-body and hydrodynamic simulations

Our observational analysis is therefore compared with the numerical results obtained by the HYDRO-SKI team of the Institut für Astrophysik in Innsbruck University (Schindler and collaborators). Combined N-body and hydrodynamic simulations are used to model the massive components of clusters, i.e. dark-matter, galaxies and ICM (van Kampen et al.1999, Domainko et al. 2004, Kapferer et al. 2004).

The comparison of observational results with these simulations is first of all essential to refine the merging scenario reconstructed from the multi-wavelength observations. Secondly it allows to analyse in detail a) which physical mechanisms related to the merging event can affect the SF properties of cluster members, and b) one of the main effects of SF, that is the metal enrichment of the ICM. Since X-ray spectra reveal that the ICM contains metals (Sarazin 1988) and heavy elements are only produced in stars, the processed material must have been ejected by cluster galaxies into the ICM. After SF and subsequent SNs explosions have taken place in cluster members transferring metals to the ISM, the enriched material has to be transported into the ICM. This can happen through different physical mechanisms, triggered either by the environment, e.g. ram-pressure stripping, or by violent internal processes, e.g. galactic winds developed by massive stars (De Young 1978). The effect of ram-pressure stripping according to the local properties of the ICM, as well as the effect of galactic winds are included in the simulations by Schindler and collaborators. Metallicity is used as a tracer to follow the enriched material, and the simulated metallicity maps can of course be compared to those reconstructed from our X-ray observations.

# 3 A case study: the merging cluster Abell 3921

In the following I will present our study of the dynamical and SF properties of the merging cluster Abell 3921. These results are based on the analysis of multi-object spectroscopy (EFOSC2@ESO 3.6m) and multi-band (V,R,I) deep imaging (WFI@ESO 2.2m) observations, and are presented in detail in Ferrari et al. (2005). Our optical analysis is compared to the X-ray results of Belsole et al. (2005), based on the XMM observation of A3921.

## 3.1 Dynamical state of A3921

### 3.1.1 Optical morphology of A3921

We have investigated the projected spatial morphology of A3921 through several density maps of the galaxy distribution built on the basis of a multi-scale approach. The adopted algorithm is a 2D generalisation of the algorithm presented in Fadda et al. (1998). It involves a wavelet decomposition of the galaxy catalogue performed on five successive scales from which the significant structures are recombined into the final map (following the Eq. [C7] of Fadda et al. 1998). These significant structures are obtained by thresholding each wavelet plane at a level of three times the variance of the coefficients of each plane except for the two smallest scales for which the threshold is increased to four and five times the variance in order to reduce false detections due to the very low mean density of the Poisson process at these scales (0.01 for a chosen grid of $128\times128$ pixel$^2$).

In order to avoid possible projection effects, we have isolated galaxies likely to be early types at the cluster redshift on the basis of their colour properties. Indeed, one can notice in the colour magnitude diagram (CMD) of Fig. 1 a well defined red sequence, the characteristic linear structure defined by the bulk of early-type galaxies in a cluster. Finally, from the sample of red sequence galaxies we have additionally excluded galaxies known from spectroscopy not to be cluster members. Fig. 2 shows the resulting red sequence density map at different magnitude cuts (using only galaxies at $\pm1\sigma$ around the red sequence). We can notice that the optical morphology of A3921 is characterised by: a) the presence of several substructures, b) an overall bimodal morphology, with two main clumps (A3921-A and A3921-B), c) an eastern extension of galaxies stronger at faint magnitude cuts, and d) an offset of the Brightest Cluster Galaxy (BCG) from the main density peak of clump A. These results suggest that this system is out of dynamical equilibrium and that it is probably composed of a main cluster interacting with at least two groups, one to the North-East (clump B) and one to the East, the latter being significantly less luminous than the former.

### 3.1.2 Dynamical and kinematical properties of A3921

The analysis of the velocity distribution of the cluster members can help us to understand which phase of the merging process we are witnessing. We have therefore performed a kinematical and dynamical analysis of A3921, considering firstly all the confirmed cluster members as a whole dataset (104 galaxies), and secondly the galaxies in the central region of the two main clumps separately (see Fig. 3).

We have applied the statistical indicators of Beers et al. (1990) that give the best estimation of velocity location ("mean") and scale ("dispersion") of a dataset depending on the number of its points. The results are summarised in Table 1. A3921-A and A3921-B show a mean velocity very close to each other and to the whole cluster value, with a velocity offset $\Delta v$[2] between the mean velocities of the two clumps of only $89^{+155}_{-177}$ km/s. Clump A is characterised by the highest velocity dispersion.

Table 1: Properties of the $cz$ distribution for various subsamples of A3921 (Ferrari et al. 2005). $C_{\rm BI}$ and $S_{\rm BI}$ are the mean velocity and the velocity dispersion of the different distributions (biweight estimators for location and scale, Beers et al. 1990)

| Subsample | $N_{\rm gal}$ | $C_{\rm BI}$ [km/s] | $S_{\rm BI}$ [km/s] |
|---|---|---|---|
| Whole sample | 104 | $28047^{+76}_{-77}$ | $831^{+100}_{-76}$ |
| A3921-A | 41 | $28017^{+145}_{-173}$ | $1008^{+156}_{-106}$ |
| A3921-B | 20 | $27920^{+88}_{-86}$ | $451^{+215}_{-80}$ |

In dissipationless systems, gravitational interactions of cluster galaxies over a relaxation time generate a Gaussian distribution of their radial velocities; possible deviations from Gaussianity could provide important indications of on-going dynamical processes. We have therefore analysed the velocity distributions of the three sub-samples of Table 1 in order to test the null Gaussian hypothesis. In particular, we have analysed: a) the classical shape estimators of a distribution (skewness and kurtosis, and asymmetry and tail indexes by Bird and Beers 1993), b) 13 1-D statistical tests of Gaussianity included in the ROSTAT package (Beers et al. 1990), c) the improvement in fitting a multiple-component model over a single-one on the velocity distributions of the three samples through the KMM algorithm of McLachlan & Basford (1988), and d) the mean-velocity and velocity-dispersion profiles of

[2] Cosmologically and relativistically corrected from the mean cluster redshift: $\Delta v = c(\bar{z}_A - \bar{z}_B)/(1+\bar{z})$

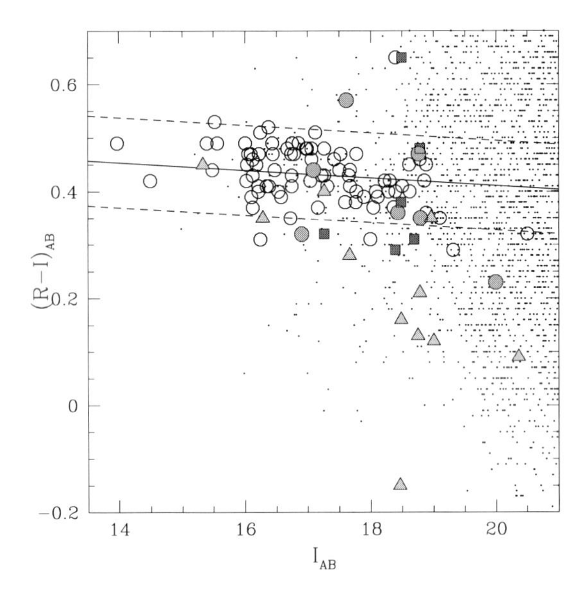


Figure 1: $(R-I)_{\rm AB}$ *vs* $I_{\rm AB}$ colour-magnitude diagrams. All galaxies within $34 \times 34$ arcmin$^2$ are shown. Big symbols correspond to confirmed cluster members whereas dots correspond to galaxies without spectroscopic information. Triangles correspond to emission line galaxies, squares to k+a type, circles to galaxies presenting an H-K inversion following the classification described in the text (Sect. 3.2). The solid line is the best linear-fit to the red sequence of the cluster ($(R-I)_{\rm AB} = -0.0071 I_{\rm AB} + 0.5531$, width of $\sigma_{\rm RS}$=0.0837) . The dotted lines are at $\pm 1\sigma_{\rm RS}$ (Ferrari et al. 2005).

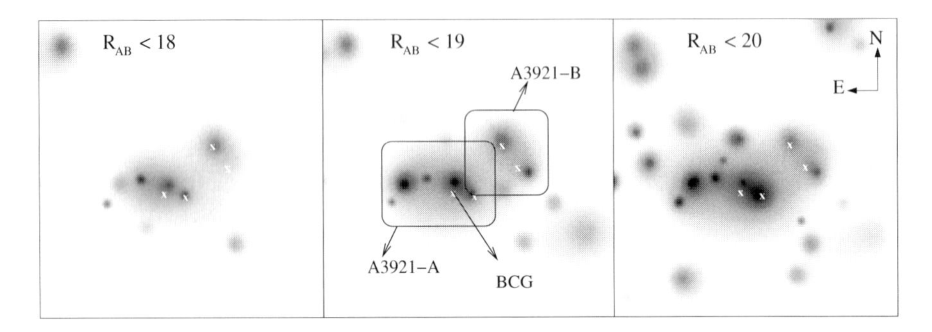


Figure 2: Projected galaxy density maps (using red sequence galaxies without galaxies known not to be cluster members from spectroscopy) on a $34 \times 34$ arcmin$^2$ field centred on A3921 and for three magnitude cuts. The white crosses indicate the positions of the four brightest cluster galaxies (Ferrari et al. 2005).

the whole cluster members. The main results of this detailed study suggest that the kinematical properties of the whole sample and of the two sub-clusters do not show strong and clear signatures of merging (see Ferrari et al. 2005 for more details). In particular, the dynamics of the central regions of the two clumps appears to be relatively unaffected by the merging event.

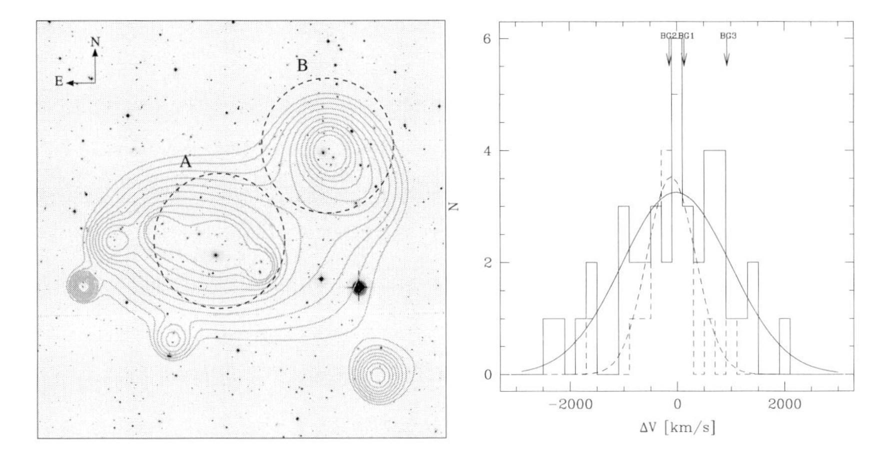


Figure 3: **Left:** Iso-density contours of the projected distribution of the red-sequence galaxies (with $R_{\rm AB} \leq 18$ and after removing galaxies known not to be cluster members from spectroscopic data) superimposed on the central $22 \times 22$ arcmin$^2$ of the R–band image of A3921. The dynamical analysis of the two clumps A and B is performed based on the galaxies selected inside the two dashed circles (R$\simeq$ 0.34 Mpc). **Right:** Velocity histogram, with a binning of 200 km/s, of the galaxies in A3921-A (solid line) and in A3921-B (dotted line), according to the division shown in the left panel. The relative Gaussian best–fits to the velocity distributions are superimposed. Arrows show the radial velocities of the three brightest cluster galaxies (Ferrari et al. 2005).

On the basis of the previous results, we have assumed that each sub-cluster is virialized, and we have calculated the mass of A3921-A and A3921-B with the classic virial equation: $M_{\rm vir} = \frac{r_{\rm vir}\sigma_{\rm vir}^2}{G}$, where $\sigma_{\rm vir}$ is the three–dimensional velocity dispersion of the system, and $r_{\rm vir}$ is the virial radius. The results obtained using the pairwise estimator for the viral radius are shown in Table 2 (see Ferrari et al. 2005 for more details).

### 3.1.3 Conclusions about the dynamical state of A3921

To summarise, in spite of clear merging signatures in the density distribution, the kinematical and dynamical properties both of the whole cluster and of the two sub-clusters do not show strong signatures of merging. Moreover, the two sub-clusters show very similar mean projected velocities.

Using the observed values of the mass of the two-clumps, their projected spatial separation, and their radial velocity offset, we have applied a two-body dynamical

Table 2: Columns 1 & 2: projected virial radii of A3921-A and A3921-B; columns 3 & 4: virial mass estimates for A3921-A and A3921-B; column 5: mass ratio. Masses are in $10^{14} M_{\odot}$ units and radiii in $h^{-1}$ Mpc units.

| $R_{\rm vir}$ (A) | $R_{\rm vir}$ (B) | $M_{\rm vir}$ (A) | $M_{\rm vir}$ (B) | $M_{\rm vir}$ (A) / $M_{\rm vir}$ (B) |
|---|---|---|---|---|
| 0.39±0.02 | 0.38±0.02 | $4.3^{+1.4}_{-1.0}$ | $0.8^{+0.7}_{-0.3}$ | $5.2^{+4.4}_{-2.5}$ |

Table 3: Scenarios that could explain the observed dynamical properties of the cluster on the basis of a two-body approach. Col.1: name of the scenario in the text – Col.2: time since last interaction of the two clumps – Col.3: angle between the plane of the sky and the line connecting the centres of the two clumps – Col.4: relative velocity between the two clumps – Col.5: spatial separation between the two systems – Col.6: state of the systems for the possible solutions (Ferrari et al. 2005).

| Scenario | $T_0$ [Gyr] | $\alpha$ [deg] | V [km/s] | R [Mpc] | Solutions |
|---|---|---|---|---|---|
| a1 | 12.6 | 84.3 | 89.4 | 7.5 | outgoing |
| a2 | 12.6 | 83.5 | 89.6 | 6.5 | infalling |
| a3 | 12.6 | 2.2 | 2318.4 | 0.7 | infalling |
| b | 0.3 | 4.9 | 1041.9 | 0.7 | outgoing |
| c | 0.5 | 27.8 | 190.8 | 0.8 | outgoing |
| d1 | 1.0 | 55.2 | 108.4 | 1.3 | outgoing |
| d2 | 1.0 | 50.6 | 115.2 | 1.2 | infalling |
| d3 | 1.0 | 4.7 | 1086.2 | 0.7 | infalling |

model (Gregory & Thomson 1984, Beers et al. 1992). Several solutions could explain the observed dynamics of A3921 allowing both the pre-merging and the post-merging cases (Ferrari et al. 2005); they are summarised in Table 3.

A comparison with the X-ray properties of the cluster is at this point essential to discriminate between the possible merging scenarios. The analysis of XMM-Newton observations by Belsole et al. (2005) reveals that the X-ray emission of A3921-A can be modelled with a 2D$\beta$-model, leaving a distorted residual structure toward the NW, coincident with A3921-B (see Fig. 4). The main cluster detected in X-rays is centred on the BCG position (BG1), while the X-ray peak of the NW clump is offset from the brightest galaxy of A3921-B (BG2) (see Belsole et al. 2005). The temperature map of the cluster shows an extended hot region oriented parallel to the line joining the centres of the two sub-clusters. A comparison of this image with numerical simulations by Ricker & Sarazin (2001) suggests that we are observing the central phases of an off-axis merger between two unequal mass objects, with clump A being the more massive component (Belsole et al. 2005), consistent with optical results.

By combining the signatures of merging derived both from the optical iso-density map and from X-ray results, we have then reconsidered the solutions of the two-body dynamical model summarised in Table 3 (Ferrari et al. 2005). In the pre-merger case, the high-angle solutions of cases (a1) and (a2) would imply a very large real separation of the two sub-clusters ($\sim 6 - 7$ Mpc), which is very unlikely taking into account the clear signs of interaction between the two clumps observed both in the optical and in X-rays. In the "recent" post-merger case, we can also exclude the solution (c), as we expect a higher value of the relative velocity ($\geq$1000 km/s) between the two clumps for obtaining so clearly a hot bar in the temperature map. Finally, the comparison of observed and simulated galaxy density and temperature maps (e.g. Schindler & Böhringer 1993, Ricker & Sarazin 2001) clearly exclude an older merger (e.g. $t_0$=1 Gyr), as we would not expect to observe a clear bimodal morphology in the optical any longer and we should not detect such obvious structure in the temperature map. Therefore, only the solutions corresponding to the very central phases of merging ($t_0 \approx \pm 0.3$ Gyr) can explain all our observational results, implying a collision axis nearly perpendicular to the line of sight. This is consistent with the absence of strong merging signatures in the observed projected velocity distribution of cluster members.

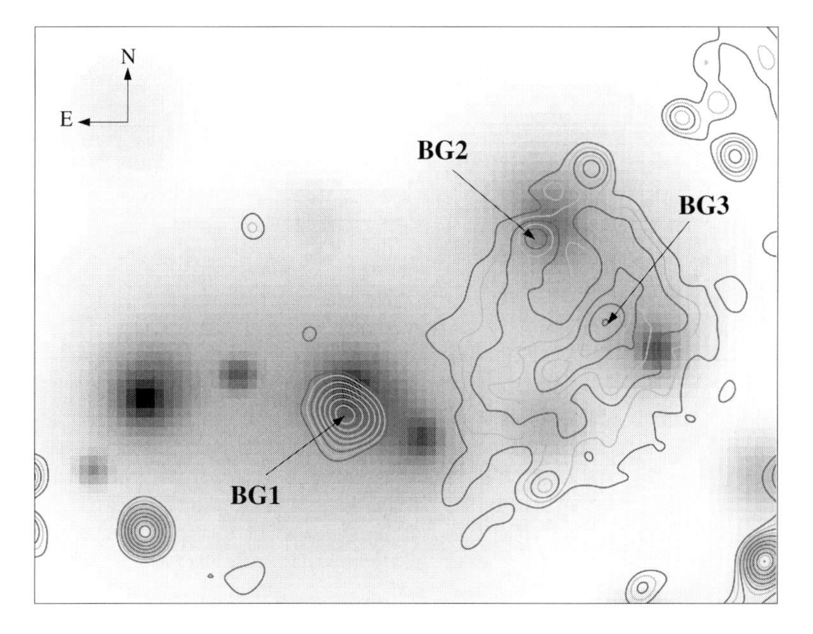


Figure 4: X-ray residuals after subtraction of a 2D-$\beta$ model (Belsole et al. 2005) overlaid on the red sequence galaxy density map of the central part of A3921 ($22 \times 18 \text{arcmin}^2$, Ferrari et al. 2005).

The superposition of the X-ray residuals onto the optical iso-density map (Fig. 4) shows that the bulk of X-ray emission in A3921-B is offset towards SW from the main concentration of galaxies. As numerical simulations show that the non-collisional component is much less affected by the collision than the gas distribution (e.g. Röttiger et al. 1993), this offset suggests that A3921-B is tangentially traversing A3921-A in the SW/NE direction, with its galaxies in advance with respect to the gaseous component. The off-axis collision geometry has probably prevented total assimilation of the B group into the main cluster A. This off-axis collision scenario is also consistent with the shape of the feature in the temperature map (Belsole et al. 2005).

## 3.2 The effects of the merging event on star formation

### 3.2.1 Identification of the active galaxies in A3921

The next step of our work has been the identification of the active cluster members to study their possible link with the dynamical state of A3921. Various methods have been suggested for this purpose, using generally the presence and strength of the [OII] ($\lambda$=3727 Å) line and of Balmer lines (typically one or a combination of $H_\delta$, $H_\gamma$, and $H_\beta$ lines). In principle, as shown by Newberry et al. (1990), the most robust approach to detect post-star-burst galaxies would make full use of all three Balmer lines. However, our limited spectral range does not allow to include $H_\beta$ in numerous cases, and the S/N ratio of the $H_\gamma$ line is generally poor. Therefore, we have used the combination of [OII] and of the $H_\delta$ line to establish our spectral classification, as in Dressler et al. (1999).

For measuring the equivalent widths of these lines, we have used a Gaussian fitting technique through the task "splot" in IRAF. The equivalent widths of absorption and emission features are defined as positive and negative respectively, and the minimum measurable EW of each spectrum, estimated as in Barrena et al. (2002), is of $\sim 2.8$ Å.

We have identified the following active galaxies:

- 11 star-forming galaxies, characterised by the presence of emission lines, that have been divided into three categories (spiral-like spectra, star-bursts and probable dusty star-bursts), as described in section 2.1.1;

- 6 k+a's or post-star-burst galaxies, characterised, following the definition of Dressler et al. (1999), by absent [OII] emission and moderately strong Balmer absorption (3 Å$<$ EW($H_\delta$) $<$ 8 Å);

- 7 k+a candidates (k+a?), that do not strictly follow the criterion of Dressler et al. (1999), but present a clear inversion of the intensities of the K and H calcium lines. Since this is due to the presence of a blend of the H line with the Balmer line $H\epsilon$, these objects have probably undergone a recent star-formation activity (Rose 1985).

The percentages of star-forming and post-star-forming galaxies (13% and 16% of the identified cluster members) are comparable to higher redshift clusters. Since several studies do not detect significant recent SF in low-redshift clusters (e.g. Dressler et al. 2004), the high fraction of active galaxies detected in A3921 is already an important results in itself, which suggests a possible link between the SF properties of this cluster and its dynamical state.

### 3.2.2 Is the recent SF due to the merging event?

In order to understand if the presence of recent or on-going star-forming galaxies is related to the merging event detected in A3921, we have compared the spatial and velocity distributions of the active and passive cluster members (Figs. 5 and 6). The colour and spectral properties of active galaxies have also been taken into account.

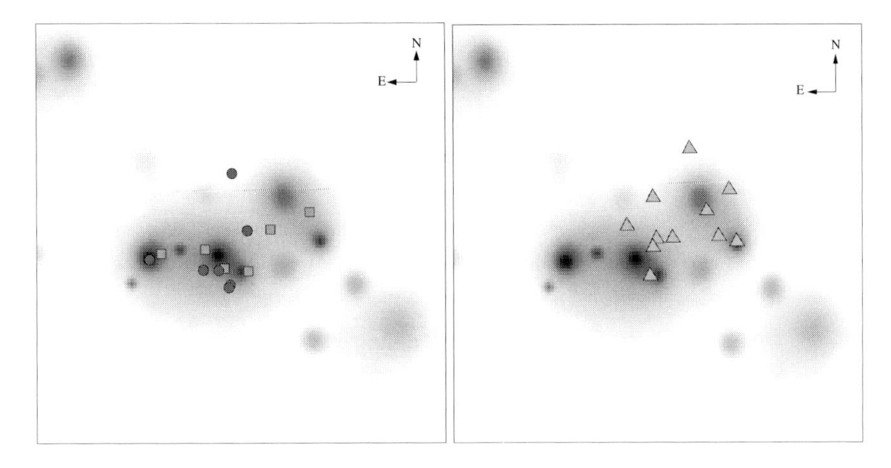


Figure 5: *Projected galaxy density map (34×34* $\mathrm{arcmin}^2$*) of the red-sequence galaxies with* $R_{\mathrm{AB}} < 19$. **Left:** *the squares represent galaxies that are members of A3921 and classified as k+a and the circles as "k+a?" (see text).* **Right:** *the triangles show the location of emission line cluster member galaxies (Ferrari et al. 2005).*

The typical EW(H$\delta$) of the k+a/k+a? galaxies detected in A3921 is moderate and most of the objects have red colours, indistinguishable from red-sequence objects. We fail to detect the population of blue k+a with strong Balmer lines as detected in Coma by Poggianti et al. (2004), which can only be reproduced by a star-burst in the recent past. In contrast, our objects reflect the evolution of galaxies having undergone star-burst or star-forming activity which has been suppressed by some physical process, and now are in the second half of their lifetime (typically $< 1.5$ Gyr), with redder colours, and fainter Balmer lines, before reaching a k-type spectrum. This population can be reproduced by simply halting continuous star formation, without evoking a strong star burst. The k+a? objects are probably galaxies that have undergone a still older and fainter last episode of star formation, as they do not have Balmer lines strong enough to enter the k+a sample, but show clear signatures of past activity. Moreover, k+a/k+a? galaxies are mostly distributed over the main cluster A, and they do not show a spatial correlation with the region of the cluster mostly affected by the merging event.

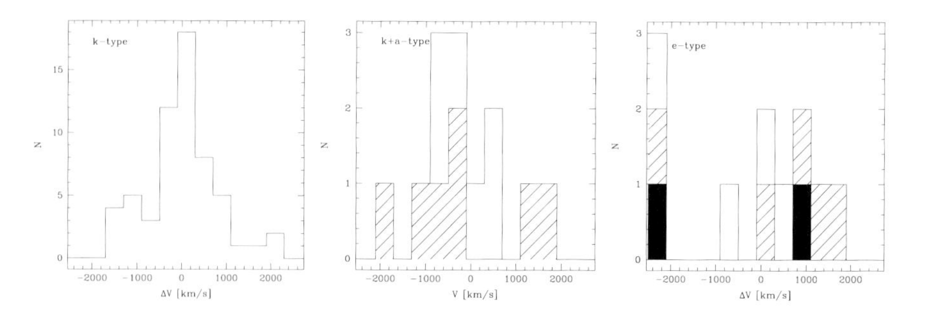


Figure 6: From left to right, velocity distribution (with a binning of 400 km/s) of k, k+a and emission line galaxies. In the central figure, the white component corresponds to secure k+a's and the shaded part to k+a candidates. In the right panel, the shaded component of the histogram corresponds to e(a)'s, the black component to e(b)'s, the white boxes to e(c) type objects (Ferrari et al. 2005).

On the contrary, most of the emission line galaxies lie in the region of the sub-cluster B, and in the region in between A and B. A similar spatial distribution is shown by blue galaxies, which are more clustered in the central region of A3921-B than in the whole field and, in particular, than in the centre of the more massive clump A (see Fig. 7). The comparison of the observed distributions of blue and emission line objects with the merging scenario presented previously ($\pm$ 0.3 Gyr) suggests that the interaction with the ICM during the passage of the sub-cluster B on the edge of cluster A may have triggered star-bursting. This hypothesis is supported by the significant difference in the radial velocity dispersions of emission-line and passive galaxies, suggesting that emission-line objects are a dynamically younger population than the general cluster members. In contrast, the k+a/k+a? population shows the signature of older star formation activity which can hardly be related to the on-going merger, but may be understood either as previously infalling galaxies, or is a relic from another older merging process, but is now at rest within the main cluster, as shown from the velocity distribution.

We therefore conclude that, in the case of A3921, the on-going merger may have triggered a star-formation episode in at least a fraction of the observed emission-line galaxies.

## 3.3 On-going follow-up

We are currently going on with the programme by observational and numerical analysis of A3921. Combined narrow-band $H_\alpha$ imaging (WFI@ESO 2.2m), high sensitivity radio observations (ATCA) and 2dF spectroscopy including the $H_\alpha$ region of the spectrum will reveal in a complementary way the active population in the central field of A3921 ($\sim 30' \times 30'$), identifying in addition the dusty star-burst galaxies previously missed due to [OII] extinction (see Sect. 2.1.1). Moreover, since both the 2dF and the ATCA observations cover a wider region of the cluster ($1.5 \times 1.5$ deg$^2$),

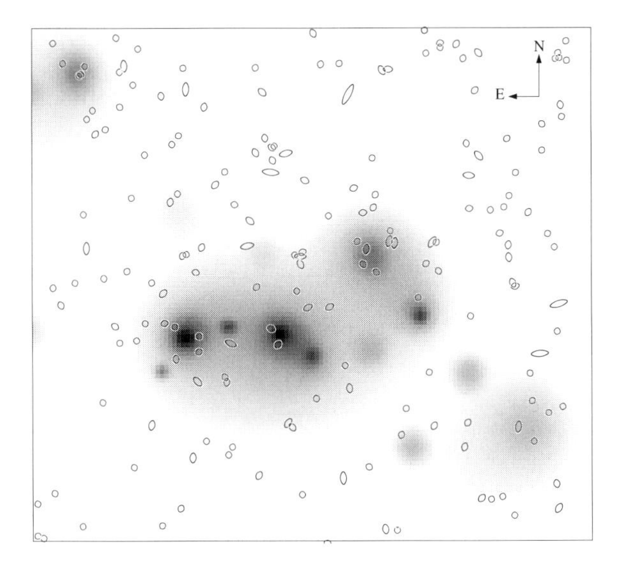


Figure 7: Projected galaxy density map (34×34 arcmin$^2$) of the red-sequence galaxies. The symbols represent the galaxies with $R_{\rm AB} < 20$ and bluer than the red sequence galaxies.

it will be possible to study the dynamics and the SF properties of the cluster out to its virial radius. Finally, we are currently analysing a Chandra observation of this cluster, which will give us high resolution density, temperature and metallicity maps of the ICM, essential to understand the link between the gas compression and the SF episode in active galaxies. These observations will allow us to test and eventually constrain better the previously found link between the merging event and the SFR of cluster members.

Our observational analysis will be finally compared with the numerical simulations performed by Schindler and collaborators. At present, we are working on modelling A3921, in order to refine the merging scenario reconstructed from the observational results. Our second step will be the comparison of the observed SF properties and metallicity maps of A3921 to the results of simulations. This will be an essential step to study in detail the physical effects acting on cluster members and affecting their SFR.

**Acknowledgements.** I warmly thank all my collaborators working on a) the optical (Christophe Benoist, Alberto Cappi, Sophie Maurogordato and Eric Slezak), X-ray (Elena Belsole, Hervé Bourdin, Gabriel W. Pratt and Jean-Luc Sauvageot) and radio (Luigina Feretti and Dick Hunstead) analysis of A3921, and b) the numerical simulations of merging clusters (Willfreid Domainko, Wolfgang Kapferer, Stefan Kimerswenger, Thomas Kronberger, Magdalena Mair, Sabine Schindler, Eelco van Kampen). This research is supported by a Marie Curie individual fellowship MEIF-CT-2003-900773.

## References

Arnaud, M., Maurogordato, S., Slezak, E., Rho, J. 2000, A&A, 355, 461

Baldi, A., Bardelli, S., Zucca, E. 2001, MNRAS, 324, 509

Barrena, R., Biviano, A., Ramella, M., Falco, E.E., Seitz, S. 2002, A&A, 386, 861

Bartholomew, L. J., Rose, J. A., Gaba, A. E., Caldwell, N. 2001, AJ, 122, 2913

Bekki, K., Couch, W. J., Shioya, Y. 2002, ApJ, 577, 651

Bekki, K., Shioya, Y, Couch, W. J. 2001, ApJL, 547, 17

Bekki, K. 1999, ApJL, 510, 15

Belsole, E., Sauvageot, J.-L., Pratt, G. W. , Bourdin, H. 2005, A&A, 430, 385

Beers, T. C., Gebhardt, K., Huchra, J.P., Forman, W., Jones, C., Bothun, G. D. 1992, ApJ, 400, 410

Beers, T. C., Flynn, K., Gebhardt, K. 1990, AJ, 100, 32

Bird, C. M., & Beers, T. C. 1993, AJ, 105, 1596

Blake, C., Pracy, M., Couch, W. et al. 2005, MNRAS, 355, 713

Borgani, S., Murante, G., Springel, V., Diaferio, A., Dolag, K., Moscardini, L., Tormen, G., Tornatore, L., Tozzi, P. 2004, MNRAS, 348, 1078

Butcher, H., Oemler, A. 1978, ApJ, 219, 18

Davis, D. S., Bird, C. M., Mushotzky, R. F., Odewahn, S. C. 1995, ApJ, 440, 48

De Young, D.S. 1978, ApJ, 223, 47

Domainko, W., Kapferer, W., Schindler, S., van Kampen, E., Kimeswenger, S., Mair, M., Ruffert, M. 2004, proceeding of the XXXIXth Rencontres de Moriond "Exploring the Universe", La Thuile (28.03.-04.04.2004), astro-ph/0405577

Donnelly, R., Forman, W., Jones, C., Quintana, H., Ramirez, A., Churazov, E., Gilfanov, M. 2001, ApJ, 562, 254

Dressler, A., Oemler, A., Poggianti, B., Smail, I., Trager, S., Shectman, S. A., Couch, W., Ellis, R. 2004, ApJ, 617, 867

Dressler, A., Gunn, J. E. 1983, ApJ, 270, 7

Dressler A., Smail, I., Poggianti, B. M., Butcher, H., Couch, W. J., Ellis, R. S., Oemler, A. 1999, ApJ, 122, 51

Durret, F., Forman, W., Gerbal, D., Jones, C., Vikhlinin, A. 1998, A&A, 335, 41

Evrard, A. E. 1991, MNRAS, 248, 8

Fadda, D., Slezak, E., Bijaoui, A. 1998, A&AS, 127, 335

Ferrari, C., Benoist, C., Maurogordato, S., Cappi, A., Slezak, E. 2005, A&A, 430, 19

Ferrari, C., Maurogordato, S., Cappi, A., Benoist, C. 2003, A&A, 399, 813

Fujita, Y., Takizawa, M., Nagashima, M., Enoki, M. 1999, PASJ, 51, L1

Gomez, P., Nichol, R., Miller, C., Goto, T. 2001, AAS, 199.7801

Gregory, S. A., Thompson, L. A. 1984, ApJ, 286, 422

Kapferer, W., Breitschwerdt, D., Domainko, W., Schindler, S., van Kampen, E., Kimeswenger, S., Mair, M. 2004, proceeding of the XXXIXth Rencontres de Moriond "Exploring the Universe", La Thuile (28.03.-04.04.2004), astro-ph/0405577

Larson, R. B., Tinsley, B. M., Caldwell, C. N. 1980, ApJ, 237, 692

Lavery, R. J., Henry, J. P. 1988 ApJ, 330, 596

Lemonon, L., Pierre, M., Hunstead, R., Reid, A., Mellier, Y., Böhringer, H. 1997, A&A, 326, 34

Maurogordato, S., Proust, D., Beers, T. C., Arnaud, M., Pellò, R., Cappi, A., Slezak, E., Kriessler, J. R. 2000, A&A, 355, 848

McLachlan, G. J., Basford, K. E. 1988, Mixture Models (Marcel Dekker, New York)

Miller, N. A., Owen, F. N. 2001, ApJ, 554, 25

Moore, B., Katz, N., Lake, G. 1996, ApJ, 457, 455

Newberry, M. V., Boronson, T. A., Kirshner, R. P. 1990, ApJ, 350, 585

Poggianti, B. M., Bridges, T. J., Komiyama, Y., Yagi, M., Carter, D., Mobasher, B., Okamura, S., Kashikawa, N. 2004, ApJ, 601, 197

Poggianti, B. M., Smail, I., Dressler, A., Couch, W. J., Barger, A. J., Butcher, H., Ellis, R. S., Oemler, A. 1999, ApJ, 518, 576

Quilis, V., Moore, B., Bower, R. 2000, SCIENCE, 288, 1617

Ricker, P. M., Sarazin, C. L. 2001, ApJ, 561, 621

Rose, J. A. 1985, AJ, 90, 1927

Röttiger, K., Stone, J. M., Mushotzky, R. F. 1998, ApJ, 493, 62

Röttiger, K., Burns, J., Loken, C. 1993, ApJL, 407, 53

Sarazin, C. L. 2003, Physics of Plasmas, 10, 1992

Sarazin, C. L. 2001, ApJ, 561, 621

Sarazin, C. L. 1988, "X-ray Emission from Clusters of Galaxies," (Cambridge: Cambridge University Press)

Schindler, S., Böhringer, H. 1993, A&A, 269, 83

Schindler, S., Müller, E. 1993, A&A, 272, 137

Struck, C. 1999, Physics Report, 1999, 321,1

Tomita, A., Nakamura, F. E., Takata, T., Nakanishi, K., Takeuchi, T., Ohta, K., Yamada, T. 1996, AJ, 111, 42

van Kampen, E., Jimenez, R., Peacock, J. A. 1999, MNRAS, 310, 43

Zabludoff, A. I., Zaritsky, D., Lin, H., Tucker, D., Hashimoto, Y., Shectman, S. A., Oemler, A., Kirshner, R. P. 1996, ApJ, 466, 104

# Continuous star formation in gas-rich dwarf galaxies

Simone Recchi and Gerhard Hensler

Institute of Astronomy, Vienna University
Türkenschanzstrasse 17, A-1180 Vienna, Austria
recchi@astro.univie.ac.at,
hensler@astro.univie.ac.at

**Abstract**

*Blue Compact Dwarf and Dwarf Irregular galaxies are generally believed to be unevolved objects, due to their blue colors, compact appearance and large gas fractions. Many of these objects show an ongoing intense burst of star formation or have experienced it in the recent past. By means of 2-D hydrodynamical simulations, coupled with detailed chemical yields originating from SNeII, SNeIa and intermediate-mass stars, we study the dynamical and chemical evolution of model galaxies with structural parameters similar to IZw18 and NGC1569. Bursts of star formation with short duration are not able to account for the chemical and morphological properties of these galaxies. The best way to reproduce the chemical composition of these objects is by assuming long-lasting episodes of star formation and a more recent burst, separated from the previous episodes by a short quiescent period. The last burst of star formation, in most of the explored cases, does not affect the chemical composition of the galaxy observable in H* II *regions, since the enriched gas produced by young stars is in a too hot phase to be detectable with the optical spectroscopy.*

## 1 Introduction

Among dwarf galaxies, Blue Compact Dwarfs (BCDs) and Dwarf Irregulars (dIrrs) are characterized by large gas content and often an active star formation (SF). They also show very blue colors and low metallicities and are therefore commonly believed to be poorly evolved systems. They are consequently ideal targets to study the feedback between star formation and interstellar medium. They have also been suggested to be the local counterparts of faint blue objects detected in excess at $z \sim 1$ (Babul & Rees 1992; Lilly et al. 1995).

It has recently become clear that most of these objects show the presence of stars of intermediate-old age (Kunth & Östlin 2000), but their importance for the global metallicity and energy budget of the galaxy is still unknown. It is interesting to

*Reviews in Modern Astronomy 18.* Edited by S. Röser
 ISBN 3-527-40608-5

simulate galaxies whose light and colors are dominated by young stars (like IZw18 and NGC1569) and to see whether their chemical and morphological properties are dominated by a recent burst of star formation or whether older episodes of SF are required in order to explain some of their characteristics.

In general, the SF in BCDs is described as a *bursting* process (Searle et al. 1973), namely, short, intense episodes of SF are separated by long inactivity periods. A *gasping* mode of SF (long episodes of SF of moderate intensity separated by short quiescent periods) is instead often used to describe the star formation in dIrrs (Aparicio & Gallart 1995). Good galaxy candidates experiencing gasping star formation are for instance NGC6822 (Marconi et al. 1995), Sextans B (Tosi et al. 1991) and the LMC (Gallagher et al. 1996). These two different SF regimes have been tested in the framework of chemical evolution models (Bradamante et al. 1998; Chiappini et al. 2003a; Romano et al. 2004), producing similar results, therefore is it not easy to discriminate between these two different SF scenarios on the basis of chemical evolution models alone. Our aim is to simulate the dynamical and chemical evolution of model galaxies by means of a 2-D hydrodynamical code in cylindrical coordinates, coupled with detailed chemical yields. Since the largest set of parameters is derived from various observations for IZw18 and NGC1569, we intend to see whether is it possible to put constraints on their past SF history.

Despite the different classification (IZw18 is a BCD galaxy, whereas NGC1569 is often classified as dIrr), these two objects show similar properties: both of these objects are in the aftermath of an intense burst of SF, are very metal poor (0.02 $Z_\odot$ for IZw18, Izotov & Thuan 1999; 0.23 $Z_\odot$ for NGC1569, González Delgado et al. 1997) and have an extremely large gas content.

In spite of their simplicity, the chemical composition of gas-rich dwarf galaxies is often peculiar and hardly understandable in terms of closed-box models. In particular, the N/O ratios are puzzling. For metallicities larger than 12 + log (O/H) $\sim$ 7.8, the log (N/O) is linearly increasing with the metallicity, although with a large scatter. This is consistent with a secondary production of nitrogen (N synthesized from the original C and O present in the star at birth). At lower metallicities, all the galaxies seem to show a constant log (N/O) (of the order of -1.55/-1.6), with almost no scatter (Izotov & Thuan 1999; hereafter IT99). This is a typical behaviour of element produced in a primary way (starting from the C and O newly formed in the star). Although some specific metal-poor objects seem to contradict the existence of the plateau (see e.g. Skillman et al. 2003; Pustilnik et al. 2004), an explanation is needed in order to understand the behavior of most dwarf galaxies in this range of metallicities.

This problem has been investigated in several papers and various ideas have been proposed to solve this puzzle. Izotov & Thuan (1999) proposed a significant primary production of nitrogen in massive stars, whereas the models of Henry et al. (2000) were able to explain the low log (N/O) at low metallicities with a very weak and constant star formation rate. Recently, Köppen and Hensler (2004) proposed the infall of metal-poor gas as a mechanism able to reduce the oxygen abundance of the galaxy, keeping the N/O ratio constant. For none of these models, however, a complete investigation of the structural and energetic effects by means of hydrodynamical simulations have been performed.

Moreover, IZw18 and NGC1569 have been carefully studied in the past and now we know, with reasonable accuracy, the chemical abundances not only in the H II medium, but also in other gas phases. In particular, recent *FUSE* (Far Ultraviolet Spectroscopic Explorer) data allowed two groups of astronomers (Aloisi et al. 2003; Lecavelier des Etangs et al. 2004) to calculate the H I abundances in IZw18. There is also an attempt to evaluate the chemical composition of the hot medium in the galactic wind of NGC1569, (Martin, Kobulnicky & Heckman 2002). It is challenging to compare the results of our simulations with these observations, in order to see whether is possible to put additional constraints on the past SF activity of IZw18 and NGC1569.

## 2 Chemical and dynamical evolution of IZw18

We first describe the evolution of a model galaxy resembling IZw18. This object, the most metal-poor galaxy locally known, has been considered in the past as a truly "young" galaxy, experiencing star formation for the very first time, since old stars could not be observed. Moreover, the spectral energy distribution is well described assuming a single, recent burst of SF (Mas-Hesse & Kunth 1999; Takeuchi et al. 2003). An underlying old population of stars has been first observed by Aloisi et al. (1999) in the optical and Östlin (2000) in the infrared, the age of which being however still disputed. This age ranges from some hundred Myrs (Aloisi et al. 1999) to a few Gyrs (Östlin 2000).

In the attempt of reproducing the characteristics of IZw18, we consider both a bursting and a gasping SF scenario. We first assume a couple of instantaneous bursts, separated by a quiescent period of 300-500 Myr. In the second subsection, we will consider a gasping SF with an old episode of SF lasting 270 Myr at a SF rate of $6 \times 10^{-3}$ $M_{\odot}$ $yr^{-1}$, a gap of 10 Myr and a recent burst lasting only 5 Myr and being 5 times more intense then the long-lasting episode. This SF history has been suggested by Aloisi et al. (1999) by fitting the observed Color-Magnitude diagram of IZw18 with synthetic ones. In this attempt at reproducing the main characteristics of IZw18, we also vary the slope of the IMF and the adopted nucleosynthetic yields, both for massive and for intermediate-mass stars (IMS). The model parameters are summarized in Table 1.

We use a 2-D hydrodynamical code with source terms. The input of energy and chemical elements into the systems is provided by SNeII (mainly responsible for the production of $\alpha$-elements), SNeIa (source of most of the iron-peak elements) and winds from low- and intermediate-mass stars (responsible for the bulk of nitrogen production and for a significant fraction of carbon). In all the considered models we will assume that SNeIa are more effective than SNeII in thermalizing the interstellar medium. This is due to the fact that SNeIa explode in a warmer and more diluted medium, owing to the previous activity of SNeII. The details about the code can be found in Recchi et al. (2001; 2002).

## 2.1 Bursting mode of star formation

The first (instantaneous) episode of SF produces $10^5$ $M_\odot$ of stars and is separated from the second one by a quiescent period of 300-500 Myr. We follow the evolution of the ISM after the first burst. After the assigned inactivity interval, we calculate how much cold gas is remained in the central part of the galaxy and we convert 10% of this gas into stars. The metallicity of this new stellar population is given by the metallicity of the gas which the starts are formed from. We obtain a second burst of SF with a mass of $\sim 5 \times 10^5$ $M_\odot$ and a metallicity of 1/50 $Z_\odot$.

The first burst is not able to account for the metallicity of the gas in IZw18. At the onset of the second burst, there is a sudden increase of the oxygen content (and, consequently, a sudden decrease of C/O and N/O abundance ratios). This mode of SF is therefore characterized by huge variations of the chemical composition of the galaxy on very short timescales. A few tens Myrs after the onset of the second burst, N/O and C/O abundance ratios begin to grow, due to the release of chemical elements from IMS. The results of our simulations match the abundance ratios found in the literature for two age intervals of the last burst: between 4 and 7 Myr and between $\sim 40$ and $\sim 80$ Myr. The second solution does not fit neither the morphology nor the spectral energy distribution of IZw18 and has to be rejected. The favoured age of the last burst is then, in the framework of a bursting scenario of SF, between 4 and 7 Myr. This solution has a very short duration, since the oxygen abundance is increasing very rapidly in this phase. Models with a SF gap of 300 or 500 Myr show approximately the same behaviour.

## 2.2 Gasping mode of star formation

As we have seen in the previous section, the bursting mode of SF can fit the observed abundances and abundance ratios found in literature for IZw18 only in tiny intervals of time. This can also be seen in Fig. 1, where we have plotted the evolution of a bursting model with 300 Myr of inactivity between two instantaneous bursts (model IZw – 1; long-dashed line). This model crosses the values of log (N/O) and log (O/H) inferred by IT99, but, as stated in the previous section, the N/O abundance ratio in particular shows large variations on short time-scales and remains within the allowed range of values only very briefly. In this section we therefore study what happens if we relax the hypothesis of instantaneous bursts of SF. The way in which continuous episodes of SF can be treated by our code is described in Recchi et al. (2004).

Models with Salpeter IMF, yields from massive stars coming from Woosley & Weaver (1995) and IMS yields coming from the most cited papers (van den Hoek & Groenewegen 1997; Renzini & Voli 1981) overestimate the nitrogen content of the galaxy by 0.4 – 0.6 dex and some of them underestimate O/H (see Recchi et al. 2004). The best fit between the observations and the results of the model is obtained when implementing the yields of both massive and IMS from Meynet & Maeder (2002). In their models, nitrogen is mainly produced in a primary way through rotational diffusion of carbon in the hydrogen-burning shell. This set of nucleosynthetic yields has given good results in chemical evolution models (Chiappini et al. 2003b) and are therefore worth testing in our simulations. Is it, however, necessary to point

Table 1: Parameters for the IZw18 models

| Model | SF mode[a] | x (IMF slope) | IMS yields | Massive stars yields |
|---|---|---|---|---|
| IZw – 1 | bursting | 1.35 | RV81[b] | WW95[c] |
| IZw – 2 | gasping | 1.10 | VG97[d] | WW95 |
| IZw – 3 | gasping | 1.35 | MM02[e] | MM02 |
| IZw – 4 | gasping | 1.35 | VG97 | WW95 |

[a] Bursting (2 instantaneous bursts of SF) or gasping (long episode of SF plus a recent burst), as described in Sect. 2.
[b] Renzini & Voli (1981)
[c] Woosley & Weaver (1995)
[d] van den Hoek & Groenewegen (1997)
[e] Meynet & Maeder (2002)

out that these models do not take into consideration the last phases of the stellar evolution (in particular the third dredge-up) and may therefore underestimate the total amount of nitrogen produced in IMS.

The results of this model are shown in Fig. 1 (model IZw – 2; dotted line). This model nicely reproduces the log (N/O) of IZw18 and slightly overestimates the observed oxygen content. Another possibility to get results closer to the observations is by using a flatter IMF (with a slope of $x = 1.1$) in order to produce more oxygen in massive stars. This model is also shown in Fig. 1 (model IZw – 3; short-dashed line). The log (N/O) is $\sim 0.2$ dex larger than the observations, therefore it better reproduces the observations compared with models adopting Salpeter IMF and van den Hoek & Groenewegen (1997) IMS yields.

It is also worth noticing in Fig. 1 that, after the onset of the second burst of SF (i.e. at 280 Myr), the metallicity of the galaxy for the gasping models (IZw – 2 and IZw – 3) does not show any sudden increase, at variance with what happens in the framework of a bursting scenario of SF (model IZw – 1; long-dashed line). This is due to the fact that the first SF episode is sufficiently energetic to create an outflow. The metals produced by the second generation of stars are released in a hot medium or directly channelled along the outflow. Consequently, they do not have the chance to cool down to temperatures detectable with the optical spectroscopy. This means that, if the SF in a galaxy has been active long enough and has been energetic enough, the last burst of SF does not affect at all the metallicity of the gas. The so-called *self-pollution* of galaxies by freshly produced metals (Kunth & Sargent 1986) can hold only under particular conditions, i.e. for the very first bursts of SF or if the gap between subsequent episodes is long enough.

Another way to see the different chemical evolution of bursting and gasping models is by plotting the log (N/O) vs. log (O/H) diagram (Fig. 2). The solid line is the evolution of the IZw – 3 model, whereas the behaviour of the IZw – 4 model is shown

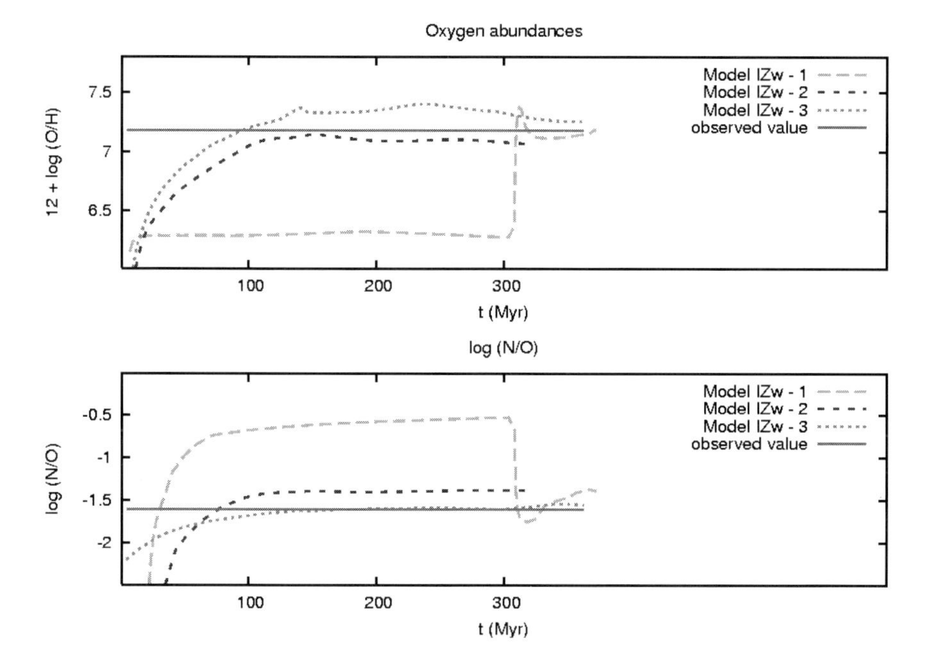


Figure 1: Evolution of 12 + log (O/H) (upper panel) and log (N/O) (lower panel) for IZw18 models. The solid line represents the observed values found for IZw18 (IT99). Model parameters are summarized in Table 1.

as dashed line. The bursting model IZw – 1 (dotted line) is plotted starting from the onset of the last burst. In Fig. 2 are also plotted the most recent data available in literature about O and N abundances in very metal-poor galaxies. As stated in the introduction, according to the results of IT99, these galaxies form a "plateau" in this diagram, the (N/O) ratios being very similar at different metallicities. Indeed, isolating the data of IT99 (filled squares and filled triangle), this plateau is pretty evident, whereas, when adding observations coming from other authors, the scatter seems to increase. We probably need more statistics and better measurements before drawing firm conclusions.

Model IZw – 4 exceeds the nitrogen of the most metal-poor galaxies (in particular the nitrogen of IZw18), whereas model IZw – 3 (solid line) matches the low log (N/O). It is also worth noticing that for this model it takes $\sim$ 100 Myr to reach the O abundance of the most metal-poor galaxies (see Fig. 1). After that, the chemical evolution tracks span a tiny region of the diagram for the remaining $\sim$ 200 Myr of the evolution of the galaxy. The bursting model IZw – 1 (dotted line) shows instead large abundance variations in very short time-scales. It takes only $\sim$ 80 Myr for this model to reach the final point. This kind of SF regime would produce a large scatter in the log (N/O) vs. log (O/H) diagram even at low metallicities. Under the hypothesis of a gasping SF regime the abundance ratios are stable for long time-scales, justifying the lack of scatter and the apparent plateau in the log (N/O) vs. log (O/H) diagram at low metallicities. Two recent papers (Aloisi et al. 2003; Lecavelier des

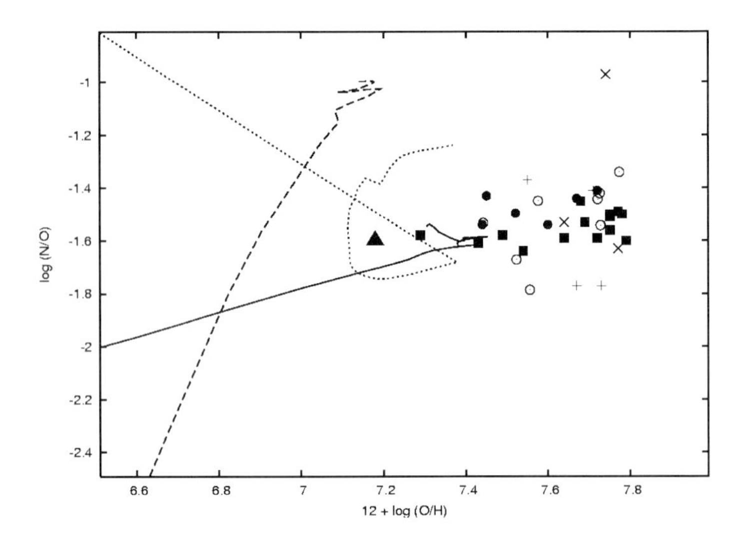


Figure 2: log (N/O) vs. 12 + log (O/H) in metal-poor BCD galaxies. The big filled triangle is the IZw18 value calculated by IT99. Filled squares are other galaxies measured by IT99. The collection of data by van Zee et al. (1997) is shown with pluses. The measurements by Kobulnicky & Skillman (1996) are indicated by suns. Crosses represent the values tabulated by Vílchez & Iglesias-Páramo (2003). The other data points (filled circles) are taken from different sources. Also shown is the evolution in the N/O vs. O/H plane of two gasping models: model IZw – 3 (solid line) and model IZw – 4 (dashed line). The dotted line represent the evolution of model IZw – 1 (bursting model), after the onset of the second burst of SF. Model parameters are summarized in Table 1

Etangs et al. 2004) tried to derive the chemical composition of the H I medium of IZw18. Even starting from the same *FUSE* data, the two papers differ significantly in the final abundance determinations. In particular, Lecavelier des Etangs et al. (2004) found an oxygen abundance in the neutral medium similar to the O/H derived in the H II regions, whereas Aloisi et al. (2003) obtained an O abundance a factor $\sim$ 3–4 lower than in the ionized gas. Due to the larger oxygen, the N/O ratio calculated by Lecavelier des Etangs et al. (2004) is much below the observations in the H II regions. The determination of Aloisi et al. (2003) is instead similar to the N/O found the in the ionized phase. In our models, we find a slight underabundance of O in the neutral medium, but not enough to justify the observations of Aloisi et al. (2003). The calculated log (N/O) is instead more consistent with the determinations of Lecavelier des Etangs et al. (2004) (see Recchi et al. 2004). Due to the differences in the results of these two groups, no robust constraints can be imposed by means of this comparison. A more careful parametrical study of gasping models for IZw18 will be presented in a forthcoming paper (Recchi et al. 2005, in prep.)

# 3 Chemical and dynamical evolution of NGC1569

We adopt the same hydrodynamical code described in the previous section to model a galaxy resembling NGC1569, a prototypical starburst galaxy. This galaxy is particularly valuable as a case study due to its proximity (2.2 Mpc according to Israel (1988); 1.95 or 2.8 Mpc according to Makarova & Karachentsev (2003)). It consists of two super star clusters (SSCs), with an absolute separation of 80–85 pc. The IMF slope of the SSCs is well constrained by the luminosity/mass ratio and is close to the Salpeter slope (Sternberg 1998). The stellar population in NGC1569 is dominated by stars younger than a few tens Myrs, the majority of which are found in two prominent super star clusters (Anders et al. 2004), although the presence of older stars have been inferred (Vallenari & Bomans 1996; Greggio et al. 1998).

We therefore adopted two possible SF histories for NGC1569. The first is a single burst of star formation, lasting for 25 Myr. The SF rate inferred by Greggio et al. (1998), assuming a Salpeter IMF, is 0.5 $M_\odot$ $yr^{-1}$. A weaker SF rate for the present burst (0.13 $M_\odot$ $yr^{-1}$) has been found in more recent studies of the CMD diagram of NGC1569 (Angeretti et al. 2005). Martin et al. (2002), by fitting the H$\alpha$ luminosity, found a SF rate of 0.16 $M_\odot$ $yr^{-1}$ for the last burst. Given the uncertainties of this value, we keep the SF rate as a free parameter.

A more complex episode of SF has been recently discovered by Angeretti et al. (2005) and is characterized by 3 episodes of SF. The first happened between 600 and 300 Myr ago at a rate of 0.05 $M_\odot$ $yr^{-1}$. This episode is followed by a period of inactivity of 150 Myr and then by a second episode lasting 110 Myr at a SF rate of 0.04 $M_\odot$ $yr^{-1}$. After a short quiescent period (3 Myr) the last episode of SF started. The onset of this episode is therefore 37 Myr ago, lasting until 13 Myr ago (24 Myr of duration in total, consistent with the estimates of Anders et al. 2004) at a rate of 0.13 $M_\odot$ $yr^{-1}$. It is worth pointing out that, in this work, the stars in a field of 200 $\times$ 200 pc have been analyzed and, therefore, the gaps in the SF process might be spurious. Table 2 summarizes the parameters adopted to model NGC1569.

Table 2: Parameters for the NGC1569 models

| Model | SF episodes | SF rate ($M_\odot$ $yr^{-1}$) | $M_{gas}$ ($M_\odot$) |
|---|---|---|---|
| NGC – 1 | 1 | 0.1 | $10^8$ |
| NGC – 2 | 1 | 0.5 | $10^8$ |
| NGC – 3 | 3 | 0.05; 0.04; 0.13 | $10^8$ |
| NGC – 4 | 3 | 0.05; 0.04; 0.13 | $1.8 \times 10^8$ |

### 3.1 Single episode of star formation

In the models described in this section, the SF lasts for 25 Myr and the SF rate is a free parameter. We have simulated two model galaxies: one with a SF rate of 0.1 $M_\odot$ $yr^{-1}$ (model NGC – 1) and one with a rate of 0.5 $M_\odot$ $yr^{-1}$ (model NGC – 2; see Table 2). It is hard to fine-tune the SF rate: a large rate (model NGC – 1) injects a large amount of energy into the ISM. This energy drives a very powerful galactic wind, able to push away from the galaxy most of the pristine gas at variance with what is observed in NGC1569. On the other hand, in the model with a lower SF rate (model NGC – 2), the oxygen produced by massive stars does not reach the observed abundance of NGC1569 of 12 + log (O/H) = 8.19 (Kobulnicky & Skillman 1997). In Fig. 3 we plot the evolution of oxygen and N/O for the model NGC – 1. As we can see, this model is neither able to explain the amount of oxygen present in the galaxy nor the (N/O) abundance ratio.

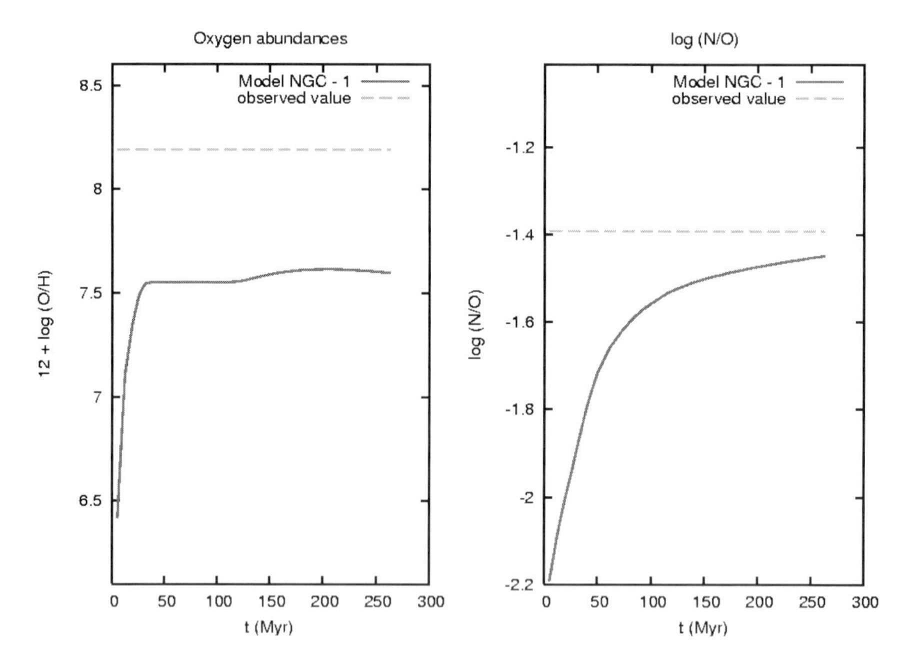


Figure 3: Evolution of 12 + log (O/H) (left panel) and log (N/O) (right panel) for a NGC1569 model with a single episode of SF at a rate of 0.1 $M_\odot$ $yr^{-1}$ (model NGC – 1; solid lines). The dashed lines are the observed values found by Kobulnicky & Skillman (1997).

As anticipated in the introduction, Martin et al. (2002) were able to give an estimate of the metallicity of the hot X-ray emitting gas in the galactic wind. This information completes the puzzle of understanding the metal enrichment. Even if the single-burst models are not able to account for the chemical and morphological properties of NGC1569, it is none the less interesting to calculate the metallicity of the hot gas (i.e. of the gas with temperatures larger than 0.3 keV) and compare it with the estimates of Martin et al. (2002). This comparison is shown in Fig. 4 for the NGC – 2 model. At the moment of the onset of a galactic wind, the oxygen abundance of the hot phase is already 1/4 of solar. It increases up to 2 times solar after $\sim$ 50 Myr from the beginning of the burst. The arrows drawn in the plot represent the estimates of the oxygen content of the galactic wind of NGC1569 (best fit; upper and lower limits). The oxygen composition of the hot gas increases continuously in this phase since, after 50 Myr, massive stars are still exploding and releasing oxygen into the interstellar medium. At later times however, the oxygen composition begins to decrease due to the larger fraction of pristine gas ablated from the supershell and entrained in the galactic wind. The [O/Fe] ratio is initially larger than solar due to the fact that the break-out occurs when SNeIa are not yet releasing their energy and metals into the ISM. At later times, since the SNeIa eject their products from the galaxy very easily (Recchi et al. 2001), the [O/Fe] ratio decreases. The observations of Martin et al. (2002) point toward a galactic wind dominated by $\alpha$-elements, therefore the outflow is probably still triggered by SNeII.

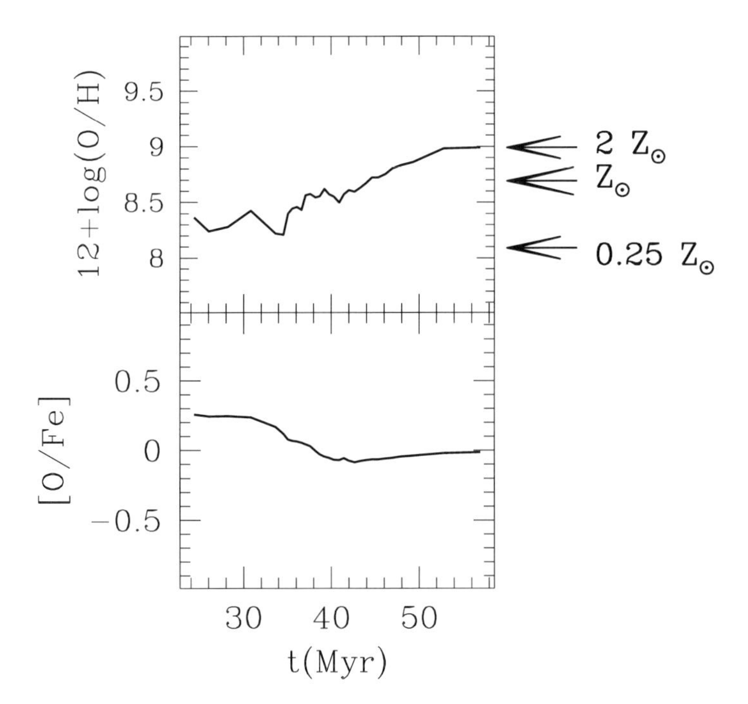


Figure 4: Evolution of 12 + log (O/H) (upper panel panel) and [O/Fe] abundance ratio (lower panel) for the hot gas entrained in the galactic wind for the model NGC – 2 (see Table 2). The arrows indicate the oxygen abundance of the galactic wind inferred by Martin et al. (2002).

## 3.2 Three episodes of star formation

As shown in the previous section, single SF bursts of short durations are not able to account for the global properties of NGC1569. We therefore describe in this section the evolution of models in which the SF is a gasping process, occurring in 3 different episodes as explained in Sect. 3. Since in this case the SF rate is no longer a free parameter, we decide to explore the effect of a different initial ISM distribution. In particular, we consider a "light" model (model NGC – 3), in which the total galactic H I mass at the beginning of the simulation is $\sim 10^8$ $M_\odot$ and a model with a factor of 2 more gas initially present inside the galaxy (model NGC – 4; see Table 2). Since there are still uncertainties about the total gas mass of NGC1569 (see e.g. Stil & Israel 2002; Mühle et al. 2003) one has the freedom to test different values of the initial mass and to see which one is more appropriate to reproduce the characteristics of NGC1569. We adopt hereafter the nucleosynthetic prescriptions of Meynet & Maeder (2002), since they seem to give the better description of the chemical properties of IZw18 (see Sect. 2), bearing in mind that the predicted nitrogen can be a lower limit.

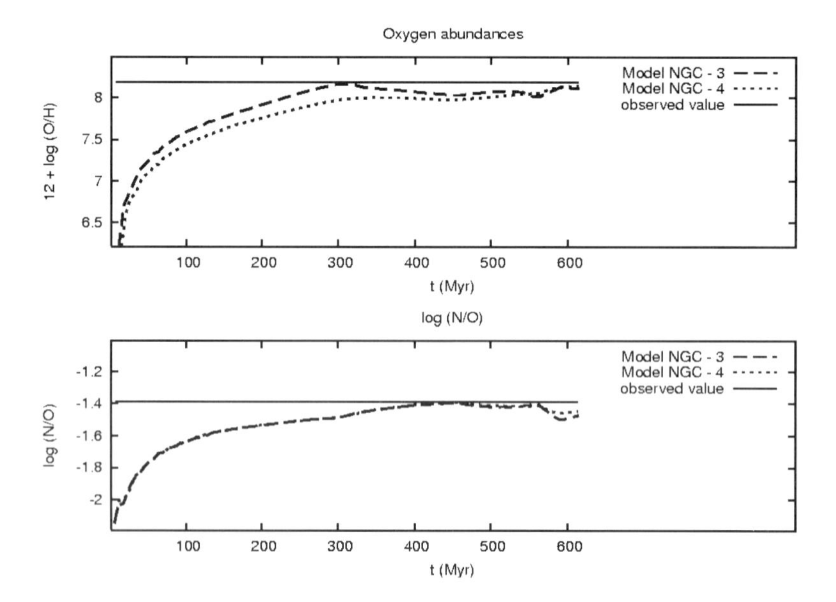


Figure 5: Evolution of 12 + log (O/H) (upper panel) and log (N/O) (lower panel) for two NGC1569 models in which the Angeretti et al. (2005) SF history is implemented. The solid line represents the observed values found for NGC1569 (Kobulnicky & Skillman 1997). The dashed line is the evolution of model NGC – 3, whereas the dotted line shows the evolution of model NGC – 4 (see Table 2). Note that the N/O evolution of the two models is almost identical until t $\sim$ 560 Myr, therefore it is diffucult, in the lower panel, to disentangle the two lines.

In Fig. 5 we show the evolution of oxygen (upper panel) and log (N/O) (lower panel) for the models NGC – 3 (dashed lines) and NGC – 4 (dotted lines) and we compare them with the abundances derived from Kobulnicky & Skillman (1997) (solid lines). At the end of the simulations (after $\sim$ 600 Myr), the oxygen is reproduced nicely by the model NGC – 4 and also model NGC – 3 is close to the observed value. It is worth noticing that the outflow created by the pressurized gas is very weak in the NGC – 4 model and of moderate intensity for the model NGC – 3. The fraction of oxygen lost through the galactic wind is larger in the light model. In the first hundreds of Myrs the oxygen abundance predicted by the model NGC – 3 is larger, since it is diluted by a smaller amount of hydrogen. At later times, however, the O abundance predicted by this model slightly decreases with time, since some oxygen is expelled from the galaxy.

The final log (N/O) predicted by these models are – 1.47 (model NGC – 3) and – 1.44 (model NGC – 4). These values slightly underestimate the observations of Kobulnicky & Skillman (1997), but are still reasonably close to it, considering the observational errors (0.05 dex).

## 3.3 Model with a big infalling cloud

Observations show the presence of extended H I clouds and complexes surrounding NGC1569 (Stil & Israel 1998). In particular, there is a series of gas clumps in the southern halo probably connected with a H I arm present in the western side of the galaxy. These H I features can be attributed to the debris of a tidally disrupted big cloud infalling towards NGC1569 (Mühle et al. 2005). The mass of this complex is difficult to asses. The lower limit given by Mühle et al. (2005) is $1.2 \times 10^7$ $M_\odot$ (the sum of the mass of all the detected groups of clouds), but some H I could have been already accreted. If this complex is similar to the high velocity clouds in the Local Group, one has to expect masses larger than a few $10^7$ $M_\odot$ (Blitz et al. 1999). As a first attempt to study the effect of a big cloud infalling towards the galaxy, we assume a mass of $2 \times 10^7$ $M_\odot$. The initial position of the cloud is 2 kpc away from the center of the galaxy along the polar axis (due to the assumed symmetry of the system, this is the only reasonable initial configuration). The infalling velocity of this cloud is 10 km $s^{-1}$, similar to the local sound speed, and its radius is 1 kpc. The other structural parameters are as in model NGC – 3 (see Table 2).

The evolution of oxygen and log (N/O) for this model is shown in Fig. 6. The development of the galactic wind is hampered by the pressure of this big infalling cloud and, consequently, no major outflow is developed during the simulation. The oxygen abundance is therefore always increasing, since only a negligible fraction of it is lost from the galaxy. The final log (N/O) is consistent with the observations, whereas this model underestimates the final oxygen content of the galaxy (by $\sim$ 0.15 dex). This is due to the fact that the big cloud on its path towards the galaxy sweeps up and drags some gas initially present in the outer regions of the galaxy. The final gas mass inside the galaxy is therefore larger than the initial one. It is worth noticing that the prominent outflow visible in NGC1569 (Martin et al. 2002) deviates from the results of this simulation. Therefore, we either have to consider a larger input of energy into the system or have to consider a different infall direction of

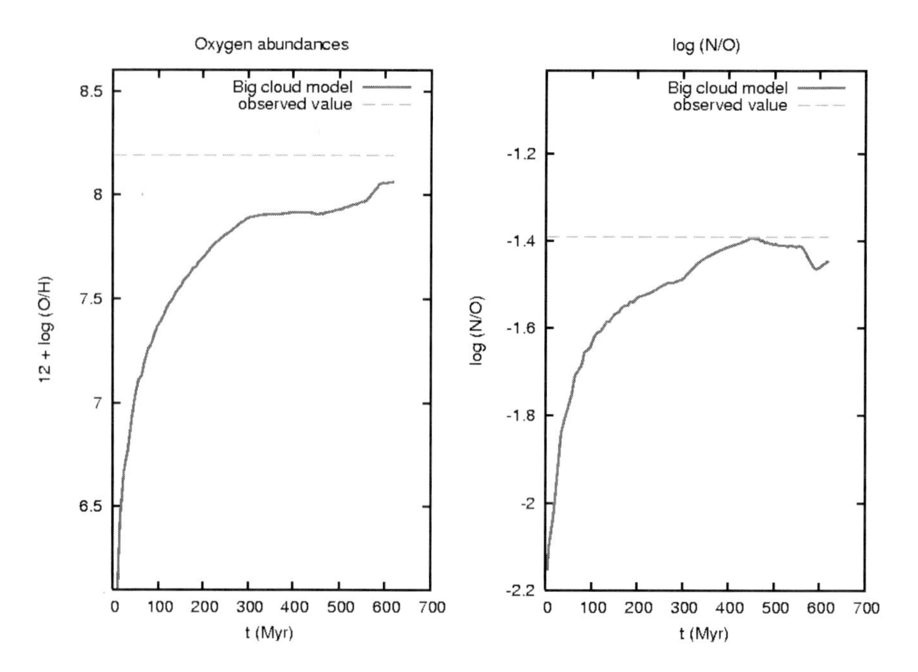


Figure 6: Evolution of 12 + log (O/H) (left panel) and log (N/O) (right panel) for a NGC1569 models in which the infall of a big cloud towards the center of the galaxy is taken into consideration.

the cloud. Indeed, observations show that this H I complex seems to wrap around the disk of NGC1569 and to approach the galaxy from the western side. A more careful parametrical study of model galaxies reproducing NGC1569 will be presented in a forthcoming paper (Recchi & Hensler 2005, in prep.)

# 4 Conclusions

By means of a 2-D hydrodynamical code, we have studied the chemical and dynamical evolution of model galaxies resembling IZw18 and NGC1569, two gas-rich dwarf galaxies in the aftermath of an intense burst of SF. We have considered in both cases either episodes of SF of short duration (bursting SF), or more complex SF behaviours, in which the galaxies have experienced in the past long-lasting episodes of SF, separated from the last more intense burst by short periods of inactivity (gasping star formation).

Models with a bursting star formation are generally unable to account for the chemical and morphological properties of these two objects. In the case of IZw18, they produce huge variations of the chemical composition of the galaxy in short time-scales. They are able to fit at the same time the C, N, O composition of IZw18, but only for very short time intervals. This pattern of the chemical tracks would presumably give rise to a large scatter in the abundance ratios. The observations available nowadays disagree with this scenario, since most metal-poor galaxies seem to share the same [N/O] abundance ratio (IT99). In the case of NGC1569, models with a single short episodes of SF either severely underproduce O or inject too much energy into the system, enough to unbind a too large fraction of the gas initially present in the galaxy.

The best way to reproduce the chemical composition of both, IZw18 and NGC1569, is therefore assuming long-lasting, continuous episodes of SF of some hundreds Myrs of age and a recent and more intense short burst. Adopting the star formation prescriptions derived from the comparison of the color-magnitude diagrams with synthetic ones (Aloisi et al. 1999 for IZw18; Angeretti et al. 2005 for NGC1569) we produce results in good agreement with the observations, if the yields of Meynet & Maeder (2002) are implemented.

For what concerns NGC1569, a model in which a big cloud is falling towards the center of the galaxy along the polar axis inhibits almost completely the formation of a galactic wind, at variance with what observed.

In most models with gasping star formation, the presently observed chemical composition of the galaxy reflects mostly the chemical enrichment from old stellar populations. In fact, if the first episodes of SF are powerful enough to create a galactic wind or to heat up a large fraction of the gas surrounding the star forming region, the metals produced by the last burst of star formation are released in a too hot medium or are directly expelled from the galaxy through the wind. They do not have the chance to pollute the surrounding medium and contribute to the chemical enrichment of the galaxy.

## Acknowledgments

We thank Stefanie Mühle and Luca Angeretti for interesting discussions. S.R. acknowledges generous financial support from the Alexander von Humboldt Foundation and Deutsche Forschungsgemeinschaft (DFG) under grant HE 1487/28-1. The Observatory of Vienna is also acknowledged for the travel support.

## References

Aloisi, A., Savaglio, S., Heckman, T.M., Hoopes, C.G., Leitherer, C., Sembach, K.R. 2003, ApJ, 595, 760

Aloisi, A., Tosi, M., Greggio, L. 1999, AJ, 118, 302

Anders, P., de Grijs, R., Fritze-v. Alvensleben, U., Bissantz, N. 2004, MNRAS, 347, 17

Angeretti, L., Tosi, M., Greggio, L., Sabbi, E., Aloisi, A., Leitherer, C. 2005, AJ, submitted

Aparicio, A., Gallart, C. 1995, AJ, 110, 2105

Babul, A., Rees, M.J. 1992, MNRAS, 255, 346

Blitz, L., Spergel, D.N., Teuben, P.J., Hartmann, D., Burton, W.B. 1999, ApJ, 514, 818

Bradamante, F., Matteucci, F., D'Ercole, A. 1998, A&A, 337, 338

Chiappini, C., Matteucci, F., Meynet, G. 2003b, A&A, 410, 257

Chiappini, C., Romano, D., Matteucci, F. 2003a, MNRAS, 339, 63

Gallagher, J.S. et al. 1996, ApJ, 466, 732

González Delgado, R.M., Leitherer, C., Heckman, T., Cerviño, M. 1997, ApJ, 483, 705

Greggio, L., Tosi, M., Clampin, M., de Marchi, G., Leitherer, C., Nota, A., Sirianni, M. 1998, ApJ, 504, 725

Henry, R.B.C., Edmunds, M.G., Köppen, J. 2000, ApJ, 541, 660

Israel, F.P. 1988, A&A, 194, 24

Izotov, Y.I., Thuan, T.X. 1999, ApJ, 511, 639 (IT99)

Kobulnicky, H.A., Skillman, E.D. 1996, ApJ, 471, 211

Kobulnicky, H.A., Skillman, E.D. 1997, ApJ, 489, 636

Köppen, J., Hensler, G. 2004, A&A, submitted

Kunth, D., Östlin, G. 2000, A&AR, 10, 1

Kunth, D., Sargent, W.L.W. 1986, ApJ, 300, 496

Lecavelier des Etangs, A., Désert, J.-M., Kunth, D., Vidal-Madjar,A., Callejo, G., Ferlet, R., Hébrard, G., Lebouteiller, V. 2004, A&A, 413, 131

Lilly, S.J., Tresse, L., Hammer, F., Crampton, D., Le Fevre, O. 1995, ApJ, 455, 108

Makarova, L.N., Karachentsev, I.D. 2003, Ap, 46, 144

Marconi, G., Tosi, M., Greggio, L., Focardi, P. 1995, AJ, 109, 173

Martin, C.L., Kobulnicky, H.A., Heckman, T.M. 2002, ApJ, 574, 663

Mas-Hesse, J.M., Kunth, D. 1999, A&A, 349, 765

Meynet, G., Maeder, A. 2002, A&A, 390, 561

Mühle, S., Klein, U., Wilcots, E.M., Hüttenmeister, S. 2003, ANS, 324, 40

Mühle, S., Klein, U., Wilcots, E.M., Hüttenmeister, S. 2005, submitted

Östlin, G. 2000, ApJ, 535, L99

Pustilnik, S., Kniazev, A., Pramskij, A., Izotov, Y., Folz, C., Brosch, N., Martin, J.-M., Ugryumov, A. 2004, A&A, 419, 469

Recchi, S., Matteucci, F., D'Ercole, A. 2001, MNRAS, 322, 800

Recchi, S., Matteucci, F., D'Ercole, A., Tosi, M. 2002, A&A, 384, 799

Recchi, S., Matteucci, F., D'Ercole, A., Tosi, M. 2004, A&A, 426, 37

Renzini, A., Voli, M. 1981, A&A, 94, 175

Romano, D., Tosi, M., Matteucci, F. 2004, to appear in the proceedings of the conference "Starbursts - From 30 Doradus to Lyman break galaxies", eds. R. de Grijs and R.M. Gonzalez Delgado (Kluwer)

Searle, L., Sargent, W.L.W., Bagnuolo, W.G. 1973, ApJ, 179, 427

Skillman, E.D., Côté, S., Miller, B.W. 2003, AJ, 125, 610

Sternberg, A. 1998, ApJ, 506, 721

Stil, J.M., Israel, F.P. 1998, A&A, 337, 64

Stil, J.M., Israel, F.P. 2002, A&A, 392, 473

Takeuchi, T.T., Hirashita, H., Ishii, T.T., Hunt, L.K., Ferrara, A. 2003, MNRAS, 343, 839

Tosi, M., Greggio, L., Marconi, G., Focardi, P. 1991, AJ, 102, 951

Vallenari, A., Bomans, D.J. 1996, A&A, 313, 713

van den Hoek, L.B., Groenewegen, M.A.T. 1997, A&AS, 123, 305

van Zee, L., Haynes, M.P., Salzer, J.J. 1997, AJ, 114, 2479

Vílchez, J.M., Iglesias-Páramo, J., 2003, ApJS, 145, 225

Woosley, S.E., Weaver, T.A. 1995, ApJS, 101, 181

# M33 – Distance and Motion

Andreas Brunthaler

Joint Institute for VLBI in Europe
Postbus 2, 7990 AA Dwingeloo, the Netherlands
brunthaler@jive.nl

**Abstract**

*Measuring the proper motions and geometric distances of galaxies within the Local Group is very important for our understanding of the history, present state and future of the Local Group. Currently, proper motion measurements using optical methods are limited only to the closest companions of the Milky Way. However, Very Long Baseline Interferometry (VLBI) provides the best angular resolution in astronomy and phase-referencing techniques yield astrometric accuracies of $\approx$ 10 micro-arcseconds. This makes a measurement of proper motions and angular rotation rates of galaxies out to a distance of $\sim$ 1 Mpc feasible.*

*This article presents results of VLBI observations of two regions of $H_2O$ maser activity in the Local Group galaxy M33. The two masing regions are on opposite sides of the galaxy. This allows a comparison of the angular rotation rate (as measured by the VLBI observations) with the known inclination and rotation speed of the HI gas disk. This gives a geometric distance of 730 $\pm$ 100 $\pm$ 135 kpc. The first error indicates the statistical error from the proper motion measurements while the second error is the systematic error from the rotation model. This distance is consistent, within the errors, with the most recent Cepheid distance to M33. Since all position measurements were made relative to an extragalactic background source, the proper motion of M33 has also been measured. This provides a three dimensional velocity vector of M33, showing that this galaxy is moving with a velocity of 190 $\pm$ 59 km $s^{-1}$ relative to the Milky Way. These measurements promise a new handle on dynamical models for the Local Group and the mass and dark matter halo of Andromeda and the Milky Way.*

# 1 Introduction

The nature of spiral nebulae like M33 was the topic of a famous debate in the 1920's. While some astronomers favored a close distance and Galactic origin, others were

*Reviews in Modern Astronomy 18.* Edited by S. Röser

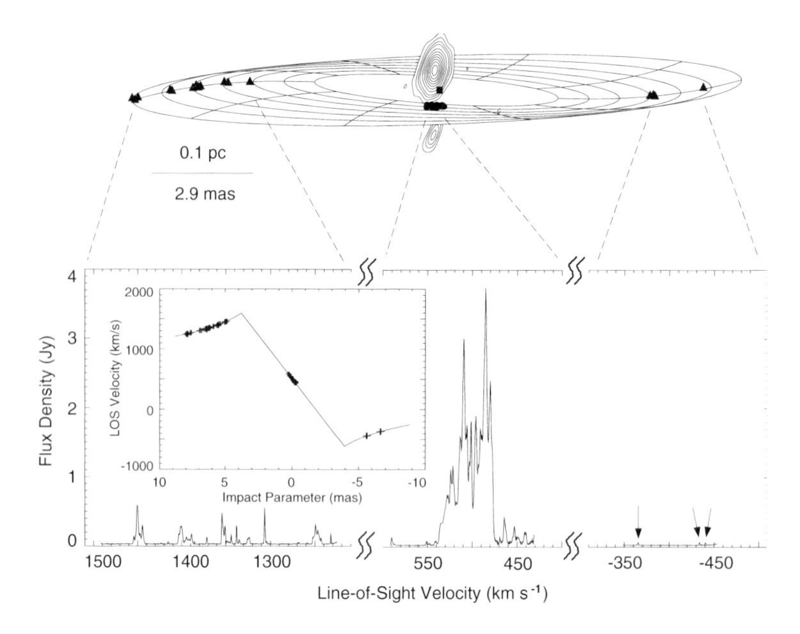


Figure 1: The warped disk model for NGC 4258.The inset shows line-of-sight (LOS) velocity versus impact parameter for the best-fitting Keplerian disk, with the maser data superposed (taken from Herrnstein et al. 1999).

convinced of the extragalactic nature. In 1923, van Maanen claimed to have measured a large proper motion and angular rotation of M33 from photographic plates separated by $\approx$ 12 years. These measurements yielded rotational motions of $\approx$ 10-30 mas yr$^{-1}$, clearly indicating a close distance to M33. However, Hubble (1926) discovered Cepheids in M33, showing a large distance to M33 and confirming that M33 is indeed an extragalactic object. The expected proper motions from the rotation of M33 are only $\approx$ 30 $\mu$as yr$^{-1}$, and 3 orders of magnitude smaller than the motions claimed by van Maanen. The source of error in van Maanens observations was never identified. After more than 80 years, the idea behind the experiment to measure the rotation and proper motions of galaxies remains interesting for our understanding of the dynamics and geometry of the Local Group. Hence, they are an important science goal of future astrometric missions, e.g. the Space Interferometry Mission (SIM, see Shaya et al. 2003).

## 1.1 Geometric Distances

Currently, the calibration of most standard candles used for extragalactic distances, are tied in one way or another to the distance of the Large Magellanic Cloud (LMC) (e.g., Freedman 2000; Mould et al. 2000). However, recently Tammann, Sandage & Reindl (2003) and Sandage, Tammann & Reindl (2004) find differences in the slopes of the Period – Luminosity (P–L) relations in Cepheids in the Galaxy, LMC and SMC. This "weakens the hope of using Cepheids to attain good accuracies in measuring galaxy distances" until the reasons for the differences in the P–L relations are understood. This has been also confirmed by Ngeow & Kanbur (2004).

Hence, geometric distances to nearby galaxies in which well understood standard candles can be studied, are needed to limit systematic errors in the extragalactic distance scale. However, geometric distances to other galaxies are very difficult to obtain. The best case is the Seyfert galaxy NGC 4258. Here, water vapor maser emission in a Keplerian circumnuclear disk that orbits a super-massive black hole in the center of the galaxy (see Fig. 1) was used to measured a geometric distance. This distance of 7.2±0.5 Mpc (Herrnstein et al. 1999) is slightly shorter than the distance estimate using Cepheids by Newman et al. (2001) who derive a value of 7.8 ± 0.3 Mpc.

To confirm and calibrate the current extragalactic distance scale it is essential to obtain geometric distances to other nearby galaxies. This involves detection of coherent motions at extragalactic distances.

## 1.2 Proper Motions

Another important astrophysical question is the nature and existence of dark matter in the universe. The existence of dark matter was first suggested in the 1930s by Fritz Zwicky from the observed peculiar motions of galaxies in clusters. It was firmly established by observations of the flat rotation curves of galaxies (e.g. Fich & Tremaine 1991). The closest place to look for dark matter halos is the Milky Way and Andromeda in the Local Group. Various attempts have been made to weigh the galaxies in the Local Group and determine size and mass of the Milky Way and its not very prominent dark matter halo (Kulessa & Lynden-Bell 1992; Kochanek 1996). Other attempts use Local Group dynamics in combination with MACHO data to constrain the universal baryonic fraction (Steigman & Tkachev 1999).

The problem when trying to derive the gravitational potential of the Local Group is that usually only radial velocities are known and hence statistical approaches have to be used. Kulessa & Lynden-Bell (1992) introduced a maximum likelihood method which requires only the line-of-sight velocities, but it is also based on some assumptions (eccentricities, equipartition).

Clearly, the most reliable way of deriving masses is using orbits, which requires the knowledge of three-dimensional velocity vectors obtained from measurements of proper motions. The usefulness of proper motions was impressively demonstrated for the Galactic Center where the presence of a dark mass concentration, presumably a super-massive black hole, has been unambiguously demonstrated by stellar proper motion measurements (Eckart & Genzel 1996; Schödel et al. 2002; Ghez et al. 2003).

However, measuring proper motions of members of the Local Group to determine its mass is difficult. For the LMC Jones, Klemola & Lin (1994) claim a proper motion of $1.2 \pm 0.28$ mas yr$^{-1}$ obtained from comparing photographic plates over a time-span of 14 years. Schweitzer et al. (1995) claim $0.56 \pm 0.25$ mas yr$^{-1}$ for the Sculptor dwarf spheroidal galaxy from plates spanning 50 years in time (Fig 2). Kochanek (1996) shows that inclusion of these marginal proper motions can already

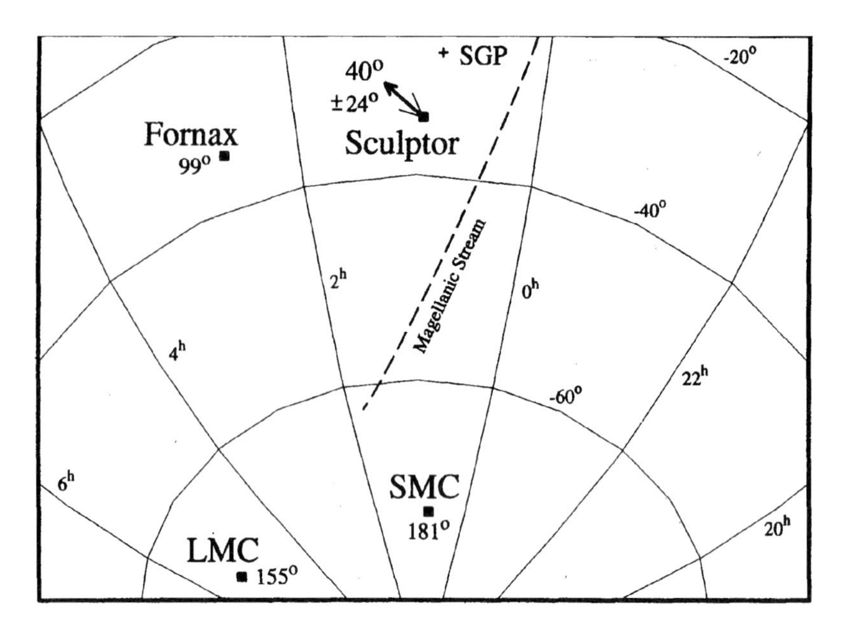


Figure 2: The proper motion of the Sculptor Dwarf Spheroidal Galaxy. Taken from Schweitzer et al. (1995).

significantly improve the estimate for the mass of the Milky Way, since it reduces the strong ambiguity caused by Leo I, which can be treated as either bound or unbound to the Milky Way. The same work also concludes that if the claimed optical proper motions are true, the models also predict a relatively large tangential velocity of the other satellites of the Galaxy.

The dynamics of nearby galaxies are also important to determine the solar motion with respect to the Local Group to help define a standard inertial reference frame (Courteau & van den Bergh 1999). And again, as Courteau and van den Bergh point out, "interpretation errors for the solar motion may linger until we obtain a better understanding of the true orbital motion of Local Group members".

Despite the promising start, the disadvantage of the available optical work is obvious: a further improvement and confirmation of these measurements requires an additional large time span of many decades and will still be limited to only the closest companions of the Milky Way.

## 1.3 Proper Motions with VLBI

On the other hand, the expected proper motions for galaxies within the Local Group, ranging from 0.02 – 1 mas yr$^{-1}$, can be seen with Very Long Baseline Interferometry (VLBI) using the phase-referencing technique. A good reference point is the motion of Sgr A* across the sky reflecting the Sun's rotation around the Galactic Center. This motion is well detected between epochs separated by only one month with the

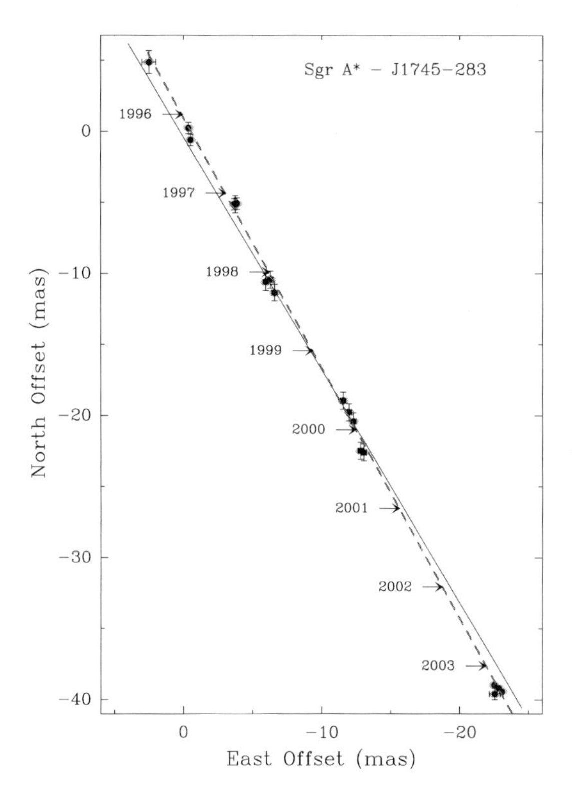


Figure 3: Apparent motion of Sgr A* relative to a background quasar. The blue line indicates the orientation of the Galactic plane and the dashed line is the variance-weighted best-fit proper motion.

VLBA (Reid et al. 1999). New observations increased the time span to 8 years. The position of Sgr A* relative to the background quasar J1745-283 is shown in Figure 3 (taken from Reid & Brunthaler 2004).

The motion of Sgr A* is 6.379 $\pm$ 0.026 mas yr$^{-1}$ and almost entirely in the plane of the galaxy. This gives, converted to Galactic coordinates a proper motion of -6.379 $\pm$ 0.026 mas yr$^{-1}$ in Galactic longitude and -0.202 $\pm$ 0.019 mas yr$^{-1}$ in Galactic latitude. If one assumes a distance to the Galactic Center ($R_0$) of 8 $\pm$ 0.5 kpc (e.g. Reid 1993; Eisenhauer et al. 2003), these motions translate to a velocity of -241 $\pm$15 km s$^{-1}$ along the Galactic plane and -7.6 $\pm$ 0.6 km s$^{-1}$ out of the plane of the Galaxy. This motion can be entirely explained by a combination of a circular rotation of the Local Standard of Rest (LSR) $\Theta_0$ and the deviation of the motion of the Sun from the motion of the LSR. Removing the Solar motion relative to the LSR as measured by Hipparcos data by Dehnen & Binney (1998) (5.25 $\pm$ 0.62 km s$^{-1}$ in longitude and -7.17 $\pm$ km s$^{-1}$ in latitude) yields an estimate of $\Theta_0$=236$\pm$15 km s$^{-1}$. Here the error is dominated by the uncertainty in the distance $R_0$ to the Galactic Center. The motion of Sgr A* out of the plane of the Galaxy is only -0.4 $\pm$ 0.8

Figure 4: Position of two regions of maser activity in M33. Also shown are predicted motions due to rotation of the H I disk. The image of M33 was provided by Travis Rector (NRAO/AUI/NSF and NOAO/AURA/NSF), David Thilker (NRAO/AUI/NSF) and Robert Braun (ASTRON).

km $s^{-1}$. This lower limit can be used to put tight constrains for the mass of Sgr A* (for details see Reid & Brunthaler 2004).

With the accuracy obtainable with VLBI one can in principle measure very accurate proper motions for most Local Group members within less than a decade. The main problem so far is finding appropriate radio sources. Useful sources would be either compact radio cores or strong maser lines associated with star forming regions. Fortunately, in a few galaxies bright masers are already known. Hence the task that lies ahead of us, if we want to significantly improve the Local Group proper motion data and mass estimate, is to make phase-referencing observations with respect to background quasars of known Local Group galaxies with strong $H_2O$ masers.

The most suitable candidates for such a VLBI phase-referencing experiment are the strong $H_2O$ masers in IC 10 ($\sim$ 10 Jy peak flux in 0.5 km $s^{-1}$ line, the brightest known extragalactic maser; Becker et al. 1993) and IC 133 in M33 ($\sim$ 2 Jy, the first extragalactic maser discovered). Both masers have been observed successfully

Table 1: Details of the observations: Observing date, observation length $t_{obs}$, beam size $\theta$ and position angle $PA$ of the beam.

| Epoch | Date | $t_{obs}$ | $\theta$ [mas] | $PA[°]$ |
|---|---|---|---|---|
| I | 2001/03/27 | 10 h | 0.88×0.41 | 164 |
| I | 2001/04/05 | 10 h | 0.86×0.39 | 169 |
| II | 2002/01/28 | 10 h | 0.62×0.33 | 176 |
| II | 2002/02/03 | 10 h | 0.71×0.33 | 175 |
| III | 2002/10/30 | 10 h | 0.87×0.38 | 171 |
| III | 2002/11/12 | 10 h | 0.84×0.36 | 165 |
| IV | 2003/12/14 | 12 h | 0.85×0.36 | 159 |
| IV | 2004/01/08 | 12 h | 1.15×0.47 | 164 |

with VLBI (e.g. Argon et al. 1994; Greenhill et al. 1990; Greenhill et al. 1993; Brunthaler et al. 2002).

The two galaxies belong to the brightest members of the Local Group and are thought to be associated with M31. In both cases a relatively bright phase-referencing source is known to exist within a degree. In addition their galactic rotation is well known from H I observations. Consequently, M33 and IC 10 seem to be the best known targets for attempting to measure Local Group proper motions with the VLBA.

The next sections are based on Brunthaler et al. (2005) and present the first results of a program to measure the geometric distance and proper motion of M33 . The results on IC 10 will be discussed in a subsequent paper.

# 2 VLBI observations of M33

## 2.1 Observations and Data Reduction

We observed two regions of $H_2O$ maser activity in M33, M33/19 (following the notation of Israel & van der Kruit 1974) and IC 133, eight times with the NRAO Very Long Baseline Array (VLBA) between March 2001 and January 2004. M33/19 is located in the south-eastern part of M33, while IC 133 is located in the north-east of M33 (see Fig. 4). Details of the observations are shown in Table 1. The observations are grouped into four epochs, each comprising two closely spaced observations to enable assessment of overall accuracy and systematic errors. The separations of the two observations within each epoch were large enough that the weather conditions are uncorrelated, but small enough that proper motions are negligible during this time.

We observed four 8 MHz bands, in dual circular polarization. The 128 spectral channels in each band yield a channel spacing of 62.5 kHz, equivalent to 0.84 km s$^{-1}$, and covered a velocity range of 107 km s$^{-1}$. The observations involved rapid switching between the phase-calibrator J0137+312, which is a compact background source with continuum emission, and the target sources IC 133 and M33/19

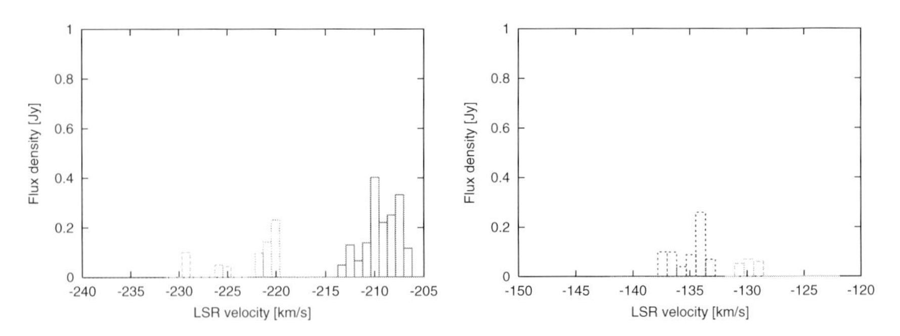


Figure 5: VLBI-spectrum of IC 133 (left) and M33/19 (right) on 2002 February 03.

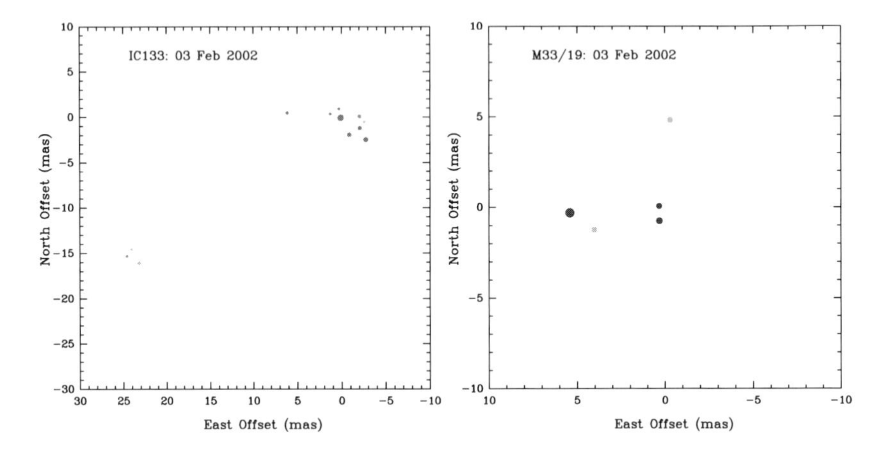


Figure 6: Composite map of the $H_2O$ masers in IC 133 (left) and M33/19 (right). The area of the circles is proportional to the flux density of the components.

(J0137+312 – IC 133 – J0137+312 – M33/19 – J0137+312). With source changes every 30 seconds, an integration time of 22 seconds was achieved. The background source was assumed to be stationary on the sky. Being separated by only $1^\circ$ on the sky from the masers, one can obtain a precise angular separation measurement for all sources. The data were edited and calibrated using standard techniques in the Astronomical Image Processing System (AIPS) as well as zenith delay corrections (see Brunthaler, Reid & Falcke 2003 for discussion) to improve the accuracy of the phase-referencing. The masers in IC 133 and M33/19 were imaged with standard techniques in AIPS.

In IC 133, 29 distinct emission features were detected. The VLBI spectra and spatial distribution of the maser components in IC 133 in the observation on 2004 February 03 are shown in Figures 5 and 6. Most maser components cluster in two regions, IC 133M (main) and IC 133SE (southeast), following the notation of Greenhill et al. (1990). These two regions are separated by $\approx$ 30 mas, which translates at an assumed distance of 790 kpc to $3.5\times10^{17}$cm. The emission in both regions is

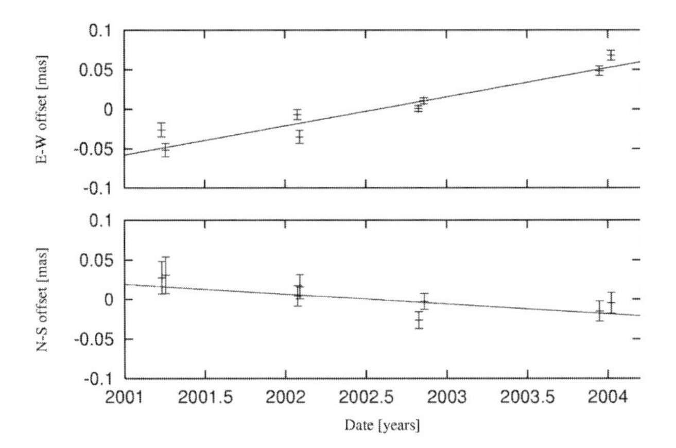


Figure 7: Average position of the masers M33/19 in right ascension (upper) and declination (lower) relative to a background source.

spread over $\approx 4$ mas, corresponding to $\approx 3200$ AU. This is similar to water maser emission in Galactic star-forming regions like W3(OH), W49 or Sgr B2 (e.g. Reid et al. 1995; Walker et al. 1977; Kobayashi et al. 1989 respectively). The spetrum and the spatial distribution is very similar to earlier VLBI observations by Argon et al. (2004), Greenhill et al. (1990), and Greenhill et al. (1993). All components were unresolved.

In M33/19, 8 maser features (see Figures 5 and 6) could be detected. The two features near the phase center were separated by less than a beam size and blend together. These two features were fit by two elliptical Gaussian components simultaneously. All other features were fit by a single elliptical Gaussian component.

## 2.2 Proper Motions of M33/19 and IC 133

The maser emission in M33/19 and IC 133 is variable on timescales less than one year. Between the epochs, new maser features appeared while others disappeared. However, the motions of 4 components in M33/19 and 6 components in IC 133 could be followed over all four epochs. The component identification was based on the positions and radial velocities of the maser emission. Each component was usually detected in several frequency channels. A rectilinear motion was fit to each maser feature in each velocity channel separately. Fits with reduced $\chi^2$ larger than 3 were discarded as they are likely affected by blending. Then, the variance weighted average of all motions was calculated. This yields an average motion of the maser components in M33/19 of $35.5 \pm 2.7$ $\mu$as yr$^{-1}$ in right ascension and $-12.5 \pm 6.3$ $\mu$as yr$^{-1}$ in declination relative to the background source J0137+312. For IC 133 one gets an average motion of $4.7 \pm 3.2$ $\mu$as yr$^{-1}$ in right ascension and $-14.1 \pm 6.4$ $\mu$as yr$^{-1}$ in declination. All motions are on the plane of the sky. The larger uncertainty in declination is caused by the elliptical beam shape of the VLBA (see also Table 1). The accuracy and number of measured motions is not adequate to model the internal

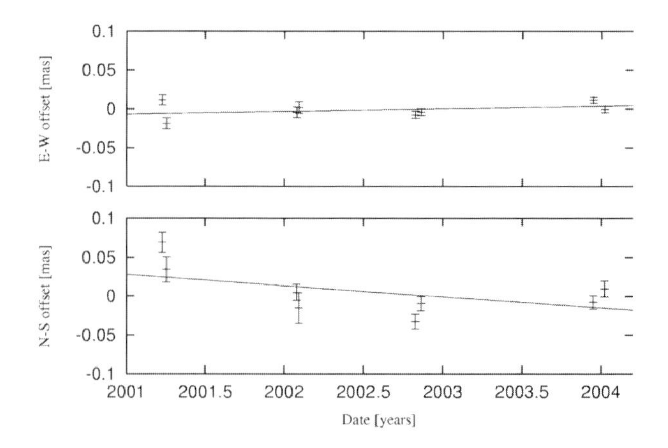


Figure 8: Average position of the masers in IC 133 in right ascension (upper) and declination (lower) relative to a background source.

dynamics of the IC133 and M33/19 regions (e.g., outflow), as was done by Greenhill et al. (1993) and Argon et al. (2004). Though for now the internal dynamics are small enough to be in the noise, future observations that are more sensitive will have to incorporate dynamical models for each region.

One can also calculate the average position of all maser components for each observation. Here, the individual fits for each maser feature were used to remove a constant position difference for each maser feature. We used the position difference at epoch 2002.627, which is in the middle of our observations. Then the variance weighted average of the positions of all detected features was calculated. The resulting average position of the masers in M33/19 and IC 133 are shown in Fig. 7 and Fig. 8 respectively. A fit of a rectilinear motion to these average positions yields motions of $37 \pm 5$ $\mu$as yr$^{-1}$ in right ascension and $-13 \pm 6$ $\mu$as yr$^{-1}$ in declination for M33/19. For IC 133 one gets a motion of $3 \pm 3$ $\mu$as yr$^{-1}$ in right ascension and $-13 \pm 10$ $\mu$as yr$^{-1}$ in declination. This is consistent with variance weighted average of all maser feature motions and suggests that the systematic internal motions within the two regions (e.g., outflow) are probably not a substantial source of bias with the current astrometric accuracy.

# 3 Geometric Distance of M33

The relative motions between M33/19 and IC 133 are independent of the proper motion of M33 and any contribution from the motion of the Sun. Since the rotation curve and inclination of the galaxy disk are known one can predict the expected relative angular motion of the two masing regions. The rotation of the H I gas in M33 has been measured by Corbelli & Schneider (1997), hereafter CS97. They fit a tilted-ring model to the measured velocities. According to their model of the rotation of M33, one can calculate the expected transverse velocities of M33/19 and IC 133.

For M33/19, one expects a motion of 42.4 km s$^{-1}$ in right ascension and $-39.6$ km s$^{-1}$ in declination. For IC 133 we expect $-64.0$ km s$^{-1}$ in right ascension and $-74.6$ km s$^{-1}$ in declination. This gives a relative motion of 106.4 km s$^{-1}$ in right ascension and 35 km s$^{-1}$ in declination between the two regions of maser activity.

The radial velocities of the $H_2O$ masers in M33/19 and IC 133 and the proximate H I gas are in very good agreement ($< 10$ km s$^{-1}$). This strongly suggests that the maser sources are rotating with the H I gas in the galaxy. However, while the agreement between the rotation model presented by CS97 and the radial velocity of the H I gas at the position of IC 133 is also very good ($< 5$ km s$^{-1}$), there is a difference of $\sim$ 15 km s$^{-1}$ at the position of M33/19. Hence, we conservatively assume a systematic error of 20 km s$^{-1}$ in each velocity component for the relative velocity of the two maser components. Comparing the measured angular motion of 30.8 $\pm$ 4 $\mu$as yr$^{-1}$ in right ascension with the expected linear motion of 106 $\pm$ 20 km s$^{-1}$, one gets a geometric distance of

$$D = 730 \pm 100 \pm 135 \text{ kpc},$$

where the first error indicates the statistical error from the VLBI proper motion measurements while the second error is the systematic error from the rotation model.

After less than three years of observations, the uncertainty in the distance estimate is already dominated by the uncertainty of the rotation model of M33. However, this can be improved in the near future by better rotation models using higher resolution (e.g., Very Large Array or Westerbork Synthesis Radio Telescope) data of H I gas in the inner parts of the disk. Also, the precision of the proper motion measurements will increase with time as $t^{3/2}$ for evenly spaced observations.

Within the current errors the geometric distance of $730 \pm 100 \pm 135$ kpc is in good agreement with recent Cepheid and Tip of the Red Giant Branch (TRGB) distances of 802$\pm$51 kpc (Lee et al. 2002) and 794$\pm$23 kpc (McConnachie et al. 2004) respectively. It also agrees with a geometric distance estimate of 800$\pm$180 kpc obtained by Argon et al. (2004), who modeled the outflow within IC 133, as traced by $H_2O$ maser proper motions estimated after a 14 year VLBI monitoring program. Other measurements using TRGB and Red Clump (RC) (Kim et al. 2002) favor a larger distance of 916 $\pm$ 55 kpc and 912 $\pm$ 59 kpc respectively.

# 4 Proper Motion of M33

The observed proper motion $\tilde{\vec{v}}_{prop}$ of a maser (e.g., M33/19 or IC 133) in M33 can be decomposed into three components $\tilde{\vec{v}}_{prop} = \vec{v}_{rot} + \vec{v}_{\odot} + \vec{v}_{prop}$. Here $\vec{v}_{rot}$ is the motion of the masers due to the internal galactic rotation in M33 and $\vec{v}_{\odot}$ is the apparent motion of M33 caused by the rotation of the Sun around the Galactic Center. The last contribution $\vec{v}_{prop}$ is the true proper motion of M33 relative to the Galaxy.

- The motion of the Sun can be decomposed into a circular motion of the Local Standard of Rest (LSR) and the peculiar motion of the Sun. Using a LSR velocity of 236$\pm$15 km s$^{-1}$ (Reid & Brunthaler 2004) and the peculiar velocity of the Sun from Dehnen & Binney (1998), the motion of the Sun causes an apparent proper motion of $\dot{\alpha}_{\odot} = 52.5 \pm 3.3$ $\mu$as yr$^{-1}$ in right ascension and $\dot{\delta}_{\odot} = -37.7 \pm 2.4$ $\mu$as yr$^{-1}$ in declination, assuming a distance of 730 kpc and the Galactic coordinates of M33 ($l = 133.6°$, $b = -31.3°$).

- Using the rotation model of CS97, the contribution from the rotation of M33 (for IC 133) is $\dot{\alpha}_{rot} = -18.5 \pm 6$ $\mu$as yr$^{-1}$ in right ascension and $\dot{\delta}_{rot} = -21.6 \pm 6$ $\mu$as yr$^{-1}$ in declination. Here, we assumed again an uncertainty of 20 km s$^{-1}$ for the rotation velocity and a distance of 730 kpc.
- Combining these velocity vectors, one gets the true proper motion of M33:

$$
\begin{aligned}
\dot{\alpha}_{prop} &= \dot{\tilde{\alpha}}_{prop} - \dot{\alpha}_{rot} - \dot{\alpha}_{\odot} \\
&= (4.7 \pm 3.2 + 18.5 \pm 6 - 52.5 \pm 3.3)\frac{\mu\mathrm{as}}{\mathrm{yr}} \\
&= -29.3 \pm 7.6\frac{\mu\mathrm{as}}{\mathrm{yr}} = -101 \pm 35\frac{km}{s} \\
\text{and} & \\
\dot{\delta}_{prop} &= \dot{\tilde{\delta}}_{prop} - \dot{\delta}_{rot} - \dot{\delta}_{\odot} \\
&= (-14.1 \pm 6.4 + 21.6 \pm 6 + 37.7 \pm 1)\frac{\mu\mathrm{as}}{\mathrm{yr}} \\
&= 45.2 \pm 9.1\frac{\mu\mathrm{as}}{\mathrm{yr}} = 156 \pm 47\frac{km}{s}.
\end{aligned}
$$

The transverse velocity changes by less than 5 km s$^{-1}$ if one uses the TRGB distance of 794$\pm$23 kpc (McConnachie et al. 2004) for this analysis. Finally, the systemic radial velocity of M33 is $-179$ km s$^{-1}$ (CS97). The radial component of the rotation of the Milky Way towards M33 is $-140 \pm 9$ km s$^{-1}$. Hence, M33 is moving with $-39 \pm 9$ km s$^{-1}$ towards the Milky Way. This gives now the three dimensional velocity vector of M33 which is plotted in Fig. 9. The total velocity of M33 relative to the Milky Way is 190 $\pm$ 59 km s$^{-1}$.

It is often argued (e.g. Kahn & Woltjer 1959) that the Milky Way is falling directly towards Andromeda (M31). This is based on the argument that there are no other large galaxies in the Local Group to generate angular momentum through tidal torques. Following this argument, we assume a zero proper motion of Andromeda. Then the angle between the velocity vector of M33 relative to Andromeda and the vector pointing from M33 towards M31 is 30°$\pm$15°. This angle strongly depends on the relative distance between Andromeda and M33. Thus, it is crucial to use distances for the two galaxies which have similar systematic errors. Thus, the TRGB distances to M33 and Andromeda from McConnachie et al. (2004) and McConnachie et al. (2005) respectively are used. For an angle of 30°, only elliptical orbits with eccentricities of e>0.88 are allowed. For the largest allowed angle of 45°, the eccentricities are restricted to e>0.7. This eccentricity limit weakens if the proper motion of Andromeda is non-negligible and, for a motion of $>$150 km s$^{-1}$, any eccentricity is allowed.

Gottesman, Hunter & Boonyasait (2002) consider the effects of dynamical friction of Andromeda on M33. They conclude that Andromeda cannot have a very massive halo unless the orbit of M33 has a low eccentricity. Otherwise, the dynamical friction would have lead to a decay of the orbit of M33. If this is correct,

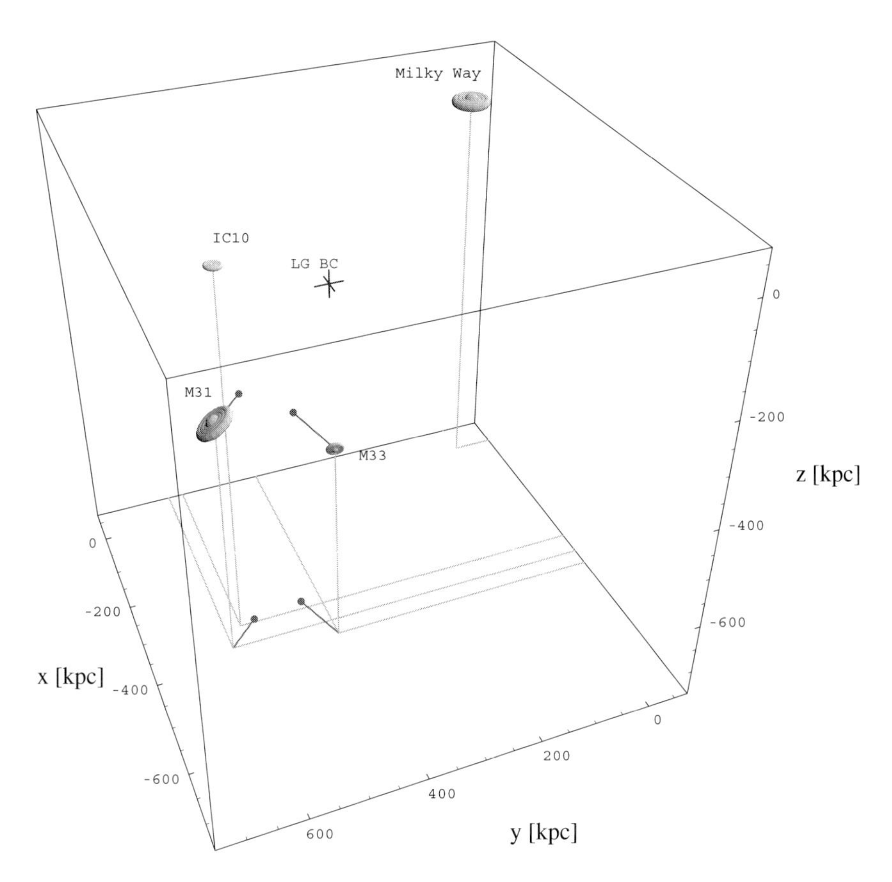


Figure 9: Schematic view of the Local Group with the space velocity of M33 and the radial velocity of Andromeda. The blue cross marks the position of the Local Group Barycenter (LG BC) according to van den Bergh (1999).

then the high eccentricity of the orbit of M33 would lead to the conclusion that the total mass of Andromeda must be less than $10^{12}$ $M_{\odot}$. This is in good agreement with recent estimates by Evans & Wilkinson (2000) and Evans et al. (2003) ($\sim 1.2\times10^{11} M_{\odot}$) derived from the three-dimensional positions and radial velocities of its satellite galaxies.

The next step towards a kinematic model of the Local Group is a model of the Andromeda subgroup. If one assumes that M33 is bound to Andromeda, a lower limit on the mass of Andromeda can be estimated by comparing the relative speed with the escape velocity. This mass limit still depends on the unknown proper motion vector of Andromeda. However, some proper motion vectors can be excluded for Andromeda, since it would lead to scenarios in which M33 would have been destroyed by the close interaction with Andromeda. Similar same arguments can be made for IC 10.

# 5 Discussion & Outlook

More than 80 years after van Maanen's observation, we have succeeded with measuring the rotation and proper motion of M33. These measurements provide a new handle on dynamical models for the Local Group and the mass and dark matter halo of Andromeda and the Milky Way.

Further VLBI observations within the next few years and an improved rotation model have the potential to improve the accuracy of the distance estimate to less than 10%. At least one additional maser source exists in M33 (M33/50, see Huchtmeier, Eckart & Zensus 1988) that will be used in the future to increase the accuracy of the measurements. A third region of maser activity will also help to check for non-circular velocities of the masers.

Unfortunately, no maser emission could be found in Andromeda despite intensive searches (e.g. Imai et al. 2001). Hence, the proper motion vector of Andromeda is still unknown. In the near future, new technical developments using higher bandwidths will increase the sensitivity of VLBI. This will allow the detection of radio emission from the central black hole of Andromeda, M31*, with flux densities of $\sim$30 $\mu$Jy (Crane, Dickel & Cowan 1992), to measure the proper motion of Andromeda. Today we are only able to study the extreme (bright) tip of the maser luminosity distribution for interstellar masers. The Square Kilometer Array (SKA), with substantial collecting area on intercontinental baselines and a frequency coverage up to 22 GHz (Fomalont & Reid 2004), will provide the necessary sensitivity to detect and measure the proper motions of a much greater number of masers in active star forming regions in the Local Group.

# Acknowledgements

The VLBA is operated by the National Radio Astronomy Observatory (NRAO). The National Radio Astronomy Observatory is a facility of the National Science Foundation operated under cooperative agreement by Associated Universities, Inc.

## References

Argon, A. L., Greenhill, L. J., Moran, J. M. et al. 1994, ApJ, 422, 586

Argon, A. L., Greenhill, L. J., Moran, J. M. et al. 2004, ApJ, 615, 702

Becker, R., Henkel, C., Wilson, T. L., Wouterloot, J. G. A. 1993, A&A, 268, 483

Brunthaler, A., Reid, M. J., Falcke, H., Greenhill, L. J., Henkel, C. 2002, Towards Proper Motions in the Local Group, in Proceedings of the 6th EVN Symposium, p. 189

Brunthaler, A., Reid, M. J., Falcke, H., 2003, Atmosphere-Corrected Phase-Referencing, in ASP conf. series "Future Directions in High Resolution Astronomy: The 10th Anniversary of the VLBA", J. D. Romney & M. J. Reid (eds.), in press, astro-ph0309575

Brunthaler, A., Reid, M., Falcke, H., Greenhill, L. J., Henkel, C. 2005, Science, 307, 1440

Corbelli, E., Schneider, S. E. 1997, ApJ, 479, 244

Courteau, S., van den Bergh, S. 1999, AJ, 118, 337

Crane, P. C., Dickel, J. R., Cowan, J. J. 1992, ApJ, 390, L9

Dehnen, W., Binney, J. J. 1998, MNRAS, 298, 387

Eckart, A., Genzel, R. 1996, Nature, 383, 415

Eisenhauer, F., Schödel, R., Genzel, R. et al. 2003, ApJ, 597, L121

Evans, N. W., Wilkinson, M. I. 2000, MNRAS, 316, 929

Evans, N. W., Wilkinson, M. I., Perrett, K. M., Bridges, T. J. 2003, ApJ, 583, 752

Fich, M., Tremaine, S. 1991, ARA&A, 29, 409

Fomalont, E., Reid, M. 2004, New Astronomy Review, 48, 1473

Freedman, W. L. 2000, Phys. Rep., 333, 13

Ghez, A. M., Duchêne, G., Matthews, K. et al. 2003, ApJ, 586, L 127

Gottesman, S. T., Hunter, J. H., Boonyasait, V. 2002, MNRAS, 337, 34

Greenhill, L. J., Moran, J. M., Reid, M. J. et al. 1990, ApJ, 364, 513

Greenhill, L. J., Moran, J. M., Reid, M. J., Menten, K. M., Hirabayashi, H. 1993, ApJ, 406, 482

Herrnstein, J. R., Moran, J. M., Greenhill, L. J. et al. 1999, Nature, 400, 539

Hubble, E. P. 1926, ApJ, 63, 236

Huchtmeier, W. K., Eckart, A., Zensus, A. J. 1988, A&A, 200, 26

Imai, H., Ishihara, Y., Kemeya, O., Nakai, N. 2001, PASJ, 53, 489

Israel, F. P., van der Kruit, P. C. 1974, A&A, 32, 363

Jones, B. F., Klemola, A. R., Lin, D. N. C. 1994, AJ, 107, 1333

Kahn, F. D., Woltjer, L. 1959, ApJ, 130, 705

Kim, M., Kim, E., Lee, M. G., Sarajedini, A., Geisler, D. 2002, AJ, 123, 244

Kobayashi, H., Ishiguro, M., Chikada, Y. et al. 1989, PASJ, 41, 141

Kochanek, C. S. 1996, ApJ, 457, 228

Kulessa, A. S., Lynden-Bell, D. 1992, MNRAS, 255, 105

Lee, M. G., Kim, M., Sarajedini, A., Geisler, D., Gieren, W. 2002, ApJ, 565, 959

McConnachie, A. W., Irwin, M. J., Ferguson, A. M. N. et al. 2004, MNRAS, 350, 243

McConnachie, A. W., Irwin, M. J., Ferguson, A. M. N. et al. 2005, MNRAS, 356, 979

Mould, J. R., Huchra, J. P., Freedman, W. L. et al. 2000, ApJ, 529, 786

Newman, J. A., Ferrarese, L., Stetson, P. B. et al. 2001, ApJ, 553, 562

Ngeow, C., Kanbur, S. M. 2004, MNRAS, 349, 1130

Reid, M. J. 1993, ARA&A, 31, 345

Reid, M. J., Argon, A. L., Masson, C. R., Menten, K. M., Moran, J. M. 1995, ApJ, 443, 238

Reid, M. J., Brunthaler, A. 2004, ApJ, 616, 872

Reid, M. J., Readhead, A. C. S., Vermeulen, R. C., Treuhaft, R. N. 1999, ApJ, 524, 816

Sandage, A., Tammann, G. A., Reindl, B. 2004, A&A, 424, 43

Schödel, R., Ott, T., Genzel, R. et al. 2002, Nature, 419, 694

Schweitzer, A. E., Cudworth, K. M., Majewski, S. R., Suntzeff, N. B. 1995, AJ, 110, 2747

Shaya, E. J., Tully, R. B., Peebles, P. J. E. et al. 2003, Space interferometry mission dynamical observations of galaxies (SIMDOG) key project, in Interferometry in Space. Edited by Shao, Michael. Proceedings of the SPIE, Volume 4852, 2003, p. 120

Steigman, G., Tkachev, I. 1999, ApJ, 522, 793
Tammann, G. A., Sandage, A., Reindl, B. 2003, A&A, 404, 423
van den Bergh, S. 1999, A&ARv, 9, 273
van Maanen, A. 1923, ApJ, 57, 264
Walker, R. C., Burke, B. F., Johnston, K. J., Spencer, J. H. 1977, ApJ, 211, L135

# NIR Observations of the Galactic Center

R. Schödel, A. Eckart, C. Straubmeier & J.-U. Pott

I.Physikalisches Institut
Universität zu Köln, Zülpicher Str.77, 50937 Köln
rainer@ph1.uni-koeln.de
www.ph1.uni-koeln.de

**Abstract**

*High-resolution near-infrared observations of the central parsec of the Milky Way are currently revolutionizing our understanding of this nearest galactic nucleus. In this contribution we review some recent results of this research. Via the observation of stellar dynamics it has by now delivered ironclad evidence that the central radio source, Sagittarius A*, is associated with a supermassive black hole. NIR monitoring of the flux from Sagittarius A* allows to study the accretion and emission mechanisms in this extremely sub-Eddington system. Coordinated observations of the near-infrared and X-ray quasi-quiescent and flaring emission from Sagittarius A* show that the emission is most likely produced via synchrotron or synchrotron-self Compton processes. There are numerous young, massive stars present in the central stellar cluster. In fact, star formation may be even a still ongoing process: A group of deeply embedded sources north of the IRS 13 complex of bright stars may represent young stars. Compact MIR sources close to the northern arm of the mini-spiral are further potential candidates for young stars, or may be, alternatively, witnesses of the close interaction of stellar winds with the ISM. Very recently. the first successful interferometric observations of a stellar source in the galactic center have been carried out at mid-infrared wavelengths.*

# 1 Introduction

The recent advent of adaptive optics instrumentation on 8-10 m-class telescopes has been a major breakthrough for near-infrared (NIR) observations of the center of the Milky Way. Due to its proximity (8 kpc, see, e.g., Reid, 1993; Eisenhauer et al., 2003) it is the only galaxy in which the central cluster can be resolved observationally and the stellar population and the dynamics of gas and dust examined in detail. Observations of the Galactic Center (GC) have delivered evidence beyond doubt for the existence of supermassive black holes. This evidence was obtained mainly through the observation of stellar proper motions and orbits near the central black hole (e.g., Eckart & Genzel, 1996; Ghez et al., 1998; Genzel et al. 2000; Ghez et al., 2000; Eckart et al. 2002; Schödel et al., 2002; Ghez et al., 2003) and through

---

*Reviews in Modern Astronomy 18.* Edited by S. Röser
 ISBN 3-527-40608-5

radio interferometric observations of the radio source Sagittarius A* (Sgr A*) that is associated with the black hole (e.g., Reid et al., 2004).

In this contribution we present a brief overview of recent research carried out with high-resolution near- and mid-infrared observations of the central parsec of the Milky Way. In particular we highlight the following results: The study of stellar dynamics near Sgr A* has delivered stringent constraints on its nature by showing that all models for its nature are extremely implausible with the exception of a massive black hole. Coordinated, simultaneous NIR/X-ray observations of Sgr A* allow us to gain insight into the accretion and emission processes near this supermassive black hole. The stellar population in the GC has been found to be characterized, surprisingly, by the presence of numerous massive, young stars. In this context, we discuss a recently discovered complex of infrared-excess sources, close to Sgr A*, that are potential candidates for young stars and report on newly discovered compact mid-infrared (MIR) excess sources. Finally, we introduce the results of the first MIR interferometric observations of a stellar source in the GC. These VLTI observations open a new era of exploration at the highest angular resolutions.

## 2 Stellar dynamics and the nature of Sgr A*

High-resolution NIR imaging with speckle interferometry (and adaptive optics since about 2000) led to the first clear evidence that the gravitational potential in the central parsec of the Milky Way is dominated by a point mass (e.g., Eckart & Genzel, 1996; Ghez et al., 1998). Continued observations revealed the first detection of stellar acceleration near Sgr A* (Ghez et al., 2000; Eckart et al., 2002). Finally, a unique solution for the orbit of the star S2 could be determined after the observation of its peri-center passage in spring 2002 (Schödel et al., 2002; Ghez et al., 2003). The Keplerian orbit of S2 indicates that a mass of $3.6 \pm 0.6 \times 10^6$ $M_\odot$ is located within a sphere of radius $0.6$ mpc around Sgr A* (Eisenhauer et al., 2003). This enormous mass density excludes the possibility that a ball of heavy, degenerate fermions ("neutrino ball") is responsible for the central mass concentration of the Milky Way. As concerns the hypothesis of a cluster of dark astrophysical objects (e.g., neutron stars), the lifetime of such a configuration with the required high density would be less than $10^5$ yrs. This leaves a supermassive black hole as the only plausible explanation for the nature of Sgr A*. In Figure 1 we show the orbits of six stars around Sgr A* as determined by Schödel et al. (2003).

Figure 2 illustrates the scales near Sgr A* that are probed by observations of stellar dynamics and of variable emission from Sgr A*. Stellar dynamics, especially the orbits of individual stars, probes the gravitational potential on the scale of several light hours down to several light days ($\approx$ 1.2 light days correspond to one mpc). On the left hand side of Fig. 2, the core radii of two hypothetical dense clusters (Plummer models) with total masses of $\sim 3.5 \times 10^6$ $M_\odot$, as derived from the orbit of S2, are indicated: A cluster of 3 $M_\odot$ black holes with a life time of $10^7$ yr, marked by a dashed line, and a dark cluster that could marginally fit the gravitational potential as constrained by the orbit of S2. The latter cluster, marked by a dotted line, would have a life time of less than $10^5$ yr. The radius of a neutrino ball, composed of

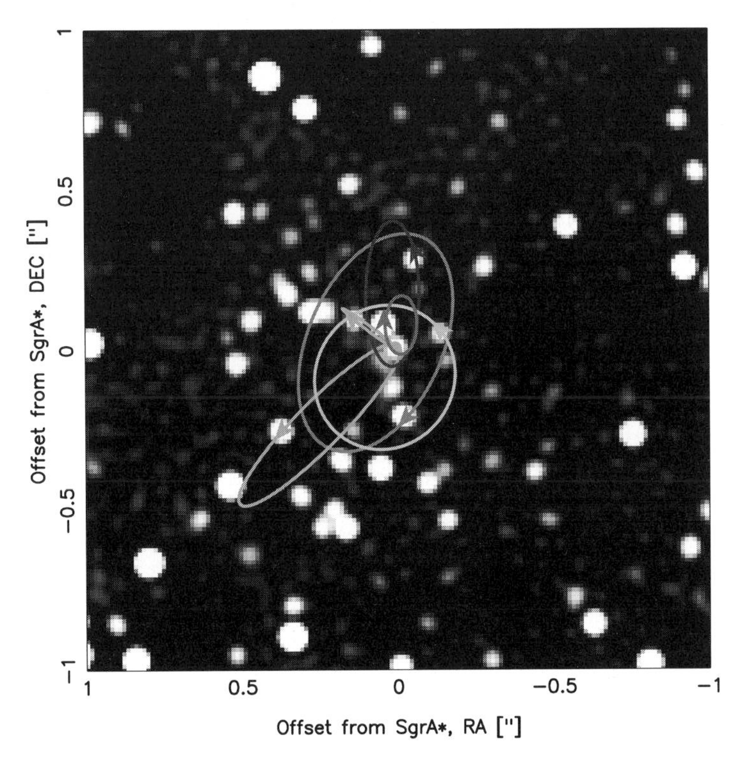


Figure 1: Stellar orbits around Sgr A* as determined by Schödel et al. (2003). The orbits are shown on the background of a Lucy-Richardson deconvolved and beam restored K-band image (NACO/VLT) of a region within $1''$ of Sgr A*.

degenerate 17 keV neutrinos, is indicated by a circle with a straight line. All these models are excluded by the observed orbit of S2 and the requirement that a given configuration should have a life time comparable to the life time of the galaxy. The right hand side of Fig. 2 illustrates a zoom into a region with the size of a few tens of Schwarzschild radii that is probed by radio/mm observations and observations of the variability of the X-ray/NIR emission (see below) from Sgr A*. Indicated are the size constraints due to the duration of the observed X-ray and NIR flares with a duration of the order 60 min (outer circle, marked by "flares:duration"), the size inferred from 7 mm interferometry (Bower et al., 2004), the size limit imposed by the variability time (rise-and-fall time, $T_{\rm Var}$) of the flares (dotted line), and the Schwarzschild radius, $R_S$, of a $3.6 \times 10^6\, M_\odot$ black hole.

# 3 Coordinated NIR/X-ray observations of Sgr A*

Sgr A* is a very weak X-ray and NIR source (see, e.g., Melia & Falcke 2001). When it could be finally detected unambiguously with instruments capable of high-

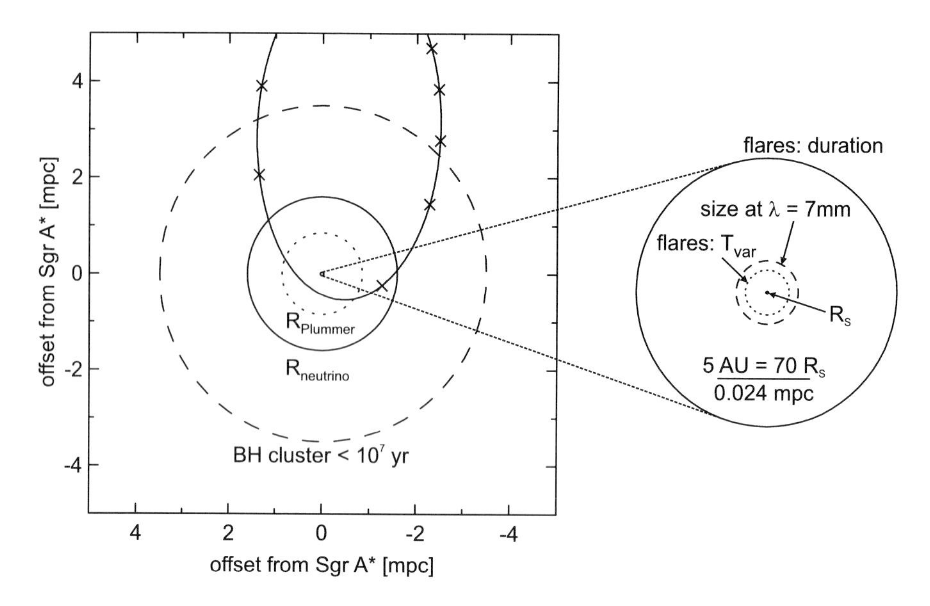


Figure 2: Illustration of the scales near Sgr A* that have been probed by observations.

resolution imaging at the corresponding wavelengths, such as *Chandra* (X-rays) and NAOS/CONICA at the ESO VLT (NIR), the emission from Sgr A* turned out to be highly variable at these wavelengths. X-ray flares, during which the emission from Sgr A* rises by factors of a few tens up to one hundred, were first detected by Baganoff et al. (2001). NIR flares were detected by Genzel et al. (2003) and Ghez et al. (2004). During NIR flares the emission from Sgr A* increases by a factor of a few in this wavelength regime. Both at X-ray and NIR wavelengths the flares show variability on time scales of just a few minutes. This points to very compact source regions of not more than a few tens of Schwarzschild radii in size.

In order to understand the nature of these flares, coordinated simultaneous observations at the different wavelengths are necessary. The first successful simultaneous near-infrared and X-ray detection of the Sgr A* counterpart, was carried out using the NACO adaptive optics (AO) instrument at the European Southern Observatory's Very Large Telescope and the ACIS-I instrument aboard the *Chandra X-ray Observatory* (Eckart et al., 2004). A flare was detected at X-rays that was covered simultaneously in its decaying part by the NIR observations (Fig. 3).

Eckart et al. (2004) find that the flaring state can be conveniently explained with a synchrotron self-Compton (SSC) model involving up-scattered sub-millimeter photons from a compact source component, possibly with modest bulk relativistic motion. The size of that component is assumed to be of the order of a few times the Schwarzschild radius. A conservative estimate of the upper limit of the time lag between the ends of the NIR and X-ray flares is of the order of 15 minutes. The simultaneity of the flares at NIR/X-ray also confirms clearly that the X-ray source seen by *Chandra* is indeed associated with Sgr A*.

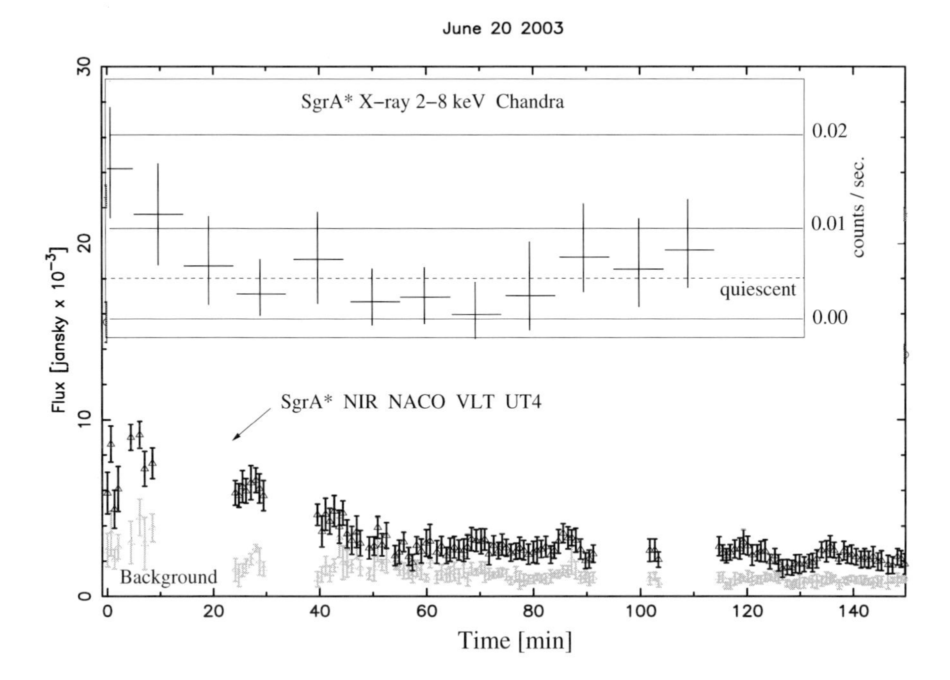


Figure 3: Sgr A* flare observed simultaneously at NIR/X-ray wavelengths (Eckart et al., 2004). The X-ray and NIR light curves are plotted with a common time axis. Straight solid lines in the inserted box represent the 0.00, 0.01, and 0.02 counts per second levels. The straight dashed line represents the X-ray quiescent-state flux density level. The NIR observations started on 19 June 2003 at 23:51:15 (AT), just 0.38 minutes before the midpoint of the highest X-ray measurement.

New coordinated NIR/X-ray observations were successful in observing several flares simultaneously at NIR/X-ray wavelengths (Eckart et al., 2005, in preparation). They show that the variability at the two wavelengths is closely related, i.e., that the time lag between the events at the two different wavelengths is close to zero. Both direct synchrotron emission and SSC emission appear to contribute in varying amounts to the emission from NIR flares, while at X-ray wavelengths only the SSC process is important. This may also be the reason why flares in the NIR regime are a factor of $\sim$2 more frequent than at X-ray wavelengths.

# 4 Embedded/MIR excess sources

Numerous young, massive stars found in the central parsec of the GC are witnesses of star formation episodes in the GC that occurred not more than a few million years ago. There are, e.g., the hot, massive so-called He-stars, mainly concentrated in the IRS 16 and IRS 13 associations (e.g., Krabbe et al. 1995; Paumard et al. 2001). The so-called S-stars, located within just a few milli-parsecs of the central black hole,

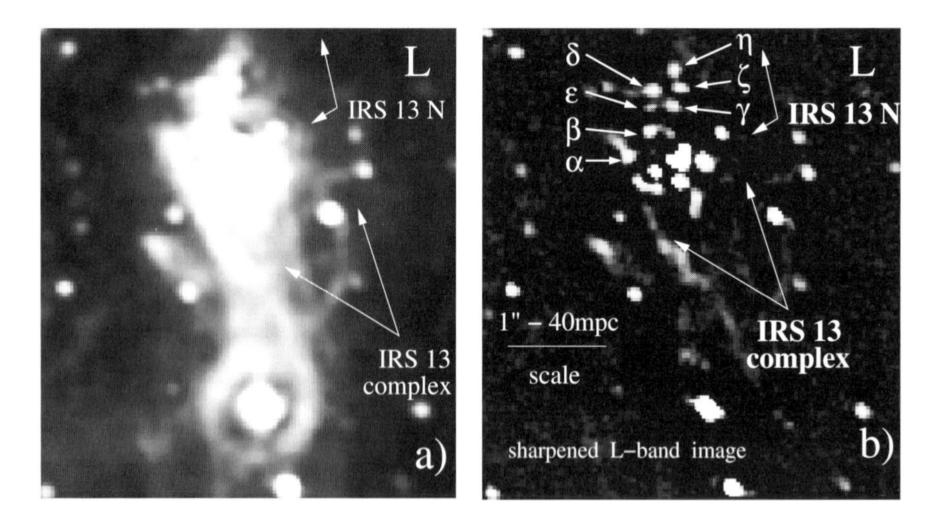


Figure 4: $L'$ band images of the IRS 13/IRS 2 region. The image scale is given in (b). Ring structures around the brighter stars in panel (a) are artifacts of the deconvolution algorithm. Panel (b) is a high pass filtered L'-image which shows the locations of individual stars, including the newly discovered $L'$-band excess sources $\alpha$ through $\eta$. The blue source $\kappa$ is located between $\zeta$ and $\delta$ (Eckart et al. 2004).

have been identified as B-type main sequence stars (e.g., Ghez et al. 2003; Eisenhauer et al. 2005). The existence of these massive, young stars so close to the central supermassive black hole is a very enigmatic finding because in the light of current astrophysical knowledge they could not have formed near the black hole nor have migrated there from larger distances within their life times (see, e.g., discussions in Ghez et al. 2003; Genzel et al. 2003a). However, the existence of these young stars raises the question whether star formation may still be an ongoing process near Sgr A*. Highly reddened sources, such as IRS 21, were previously regarded as potential young stellar objects. However, newer, higher-resolution observations show that these stars are rather massive windy stars that interact with the local ISM (Tanner et al. 2002, 2005).

New candidates for potentially very young stars were discovered by Eckart et al., (2004a), who identified a small (0.13 lt-yr diameter) cluster of compact sources about $0.5''$ north of IRS 13 with strong IR excess due to T > 500 K dust (see Fig. 4). The nature of the sources is still unclear. They may be a cluster of highly extincted stars that heat the local ISM. Eckart et al. (2004a) also consider an explanation that involves the presence of young stars at evolutionary stages between young stellar objects and Herbig Ae/Be objects with ages of about 0.1 to 1 million yr. This scenario would imply more recent star formation in the GC than previously suspected. The AO observations also resolve the central IRS 13 complex. In addition to the previously known bright stars E1 and E2, the K- and L-band images for the first time resolve object E3 into two components, E3N and E3c. The latter one is close to the 7/13 mm Very Large Array radio continuum source found at the location of the

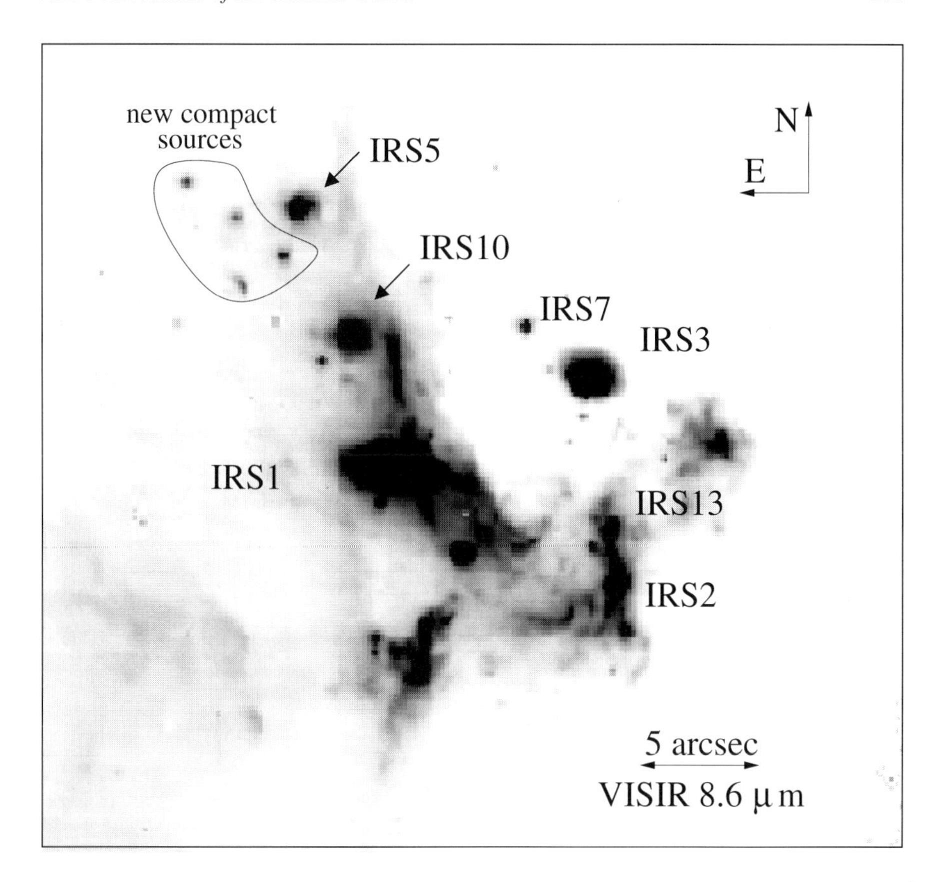


Figure 5: VISIR 8.6 μm image with the previously unknown compact sources close to the Northern Arm, indicated by the thin black line.

IRS 13 complex (Zhao & Goss 1998). E3c may therefore be associated with a dusty Wolf-Rayet like star at that location.

Recent images of the Galactic Center in the N-band (8.6 μm) with the new instrument VISIR at the VLT clearly show four compact sources located to the east of the bright Northern Arm source IRS 5 (Fig. 5). These sources are exceptional because of their compactness: All other 8.6 μm emission sources in the central parsec are extended and associated with the mini-spiral of ionized gas and warm dust, with the clearly extended luminous, dusty sources like IRS 1, 5, 10, 21, 3, or with the extended IRS 13/IRS 2 complex. Only the supergiant IRS 7 has a comparable point-like appearance on all MIR images. While the brighter 8.6 μm sources such as IRS 1, 5, 10, or 21 are now interpreted as bow-shock type interactions between hot emission-line stars with strong winds and the dusty mini-spiral ISM (Tanner et al. 2002, 2005), the nature of the four newly discovered compact sources is unclear. These sources are MIR bright and clearly discernible in M- and L-band images, and can be identified in the K-band, too. High-resolution NAOS/CONICA K- and L-

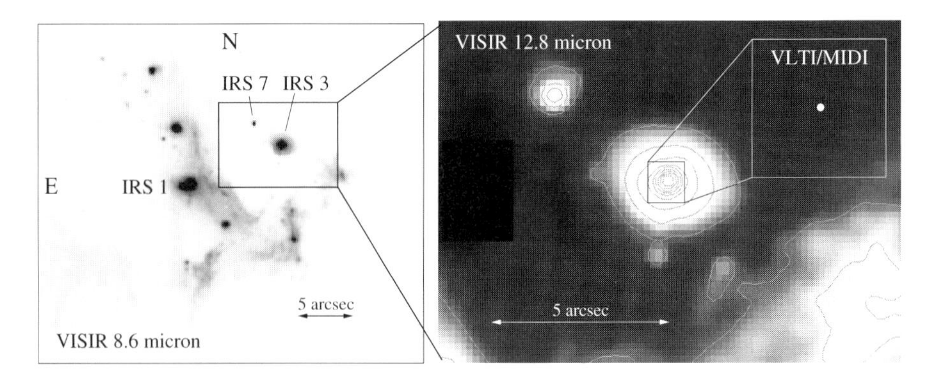


Figure 6: VISIR MIR image and an inset demonstrating the scale on which the current UT2-UT3 VLTI/MIDI data detected a compact source with a visibility of about 30%.

band images reveal that the southernmost of the four sources, which appears *slightly* extended (~0.2"; compared to IRS7) at longer wavelengths, is in fact double, consisting of a blue point source to the east and a fainter point source to the west, which shows a tail-like structure that appears to be the main source of emission at longer wavelengths. This is an intriguing case of possible interaction, either of the two sources with each other (if they lie spatially close together), or of the western component with the surrounding ISM.

There are two plausible possibilities to explain the nature of the compact MIR sources: They may be low-luminosity counterparts of the more luminous bow-shock sources, such as IRS 21. In that case they could be a more dispersed group of stars similar to the IRS13N complex described by Eckart et al. (2004a). Alternatively, they might be young stars that have been formed while falling into the central parts of the stellar cluster. These objects could be bright in the MIR since they are still surrounded by their proto-stellar dust shells or disks.

# 5 VLTI observations of IRS 3

In June 2004, IRS 3, the brightest compact MIR source in the GC was detected and partially resolved with VLTI using MIDI (N-band, 8-12 micron) on the 47 m UT1-UT2 baseline (Pott et al. 2005, in preparation). These are the first interferometric observations of an object in the GC at infrared wavelengths. About ~30% of the flux density were found to be in a compact component with a size of less than 40 mas (i.e. less than 300 AU). This agrees with the interpretation that IRS 3 is a luminous compact object in an intensive dust forming phase. Therefore, IRS 3 is the hottest and most compact bright source of MIR emission within the central stellar cluster. All other MIR sources at the GC were resolved at a 2 Jy flux level (see Figure 6, Pott et al., in preparation).

In general, the visibility amplitude of IRS 3 was found to be smaller ($0.23 \pm 0.06$) at shorter and larger ($0.28 \pm 0.04$) at longer wavelengths. Although the uncertainty of a single visibility value seems at first glance to be too large to identify such a trend, in fact the uncertainty of the slope of a visibility dataset over the N-band was found to be of the order of 1%. Thus, the estimated trend is indicating that the compact portion of the IRS 3 dust shell is extended and only partially resolved on the UT2-UT3 baseline. We also find indications for a narrower width of the 9.3 μm silicate line towards the center, indicating the presence of 'fresh', unprocessed small grains closer to the central star in IRS 3 (van Boekel et al., 2003).

## References


Baganoff, F. K., Bautz, M. W., Brandt, W. N., et al. 2001, Nature, 413, 45

van Boekel, R., Waters, L. B. F. M., Dominik, C., et al. 2003, A&A, 400, L21

Bower, G. C., Falcke, H., Herrnstein, R. M., et al. 2004, Science, 304, 704

Eckart, A. & Genzel, R. 1996, Nature, 383, 415

Eckart, A., Genzel, R., Ott, T., & Schödel, R. 2002, MNRAS, 331, 917

Eckart, A., Baganoff, F., Morris, M., et al. 2004, A&A, 427, 1

Eckart, A., Moultaka, J., Viehmann, C., et al. 2004a, ApJ, 602, 760

Eisenhauer, F., Schödel, R., Genzel, R. et al. 2003, ApJL, 597, L121

Eisenhauer, F., Genzel, R., Alexander, T. et al. 2005, ApJ, in press

Genzel, R., Pichon, C., Eckart, A., Gerhard, O. E., & Ott, T. 2000, MNRAS, 317, 348

Genzel, R., Schödel, R., Ott, T., et al. 2003, Nature, 425, 934

Genzel, R., Schödel, R., Ott, T. et al. 2003a, ApJ, 594, 812

Ghez, A. M., Klein, B. L., Morris, M., & Becklin, E. E. 1998, ApJ, 509, 678

Ghez, A. M., Morris, M., Becklin, E. E., Tanner, A., & Kremenek, T. 2000, Nature, 407, 349

Ghez, A. M., Duchêne, G., Matthews, K., et al. 2003, ApJ, 586, L127

Ghez, A. M., Wright, S. A., Matthews, K., et al. 2004, ApJL, 601, L159

Krabbe, A., Genzel, R., Eckart, A. et al. 1995, ApJ, 447, L95

Melia, F. & Falcke, H. 2001, ARA&A, 39, 309

Paumard, T., Maillard, J. P., Morris, M., & Rigaut, F. 2001, A&A, 366, 466

Reid, M. J. 1993 ARA&A, 31, 345

Reid, M. J. & Brunthaler, A. 2004, ApJ, 616, 972

Schödel, R., Ott, T., Genzel, R., et al. 2002, Nature, 419, 694

Schödel, R., Ott, T., Genzel, R., et al. 2003, ApJ, 596, 1015

Tanner, A., Ghez, A. M., Morris, M., et al. 2002, ApJ, 575, 860

Tanner, A., Ghez, A. M., Morris, M., & Becklin, E.E. 2003, AN, Supplementary Issue 1, 597

Tanner, A., Ghez, A. M., & Morris, M. R. 2005, ApJ, in press

Zhao, J.-H. & Goss, W. M. 1998, ApJ, 499, L163

# Structures in the interstellar medium

Soňa Ehlerová

Astronomical Institute, Academy of Sciences of the Czech Republic
Boční II 1401, 141 31 Prague 4, Czech Republic
sona@ig.cas.cz,
http://www.asu.cas.cz/person/ehlerova.html

**Abstract**

*The interstellar medium in our Galaxy — and any other galaxy — is far from being homogeneous. It consists of many structures. The majority of these structures are random fluctuations of density and velocity most probably driven by the interstellar turbulence. We review theoretical predictions for a statistical description of the turbulent interstellar medium and give examples of measurements in different galaxies, especially in HI. We try to show why the most commonly observed values for the ISM power spectra are around 3 ("the big power-law in the sky") and also why we should expect deviations from the "universal" value and what do they mean.*

*An important part of "non-turbulent" structures in the interstellar medium is formed by HI shells. We describe properties of automatically identified shells in the second galactic quadrant (in the Leiden-Dwingeloo survey). We argue why HI shells are not turbulent and fractal structures, even though their mass-size relation could indicate this at the first glance. Then we are interested in the size distribution of HI shells in the Milky Way, which we compare with that in other galaxies (M31, M33, SMC, LMC, HoII) and find that it is similar in all galaxies. According to the theory of Oey & Clarke (1997), the size distribution of shells is connected to the luminosity function of HII regions. This connection works well for dwarf galaxies (LMC, SMC, HoII), but not so well for spiral galaxies (M31, M33, Milky Way). We attribute this discrepancy to an influence of density gradients, though other explanations are not ruled out.*

## 1 Introduction

The interstellar medium (ISM) in galaxies is an interesting place. It is a mixture of several coexisting phases which interact and influence each other. It is a place where star formation takes place, and reversely a medium very much influenced by star formation and a stellar activity. Also other effects — e.g. spiral arms on galactic scales or galactic interactions — have a great impact on ISM.

*Reviews in Modern Astronomy 18.* Edited by S. Röser

The interstellar medium, as traced e.g. by the 21 $cm$ line of neutral hydrogen, contains two types of structures: 1) coherent, well-defined structures, and 2) random density and velocity fluctuations.

Random fluctuations of the neutral hydrogen emission (or other phases of the ISM) are usually studied using statistical methods. These studies showed that the power spectrum of the ISM had a form of a power-law with the (nearly) universal slope. This behaviour is interpreted as a consequence of the turbulence in the interstellar medium and the fractal geometry of the ISM.

There are several types of coherent structures in the ISM: shells, chimneys, clouds, streams. They are coherent both in position and velocity space. In this paper we will be interested in only one type of these structures, in HI shells.

These two kinds of structures interact. Shells evolve in a turbulent medium, aged and evolved shells blend with the surrounding. It is yet unknown what is the major energy source which drives the turbulence in the interstellar medium. One possibility is star formation, especially the activity of OB associations, which is also responsible for the creation of HI shells. From that reason it is interesting to study simultaneously properties of random fluctuations (i.e. the turbulence) and HI shells. Alternatively, the turbulence may be mainly driven by spiral arms, but this theory will not be considered in this paper.

We will describe some methods used to study statistical properties of the turbulent and hierarchical ISM, especially how to connect properties of observed two-dimensional intensity channel maps with the underlying three-dimensional physical reality represented by the density and velocity distributions. Several applications of these methods will be described, mostly on HI measurements of our Galaxy and Magellanic Clouds.

Then we will be interested in HI shells, with a special emphasis on shells in the Milky Way. We will try to find out, if shells really are the coherent structures, or if they are the "coherent tip of random fluctuations". Then we will compare size distributions of shells (as the most basic statistical property) in six galaxies (Milky Way, M31, M33, Large Magellanic Cloud, Small Magellanic Cloud and Holmberg II) and discuss if this distribution corresponds to the luminosity function of HII regions, to which shells should be connected.

# 2 The turbulent interstellar medium

When we look at images of the interstellar medium — and in principle it does not matter in which tracer — we see it is full of random fluctuations. These fluctuations appear both in intensity (i.e. density) and velocity. Their best description is using some kind of statistical approach, e.g. the spatial power spectrum. The analysis done for the ionised medium, Galactic HI, Galactic molecular gas and also for HI in the Small and Large Magellanic Clouds, shows that the spectrum of fluctuations has a form of a power-law with quite a universal logarithmic slope of $\sim -3$. Questions, which are naturally connected with this result, are: "Is this behaviour really universal and why is it so?" and "What are the underlying physical conditions?"

The power-law behaviour suggests the hierarchical structure organisation, though it need not necessarily be the case. In the case of the interstellar medium, however, it is probably true and the distribution is explained as the result of a Kolmogorov turbulent cascade. In the following we will describe several measurements of the spectrum of the interstellar medium. For references about the power-law measurements see Dickey et al. (2001).

## 2.1 The fractal structure of molecular clouds

The fractal structure of molecular clouds has been considered many times (see e.g. Elmegreen & Falgarone, 1996). Stutzki et al. (1998) showed that indices of the mass spectrum of clumps $\alpha$:

$$\frac{dN}{dM} \propto M^{-\alpha}, \tag{1}$$

of the mass-size relation $\gamma$:

$$M \propto L^{\gamma} \tag{2}$$

and the spectral power index $\beta$ are related:

$$\beta = \gamma(3 - \alpha). \tag{3}$$

The power index $\beta$ for two sample clouds was close to 2.8, other indices were $\alpha \simeq 1.8$ and $\gamma \simeq 2.3$.

The relation between $\alpha$, the fractal dimension and other fractal properties depends on the applied model of the fractal structure.

## 2.2 Lazarian & Pogosyan theory

Table 1: Results for the two-dimensional power spectrum of intensity fluctuations, Lazarian & Pogosyan (2000). See text for an explanation of the quantities $n$ and $m$ and thin and thick slices.

| slice thickness | shallow density $n < 3$ | steep density $n > 3$ |
|---|---|---|
| thin slice | $P(k) \propto k^{-n+m/2}$ | $P(k) \propto k^{-3+m/2}$ |
| thick slice | $P(k) \propto k^{-n}$ | $P(k) \propto k^{-3-m/2}$ |
| very thick slice | $P(k) \propto k^{-n}$ | $P(k) \propto k^{-n}$ |

The relation between the two-dimensional spatial power spectra of PPV (position-position-velocity) data cubes and the underlying three-dimensional density and velocity distribution was also studied by Lazarian & Pogosyan (2000). Assuming, that the density distribution has a power spectrum of the form

$$P_{den} \propto k^{-n} \tag{4}$$

and the velocity distribution is

$$P_{vel} \propto k^{-3-m}, \tag{5}$$

they calculated the resulting index of the two-dimensional spatial power spectrum (Table 1, which is taken from Stanimirović & Lazarian, 2001) as a function of the velocity thickness of the studied image. The power spectrum index is different, if the velocity slice is thick or thin. The slice is thin if its width is smaller than the velocity dispersion of the studied scale, and it is thick if the width is greater than the dispersion.

As a paradigm of the turbulent behaviour a Kolmogorov cascade is often used. The Kolmogorov spectrum has a slope of

$$P_{kolmogorov} \propto k^{-11/3} \tag{6}$$

(i.e. $m$ in Eq. 5 is $m = 2/3$). As we will see in a while, we do not observe this spectrum in reality, though the observed values are usually close. One lasting problem with the Kolmogorov spectrum (and other turbulent spectra) is that we do not observe the scale, at which energy is deposited. At this scale the self-similar behaviour should be broken, this, however, has not been observed so far.

### 2.2.1 Small Magellanic Cloud

Stanimirović & Lazarian (2001) analysed the dependence of the spatial power spectrum index of the HI emission of the SMC on the velocity thickness. In correspondence with the Lazarian & Pogosyan theory they found that the slope of the power spectrum decreased with the increasing velocity width, from about $-2.8$ to $-3.3$. This means, that the density index of HI in SMC is $n \simeq 3.3$ and the velocity index is $-3.4$, i.e. $m \simeq 0.4$.

### 2.2.2 Large Magellanic Cloud

The similar analysis as for SMC was also done by Elmegreen, Kim & Staveley-Smith (2001) for LMC. They found, contrary to SMC, that there is no difference between a power spectrum of the single velocity channel map and the velocity integrated map, which means, that the emission is thin (in the Lazarian & Pogosyan sense) even for the whole disc. However, they discovered a steepening of the power spectrum for smaller wavelengths, which they explain by the small scales being thick and large scales being thin. The observed transition between thick and thin scales should correspond to the thickness of the LMC disc. This transition occurs at a wavelength of about $\sim 80 - 100\ pc$,a value which the authors consider not unreasonable for LMC.

Elmegreen, Elmegreen & Leitner (2003) used this method — i.e. steepening of the power spectrum with wavelength — on the optical light of several face-on spiral galaxies to determine the thickness of the disc. They did detect this transition at scales of several hundred parsecs in most galaxies.

### 2.2.3 Milky Way

Two fields from the Southern Galactic Plane Survey were studied by Dickey et al. (2001). For one region they found, that the velocity term was not important and that only the density spectrum $P_{den}$ is important and has the slope of $n = 4$. For the second region they derived indices of $n = 2.9$ (the density spectrum) and $m = 0.2$ (the velocity spectrum). This is different from the SMC values ($n = 3.3$ and $m = 0.4$; Stanimirović & Lazarian, 2001). The authors explain the difference between the two regions by the fact, that the first region, positioned at a higher latitude, detects only (or primarily) the emission of the warm phase of HI, while for the second region at much lower latitude the contribution from the cool HI is not negligible. The difference between the two regions then corresponds to different turbulent regimes for the different HI phases.

# 3 HI shells

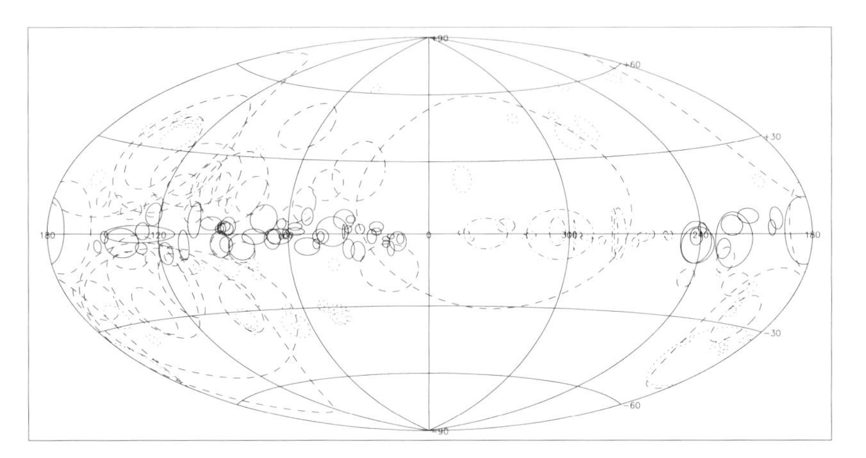


Figure 1: The galactic distribution of HI shells from the literature. Solid line: Heiles (1979), dashed: Heiles (1984), dotted: Hu (1981), dash-dotted: McClure-Griffiths et al. (2002).

HI shells are coherent structures observed in the neutral hydrogen distribution of many galaxies. They are coherent both in space and velocity: this coherence is what makes them different from random fluctuations described above. The most widely accepted explanation for the origin of shells — and also the most probable one — is the shock wave generated by winds and supernova explosions in OB associations. In individual cases other scenarios (HVC infall, hypernova explosion, ram pressure) are considered, but they certainly do not create any considerable amount of HI shells.

As already told, HI shells have been observed in many galaxies. For a review on these observations see Walter & Brinks (1999).

## 3.1 HI shells in the Milky Way

The first HI shells ever observed were found in the Milky Way. However, our position inside the Galaxy makes the overall and uniform identification of shells difficult, if not impossible. There were several efforts to detect shells in large fields on the sky.

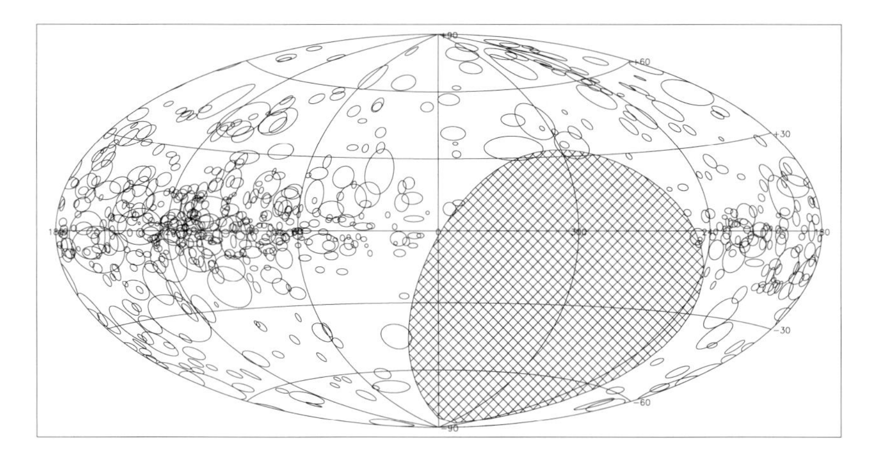


Figure 2: The galactic distribution of HI shells identified by our searching algorithm. The cross-hatched region around $l = 300°$ is a region not accessible to the Dwingeloo telescope.

Fig. 1 shows the positions of identified shells from Heiles (1979), Heiles (1984), Hu (1981) and McClure-Griffiths et al. (2002).

We have created an algorithm for the automatic identification of HI shells in three-dimensional data cubes. The algorithm is based on searching for and analysing local minima in individual velocity channels and then constructing 3D structures, whose spectra must conform to expected spectra of shells. The description of the algorithm can be found in Ehlerová & Palouš (2005) or in Ehlerová, Palouš & Wünsch (2004).

We applied our searching algorithm on the Leiden-Dwingeloo HI survey (Hartmann & Burton, 1997). The resulting shells are drawn in Fig. 2. During the analysis we found that identifications in the inner Galaxy are much influenced by a high filling factor of holes in this region and by the overlapping in velocity channels, therefore we restrict the study of HI shells properties on an analysis of structures in the second Galactic quadrant, which covers only the outer Galaxy and which is fully observed in the Leiden Dwingeloo survey. An example of an identified shell is shown in Fig. 3.

The longitudinal distribution of all HI shells in the second Galactic quadrant is shown in Fig. 4 (left) together with the distribution of HII regions (Paladini et al., 2003). These two distributions correspond quite well to each other, which is a confirmation — though certainly not a proof — that HI shells and HII regions are connected.

### 3.1.1 Properties of HI shells

A graph $r_{sh}$ versus $v_{exp}$ (Fig. 4, right) shows, that small shells ($r_{sh} \leq 100\,pc$) have low expansion velocities, while larger shells can have larger expansion velocities (the highest detected expansion velocities are around 20 $kms^{-1}$), though there is not any clear correlation between the radius and velocity. A more detailed description of the

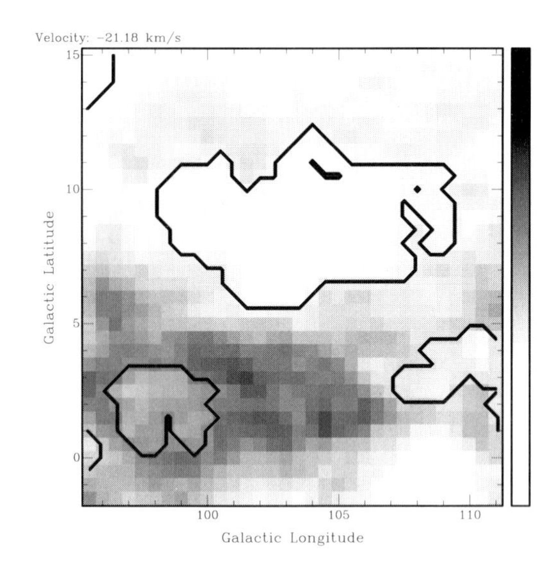


Figure 3: An example of two HI shells identified by our searching algorithm.

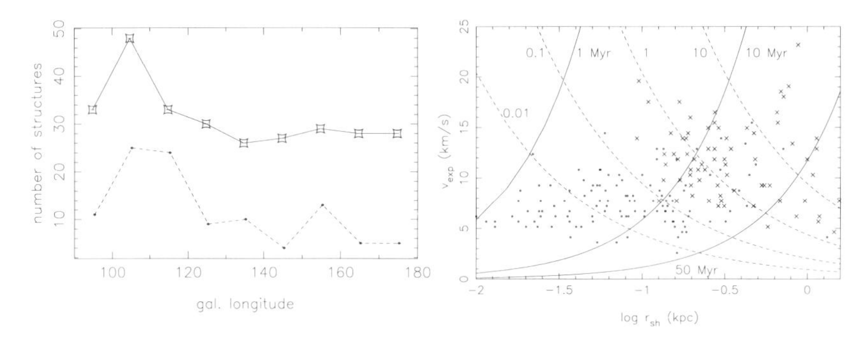


Figure 4: Left: numbers of HI shells (solid line) and HII regions (dashed line) as a function of galactic longitude. Right: shell radius versus expansion velocity for shells in the 2nd quadrant. Lines are analytical solutions after Weaver, McCray & Castor (1977). Solid lines are lines of constant age (1, 10, 50 $Myr$), dashed lines are lines of constant $L/n$ (0.01, 0.1, 1, 10 $SN\ Myr^{-1}/cm^{-3}$). Different symbols denote different energies of shells (according to Chevalier, 1974): dots are shells with $E < 10^{51} erg$, crosses are shells with $E \geq 10^{51} erg$.

Figure, together with a comparison with shells in other galaxies is given in Ehlerová & Palouš (2005).

Fig. 5 (left) shows the average density of the ambient medium in the region of the shell (this value is calculated from the spectrum of the central pixel of the structure). It seems as if the density of the ambient medium depended on the size of the shell (the least-square fit gives the dependence of the form $n \propto 1/r_{sh}$, but the correlation coefficient is not very impressive). This dependence implies that the HI column

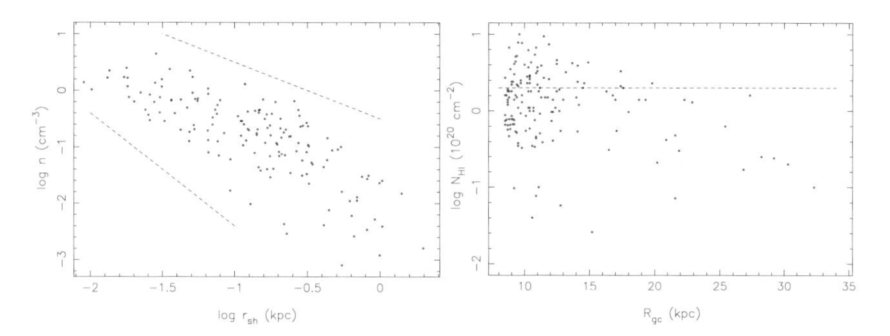


Figure 5: Left: volume density $n$ in positions of shells as the function of the shell radius $r_{sh}$. Dashed lines are power-laws with slopes of $-1$ and $-2$. Right: column density $N_{HI}$ as a function of the galactocentric distance $R_{gc}$. The dashed line is at $2 \times 10^{20} cm^{-2}$.

density of the gas swept by shells should be constant. This is roughly true. Fig. 5 (right) shows the measured column density for our identifications. The average value is $2 \times 10^{20} cm^{-2}$, however, the dispersion is high.

The dependence $n \propto 1/r_{sh}$ translates to the mass-size dependence $m_{sh} \propto r_{sh}^2$, which is a typical fractal behaviour (see also Section 2.1). However, the fractal explanation of this relation might not be the correct one, as the situation is really not straightforward, and especially the distance effects play a role.

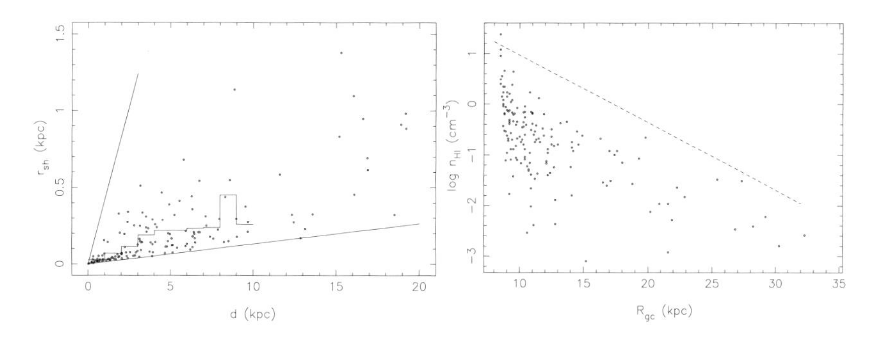


Figure 6: Left: shell radii $r_{sh}$ versus kinematic distances. Thick solid lines give the minimum and maximum radius corresponding to angular limits imposed by the searching routine. The thin line gives the average value of $r_{sh}$. Right: values of the ISM density as a function of the galactocentric distance. The dashed line shows an exponential decline with the scale of $3.3\ kpc$.

Due to constant angular limits imposed in the searching routine, we observe (on average) linearly larger shells at larger heliocentric distances and, because we study the second Galactic quadrant, it also means at larger galactocentric distances (see Fig. 6, left). The increasing galactocentric distance leads to an expected decreasing average volume density (observations in Fig. 6, right). Moreover, linearly large shells are affected by the $z$-gradient of the gaseous disc.

## 3.2 The size distribution of HI shells

Oey & Clarke (1997) predict, that the size distribution of HI shells in a galaxy should be a power-law with the index $\alpha$:

$$dN(r_{sh}) \propto r_{sh}^{-\alpha} dr_{sh} \tag{7}$$

The slope $\alpha$ is connected to the luminosity function of sources which create shells, i.e. to the luminosity function of OB associations (or HII regions):

$$\Phi(\mathcal{L}) \propto \mathcal{L}^{-\beta} \tag{8}$$

$$\alpha = 2\beta - 1 \tag{9}$$

Oey & Clarke (1997) also calculated the power-law slopes of HI shell size distributions for four galaxies, SMC, Holmberg II, M31 and M33. For the LMC galaxy the analysis was done by Kim et al. (1999). We calculate the same quantity for the Milky Way. We use only a subset of the whole catalogue of HI shells chosen in such a way, as to reduce effects of the heliocentric distance (for a description of the reduction method see Ehlerová & Palouš, 2005).

Table 2: Slopes $\beta$ of HII regions luminosity functions (Eq. 8) and $\alpha$ of HI shells size distribution (Eq. 7) for six galaxies. Values of $\alpha_{pred}$ and $\beta_{pred}$ predicted from the Eq. 7 are also given.

| | HII reg. | | HI shells | |
|---|---|---|---|---|
| | $\beta$ | $\Rightarrow \alpha_{pred}$ | $\alpha$ | $\Rightarrow \beta_{pred}$ |
| SMC | 1.9 | 2.8 | 2.7 | 1.9 |
| LMC | 1.8 | 2.5 | 2.5 | 1.8 |
| Ho II | 1.4 | 1.8 | 2.1 | 1.6 |
| M31 | 2.1 | 3.2 | 2.6 | 1.8 |
| M33 | 2.0 | 3.0 | 2.2 | 1.6 |
| MW | 2.0 | 3.0 | 2.1 | 1.6 |

Results of Oey & Clarke (1997, galaxies SMC, Ho II, M31 and M33), of Kim et al. (1999, LMC), and our results (Milky Way) for the slope $\alpha$ of the size distribution of HI shells are given in Table 2 and compared with the slope $\beta$ of the luminosity function. Slopes $\beta$ are taken from Oey & Clarke (1997), Kim et al. (1999) and Kennicutt, Edgar & Hodge (1989). The table shows that the relation between slopes $\alpha$ and $\beta$ as predicted by the Oey & Clarke theory (Eq. 9) is confirmed for dwarf galaxies (LMC, SMC and Holmberg II). For spiral galaxies (M31, M33 and Milky Way) there is a discrepancy. Spiral galaxies have a steeper luminosity function than dwarf and irregular galaxies, which predicts a steeper slope $\alpha$ (around 3) than observed (2-2.5).

One explanation of this discrepancy, supported by the fact that the observed size distribution in all studied galaxies has a similar slope, would be challenging the idea that spiral galaxies have steeper luminosity functions of HII regions (Kennicutt, Edgar & Hodge, 1989), and rather assume that the luminosity function is universal (see e.g. Thilker et al., 2002).

The alternative explanation shed some doubt on the simple connection of $\alpha$ and $\beta$ (Eq. 9) in the Oey & Clarke theory. This connection is derived under the assumption of a constant density of the ambient gas. This is not true. Even if we disregard the $z$-gradient (which could be justified by the fact, that the majority of sources and therefore shells reside in the equatorial plane), the radial gradient might be important. Indeed, we see for the Milky Way shells that there is a relation between sizes of shells and density, which may be influenced by galactic gradients of the density. If we assume that the density gradient is important, it is then evident, that the same energy source produces smaller shells in the higher density environment. Therefore, the relation between $\alpha$ and $\beta$ cannot be as simple as in (9), because not only the luminosity of the source is important, but also its ambient density. Assuming, that the major part of OB associations resides in regions with nearly constant density and a minor part is found in regions with smaller density, we would expect, that the slope $\alpha$ would be flatter than predicted by (9), which is what we observe for all three spiral galaxies (M31, M33 and Milky Way). However, the situation is quite complex (e.g. a comparison between density gradients in dwarf and spiral galaxies should be done) and we will deal with it in a later paper, as well as with a more thorough analysis of Figures 5 and 6.

# 4 Summary

In Section 2 we reviewed statistical properties of the interstellar medium. First we summarised relations among a mass spectrum, a mass-size relation and a power spectrum (Stutzki et al., 1998) and among density and velocity distributions and a power spectrum (Lazarian & Pogosyan, 2000). Then we gave typical values of these quantities from measurements for molecular clouds in the Milky Way (Stutzki et al., 1998), Small Magellanic Cloud (Stanimirović & Lazarian, 2001), Large Magellanic Cloud (Elmegreen, Kim & Staveley-Smith, 2001) and HI in the Milky Way (Dickey et al., 2001). Predictions of the theories were mostly confirmed.

Even though there are broad similarities between all mentioned cases, we see differences in a deeper look. They arise from differences between phases of the interstellar medium (molecular clouds, warm and cool HI), from different conditions in galaxies (LMC and SMC) and from different conditions for varying wavelengths (thin/thick disc of LMC).

In none of the cases the scale, at which the energy should be put into the interstellar medium to drive the turbulent cascade, was detected.

Then we studied properties of HI shells. We began by showing the catalogue of HI shells identified by our automatic searching routine applied to the Leiden-Dwingeloo HI survey of the Milky Way. For shells in the 2nd Galactic quadrant we found:

1. The average density of the ambient medium at the position of HI shells depends on the radius of the shell. This dependence is roughly $n \propto 1/r_{sh}$, which translates to the size-mass relation $m_{sh} \propto r_{sh}^2$. However, we cannot take this as a proof of the fractal nature of HI shells (or the ISM), as this be-

haviour might be a consequence of distance effects in our searching routine and of density gradients in the Milky Way.

2. The size distribution of HI shells has a power-law form with the slope $\alpha = 2.1$. According to the theory of Oey & Clarke (1997), this slope gives a slope $\beta$ of the luminosity function of HII regions, $\beta = 1.6$.

Afterwards, we compared size distributions of HI shells for six galaxies: M31, M33, Ho II, SMC (Oey & Clarke, 1997), LMC (Kim et al., 1999) and Milky Way. We found that the slopes of these distributions are similar among the galaxies (2.1-2.7). Differences lie in a comparison with the luminosity function of HII regions. For dwarf galaxies (Ho II, SMC, LMC) measured and predicted slopes are in a good agreement, for spiral galaxies (M31, M33, Milky Way) the size distribution of shells should be steeper than observed. Possible explanations are either a universality of luminosity functions regardless of the Hubble type or — more probably — a dependence of shell sizes on density gradients in galaxies, from which a more complicated relation between slopes of HII regions luminosity functions and HI shells size distributions follows.

*Acknowledgements:* I am grateful to the organisers of the conference for allowing me to present this paper during the Prague meeting. This work is supported by the grant project of the Academy of Sciences of the Czech Republic No. K1048102.

## References

Chevalier, R. 1974, ApJ 188, 501

Dickey, J. M., McClure-Griffiths, N. M., Stanimirović, S., Gaensler, B.M., Green, A. J. 2001, ApJ 561, 264

Ehlerová, S., Palouš, J. 2005, submitted to A&A

Ehlerová, S., Palouš, J., Wünsch, R. 2004, Ap&SS, 289, 279

Elmegreen, B. G., Falgarone, E. 1996, ApJ 471, 816

Elmegreen, B. G., Kim, S., Staveley-Smith, L. 2001, ApJ 548, 749

Elmegreen, B. G., Elmegreen, D. M., Leitner, S. N. 2003, ApJ 590, 271

Hartmann, D., Burton, W. B. 1997, Atlas of Galactic Neutral Hydrogen (Cambridge University Press)

Heiles, C. 1979, ApJ 229, 533

Heiles, C. 1984, ApJSS 55, 585

Hu, E., 1981, ApJ 248, 119

Kennicutt, R. C., Edgar, B. K., Hodge, P. W. 1989, ApJ 337, 761

Kim, S., Dopita, M. A., Staveley-Smith, L., Bessell, M. S. 1999, AJ 118, 2797

Lazarian, A., Pogosyan, D. 2000, ApJ 537, 173

McClure-Griffiths, N. M., Dickey, J. M., Gaensler, B. M., Green, A. J. 2002, ApJ 578, 176

Oey, M. S., Clarke, C. J. 1997, MNRAS 289, 570

Paladini, R., Burigana, C., Davies, R. D., Maino, D., Bersanelli, M., Cappellini, B., Platania, P., Smoot, G. 2003, A&A 397, 213

Stanimirović, S., Lazarian, A. 2001, ApJ 551, L53

Stutzki, J., Bensch, F., Heithausen, A., Ossenkopf, V., Zielinsky, M. 1998, A&A 336, 697

Thilker, D. A., Walterbos, R. A. M., Braun, R., Hoopes, Ch. G. 2002, AJ 124, 3118

Walter, F., Brinks, E. 1999, AJ 118, 273

Weaver, R., McCray, R., Castor, J. 1977, ApJ 218, 377

# Origins of Brown Dwarfs

Viki Joergens

Sterrewacht Leiden
PO Box 9513, 2300 RA Leiden, The Netherlands,
and
Max-Planck-Institut für Extraterrestrische Physik
Giessenbachstrasse 1, 85748 Garching, Germany
viki@strw.leidenuniv.nl

**Abstract**

*The formation of objects below or close to the hydrogen burning limit is currently vividly discussed and is one of the main open issues in the field of the origins of stars and planets. Applying various observational techniques, we explored a sample of brown dwarfs and very low-mass stars in the ChaI star forming cloud at an age of only a few million years and determined fundamental parameters for their formation and early evolution.*

*Tracking the question of how frequent are brown dwarf binaries and if brown dwarfs have planets, one of the first radial velocity (RV) surveys of brown dwarfs sensitive down to planetary masses is carried out based on high-resolution spectra taken with UVES at the VLT. The results hint at a low multiplicity fraction, which is in contrast to the situation for young low-mass stars.*

*Testing recent formation scenarios, which propose an ejection out of the birth place in the early accretion phase, we carried out a precise kinematic analysis of the brown dwarfs in our sample in comparison with T Tauri stars in the same field. This yielded the first empirical upper limit for possible ejection velocities of a homogeneous group of brown dwarfs.*

*Rotation is a fundamental parameter for objects in this early evolutionary phase. By means of studying the line broadening of spectral features in the UVES spectra as well as by tracing rotational modulation of their lightcurves due to surface spots in photometric monitoring data, one of the first rotation rates of very young brown dwarfs have been determined.*

*In the light of the presented observational results, the current scenarios for the formation of brown dwarfs are discussed.*

---

*Reviews in Modern Astronomy 18.* Edited by S. Röser

ISBN 3-527-40608-5

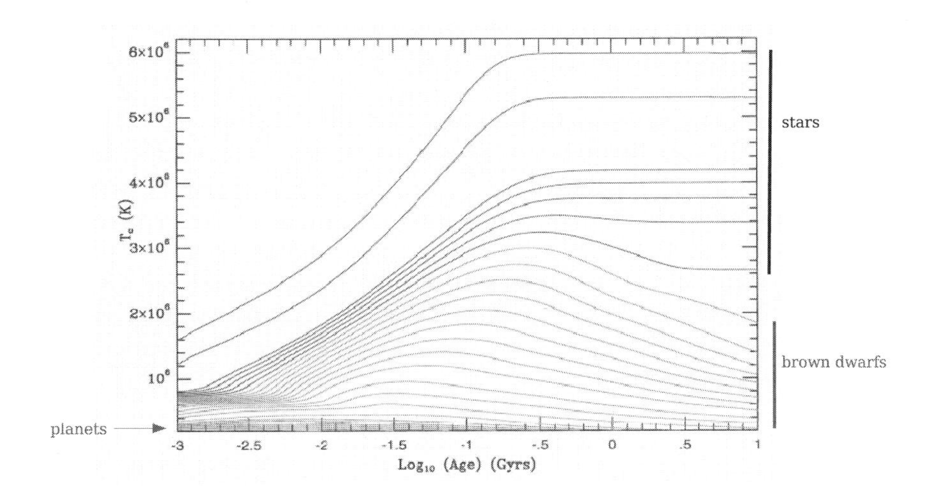


Figure 1: **Evolution of the central temperature** for stars, brown dwarfs and planets from 0.3 $M_{Jup}$ ($\sim$ Saturn) to 0.2 $M_\odot$ (211 $M_{Jup}$) versus age. From Burrows et al. (2001). The upper seven curves, which end in the plot with constant central temperatures, represent stars, the 13 curves below represent brown dwarfs and the remaining curves at the smallest temperatures mark planets with masses $\leq$ 13 $M_{Jup}$. Note the increase (classical plasma), maximum and decrease (quantum plasma) of the central temperature for brown dwarfs.

# 1 What are brown dwarfs ?

Brown dwarfs fill the gap between low-mass stars and giant planets in the mass range of about 0.08 solar masses ($M_\odot$) and about 13 Jupiter masses ($M_{Jup}$)[1], depending on metallicity. They can never fully stabilize their luminosity by hydrogen burning in contrast to stars and thus contract as they age until the electron gas in their interior is completely degenerate. At that point they have reached a final radius and become cooler and dimmer for their remaining life time. In contrast to planets brown dwarfs are able to fuse deuterium, which defines the lower mass limit of brown dwarfs. It is noted that although thermonuclear processes do not dominate the evolution of brown dwarfs, they do not only burn deuterium, but the more massive ones (masses above $\sim$0.065 $M_\odot$) burn lithium and may even burn hydrogen for a while. However, they do not burn hydrogen *at a rate sufficient to fully compensate radiative losses.*

Fig. 1 shows the evolution of the central temperature for low-mass stars (upper

[1] 1 $M_\odot$=1047 $M_{Jup}$

seven curves with constant central temperature at the end of the plot), brown dwarfs and planets (Burrows et al. 2001). The temperature for stars rises in the pre-main sequence phase due to the contraction on the Hayashi track until the interior is hot and dense enough to burn sufficient hydrogen to stabilize the central temperature and balance gravitational pressure on the hydrogen burning main sequence. For brown dwarfs, the central temperature also increases due to gravitational contraction in the first million to a few hundred million years. As the density in the interior increases part of the electron gas in the interior becomes degenerate. Electrons obey the *Pauli exclusion principle*, allowing one quantum state to be occupied by only one electron. For the compression of partly degenerate gas, energy is *needed* in order to bring the degenerate electrons closer. Therefore with onwardly contraction the temperature remains constant for a while and then starts to decrease. When the electron gas is completely degenerate, no further contraction is possible, the brown dwarf has reached its final radius and cools without compensation by compressional heating.

The existence of brown dwarfs was predicted by Kumar (1962, 1963) already in the early sixties. It followed a 30 year search for such faint objects until in 1995 almost at the same time three brown dwarfs were discovered independently: the Methane dwarf Gl 229 B, a faint companion to a nearby M-dwarf, as cool as 1000 K (Nakajima et al. 1995, Oppenheimer et al. 1995) as well as two brown dwarfs in the Pleiades, Teide 1 (Rebolo et al. 1995) and PPl 15 (Stauffer et al. 1994). The latter object turned out to be in fact a pair of two gravitationally bound brown dwarfs (Basri & Martín 1999). Up to now, more than 300 brown dwarfs have been detected in star forming regions (e.g. Béjar et al. 1999, Comerón et al. 2000), in young cluster (e.g. Pleiades, Martín et al. 2000) and in the field by the large infrared and optical surveys, DENIS (Delfosse et al. 1997), 2MASS (Kirkpatrick et al. 2000) and SLOAN (Hawley et al. 2002).

The year 1995 saw another spectacular discovery: the detection of the first extrasolar planet candidate with minimum mass around $\sim$1 $M_{\rm Jup}$ orbiting the sun-like star 51 Peg (Mayor & Queloz 1995). Since 1995, high-precision radial velocity (RV) surveys have detected more than 130 planets in orbit around stars (e.g. Mayor et al. 2003) with masses ranging from 14 Earth masses (e.g. Santos et al. 2004) to the brown dwarf border at about 13 $M_{\rm Jup}$.

Before 1995, no objects were known in the mass range between Jupiter in our own solar system and the lowest-mass stars. Today, the situation completely changed and this mass range is populated by hundreds of giant planets and brown dwarfs challenging our understanding of the formation of solar systems. Brown dwarfs play an important role in this discussion since they link the planet population to the stellar population.

# 2 Brown dwarf formation models and predictions

The substellar border defined by the hydrogen burning mass is a crucial dividing line with respect to the further evolution of an object but there is no obvious reason why it should be of any significance for the formation mechanism by which this objects was produced. Thus, by whichever process brown dwarfs are formed, it is expected to work continuously into the regime of very low-mass stars. Therefore, the term *formation of brown dwarfs* in this article stands for *the formation of brown dwarfs and very low-mass stars* even if not stated explicitly.

In the traditional picture of (low-mass) star and planet formation we assume that stars form by collapse and hierarchical fragmentation of molecular clouds (Shu et al. 1987), a process which involves the formation of a circumstellar disk due to conservation of angular momentum. The disk in turn is the birth place for planets, which form by condensation of dust and further growth by accretion of disk material. These ideas are substantially based on the situation in our own solar system. However, some properties of detected extrasolar planetary systems are difficult to explain within current theories of planet formation and it is under debate if our solar system is rather the exception than the rule. How brown dwarfs form is also an open issue but it is clear that a better knowledge of the mechanisms producing these transition objects between stars and planets will also clarify some open questions in the context of star and planet formation.

Brown dwarfs may be the high-mass extension of planets and form like giant planets in a disk around a star by either core accretion or disk instabilities (see Wuchterl et al. 2000 for a review). In the core accretion model, the formation of giant planets is initiated by the condensation of solids in a circumstellar disk, which accrete to larger bodies and form a rock/ice core. When this core reaches the critical mass for the accretion of a gas envelope ($\sim$10-15 $M_\oplus$) a so called *runaway gas accretion* is triggered, which is only slowed down when the surrounding gas reservoir is depleted by the accreting protoplanet. If brown dwarfs are formed like giant planets, we should find them in orbit around their host star. However, brown dwarfs are found in large number as free-floating objects. Furthermore, in the ongoing high-precision RV surveys for extrasolar planets, an almost complete absence of brown dwarfs in close ($<$3 AU), short-period orbits around solar-mass stars was found while these surveys detected more than 130 planets. This so-called 'brown dwarf desert' indicates that there is no continuity from planets to brown dwarfs in terms of formation.

Brown dwarfs may also form like stars by direct gravitational collapse and fragmentation of molecular clouds out of cloud cores, which are cold and dense enough to become Jeans-unstable for brown dwarf masses. Such small cloud cores have not yet been detected by radio observations but might have been missed due to insufficient detection sensitivity. However, on theoretical grounds, there is a so-called opacity limit for the fragmentation, i.e. a limiting mass, which can become Jeans-unstable: when reaching a certain density during the collapse the gas becomes optically thick and is heated up leading to an increase of the Jeans-mass (Low & Lynden-Bell 1976). That might prevent the formation of (lower mass) brown dwarfs by direct collapse. Nevertheless, assuming brown dwarfs can form in that way, they would be the low-mass extension of the stellar population and should, therefore, have similar (scaled down) properties as young low-mass stars (T Tauri stars). We know that T Tauri stars have a very high multiplicity fraction, in some star forming regions close to 100% (e.g. Leinert et al. 1993, Ghez et al. 1993, 1997, Köhler et al. 2000) indicating that the vast majority of low-mass stars is formed in binaries or higher order multiple systems. The stellar companions are thought to be formed by disk fragmentation or filament fragmentation in the circumstellar disk created by the collapse of the molecular cloud core. Therefore, we would also expect a high multiplicity fraction of brown dwarfs in this formation model. Furthermore, brown dwarfs should have circumstellar disks due to preservation of angular momentum during the collapse, harbour planets, and should have the same kinematical properties as the T Tauri stars in the same field.

Another idea is that brown dwarfs formed by direct collapse of unstable cloud cores of stellar masses and would have become stars if the accretion process was not stopped at an early stage by an external process before the object has accreted to stellar mass. It was proposed that such an external process can be the ejection of the protostar out of the dense gaseous environment due to dynamical interactions (Reipurth & Clarke 2001). It is known that the dynamical evolution of gravitationally interacting systems of three or more bodies leads to frequent close two-body encounters and to the formation of close binary pairs out of the most massive objects in the system as well as to the ejection of the lighter bodies into extended orbits or out of the system with escape velocity (e.g. Valtonen & Mikkola 1991). The escape of the lightest body is an expected outcome since the escape probability scales approximately as the inverse third power of the mass. Sterzik & Durisen (1995, 1998) and Durisen et al. (2001) considered the formation of run-away T Tauri stars by such dynamical interactions of compact clusters.

The general expectations for the properties of brown dwarfs formed by the ejection scenario are a low binary frequency (maybe 5%), no wide brown dwarf binaries and only close-in disks ($<$ 5-10 AU) since companions or disk material at larger separations will be truncated by the ejection process. Furthermore, the kinematics of ejected brown dwarfs might differ from non-ejected members of the cluster. Hydrodynamical calculations have shown that the collapse of a molecular cloud can produce brown dwarfs in this way (Bate et al. 2003, Delgado-Donate et al. 2003, Bate & Bonnell 2005), while N-body simulations of the dynamical decay allowed the prediction of statistically significant properties of ejected brown dwarfs (Sterzik & Durisen 2003, Delgado-Donate et al. 2004, Umbreit et al. 2005). However, the predictions differ significantly among the different models; this is further discussed in Sect. 4.

An external process, which prevents the stellar embryo from further growth in mass can also be a strong UV wind from a nearby hot O or B star, which ionizes and photoevaporates the surrounding gas (Kroupa & Bouvier 2003, Whitworth & Zinnecker 2004). A significant disruption of the accretion envelope by photoevaporation will also lead generally to a low multiplicity fraction and limited disk masses. Since there is no such a hot star in the Cha I cloud, the brown dwarfs in this region cannot have been formed by this mechanism.

The various ideas for the formation of brown dwarfs need to be constrained by observations of brown dwarfs, key parameters are among others the multiplicity and kinematics, which have been studied for young brown dwarfs in Cha I as described in the following. Furthermore, rotation is an important parameter for the early evolution and it might reflect the interaction with a disk due to magnetic disk braking.

# 3 Comprehensive observations of brown dwarfs and very low-mass stars in Cha I

## 3.1 Sample

The observation of very young brown dwarfs and very low-mass stars allows insights into the formation and early evolution below or close to the substellar limit. One of the best grounds for such a study is the Cha I cloud, which is part of the larger Chamaeleon complex. At a distance of 160 pc, it is one of the closest sites of active low-mass star and brown dwarf formation. Comerón and coworkers initiated here one of the first surveys for young very low-mass objects down to the substellar regime by means of an H$\alpha$ objective prism survey (Comerón et al. 1999, 2000; Neuhäuser & Comerón 1998, 1999). They found twelve very low-mass M6–M8–type objects, Cha H$\alpha$ 1 to 12, in the center of Cha I with ages of 1 to 5 Myrs, among which are four bona fide brown dwarfs and six brown dwarf candidates. Furthermore, they found or confirmed several very low-mass stars with masses smaller than 0.2 $M_\odot$ and spectral types M4.5–M5.5 in Cha I (B 34, CHXR 74, CHXR73 and CHXR 78C) as well as one 0.3 solar mass M2.5 star (Sz 23).

The membership to the Cha I cloud and therefore the youth of the objects is indicated by their H$\alpha$ emission and has been confirmed by medium-resolution spectra, the detection of lithium absorption and consistent RVs (also by the here presented observations). The substellar nature is derived from the determination of bolometric luminosities and effective temperatures (converted by means of temperature scales from spectral types) and comparison with theoretical evolutionary tracks in the Hertzsprung-Russell diagram (HRD).

Another substellar test applicable to young brown dwarfs was developed from a test for old brown dwarfs based on lithium absorption (Rebolo et al. 1992) to the age-independent statement, that *any object with spectral type M7 or later that shows lithium is substellar* (Basri 2000). Since all Cha H$\alpha$ objects show lithium absorption in their spectra (Comerón et al. 2000, Joergens & Guenther 2001), the four objects Cha H$\alpha$ 1, 7, 10 and 11 (M7.5–M8) are thus bona fide brown dwarfs, with masses in the range of about 0.03 $M_\odot$ to 0.05 $M_\odot$, Cha H$\alpha$ 2, 3, 6, 8, 9, 12 (M6.5–M7) can be classified as brown dwarf candidates and Cha H$\alpha$ 4 and 5 (M6) are most likely very low-mass stars with 0.1 $M_\odot$. An error of one subclass in the determination of the spectral type was taken into account.

We observed most of these ten brown dwarfs and brown dwarf candidates and seven (very) low-mass stars by means of high-resolution spectroscopy with UVES at the 8.2 m telescope at the VLT and by means of a photometric monitoring campaign at a 1.5 m telescope.

## 3.2 High-resolution UVES spectroscopy

High-resolution spectra have been taken for the brown dwarfs and low-mass stars Cha H$\alpha$ 1–8 and Cha H$\alpha$ 12, B34, CHXR 74 and Sz 23 between the years 2000 and 2004 with the cross-dispersed UV-Visual Echelle Spectrograph (UVES, Dekker et al. 2000) attached to the 8.2 m Kueyen telescope of the Very Large Telescope (VLT) operated by the European Southern Observatory at Paranal, Chile. The wavelength regime from 6600 Å to 10400 Å was covered with a spectral resolution of $\lambda/\Delta\lambda = 40\,000$. For each object at least two spectra separated by a few weeks have been obtained in order to monitor time dependence of the RVs. For several objects, more than two and up to twelve spectra have been taken.

After standard reduction, we have measured RVs by means of cross-correlating plenty of stellar lines of the object spectra with a template spectrum. In order to achieve a high wavelength and therefore RV precision, telluric $O_2$ lines have been used as wavelength reference. A RV precision between $40\,\mathrm{m\,s^{-1}}$ and $670\,\mathrm{m\,s^{-1}}$, depending on the S/N of the individual spectra, was achieved for the relative RVs. The errors are based on the standard deviation of two consecutive single spectra. An additional error of about $300\,\mathrm{m\,s^{-1}}$ has to be taken into account for the absolute RVs due to uncertainties in the zero point of the template. These RVs are one of the most precise ones for young brown dwarfs and very low-mass stars available up to now.

Based on the time-resolved RVs, a RV survey for planetary and brown dwarf companions to the targets was carried out (Sect. 5). The mean RVs were explored in a kinematic study of this group of brown dwarfs in Sect. 4. Additionally, projected rotational velocities $v \sin i$ were derived from line-broadening of spectral features (Sect. 6), lithium equivalent width has been measured to confirm youth and membership to the Cha I star forming region and the CaII IR-triplet has been studied as an indicator for chromospheric activity. Results have been or will be published in Joergens & Guenther (2001), Joergens (2003), Joergens et al. (2005a,b).

### 3.3 Photometric monitoring

In order to study the time-dependent photometric behaviour of the brown dwarfs and very low-mass stars, we monitored a $13' \times 13'$ region in the Cha I cloud photometrically in the Bessel R and the Gunn i filter in six consecutive half nights with the CCD camera DFOSC at the Danish 1.5 m telescope at La Silla, Chile in May and June 2000. Differential magnitudes were determined relative to a set of reference stars in the same field by means of aperture photometry. The field contained all of the targets introduced in the previous sections, however, CHXR 74 and Sz 23 were too bright and have been saturated in the images, while Cha H$\alpha$ 1, 7, 9, 10 and 11 were too faint for the size of the telescope.

Based on the lightcurves for Cha H$\alpha$ 2, 3, 4, 5, 6, 8, 12, B 34, CHXR 78C and CHXR 73, we searched for periodic variabilities and were able to determine absolute rotation periods for several brown dwarfs and very low-mass stars based on brightness modulations due to surface activity (Sect. 6, Joergens et al. 2003b).

## 4 Kinematics of brown dwarfs in Cha I and the ejection scenario

It was proposed that brown dwarfs might form in a star-like manner but have been prevented from accreting to stellar masses by the early ejection out of their birth place (Sect. 2). The ejection process might have left an observable imprint in the

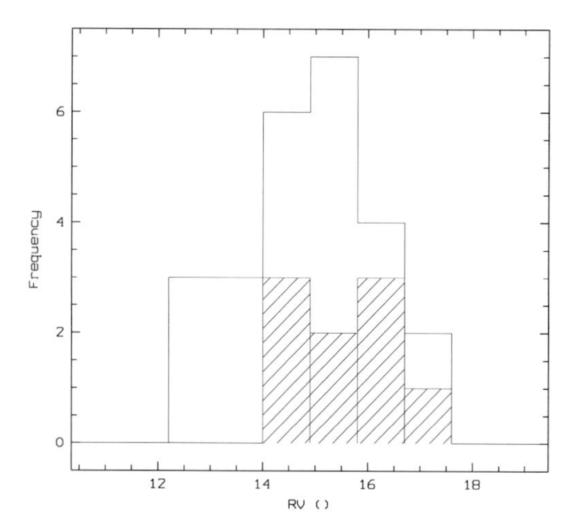


Figure 2: **Distribution of mean RVs** in km s$^{-1}$ for nine brown dwarfs and very low-mass stars with masses $\leq 0.1\,M_\odot$ (hashed) and for 25 T Tauri stars in Cha I.

kinematics of ejected members of a cluster in comparison to that of non-ejected members. Therefore, we are studying the kinematics of our target brown dwarfs in Cha I by means of precise mean RVs measured from UVES spectra. We find that the seven brown dwarfs and brown dwarf candidates and two $0.1\,M_\odot$ stars in Cha I (spectral types M6–M8) differ very less from each other in terms of their RV. The sample has a mean RV of 15.7 km s$^{-1}$ and a RV dispersion measured in terms of standard deviation of only 0.9 km s$^{-1}$. The total covered RV range is 2.6 km s$^{-1}$. We note, that in previous publications (Joergens & Guenther 2001, Joergens 2003), the dispersion was measured in terms of full width at half maximum (fwhm) of a Gaussian distribution (which is related to the standard deviation $\sigma$ of the Gaussian by fwhm $= \sigma\sqrt{8 \ln 2}$) following the procedure of radio astronomers. Based on new RV data and an improved data analysis, the here presented kinematic study (Joergens et al. 2005b) is a revised version of the previous ones. The new fwhm for the brown dwarfs is 2.1 km s$^{-1}$, which is consistent with the previous value of 2.0 km s$^{-1}$.

In order to compare the results for the brown dwarfs in Cha I with the RV distribution of higher-mass stellar objects in this cluster, we compiled all T Tauri stars confined to the same region for which RVs have been measured with a precision of 2 km s$^{-1}$ or better from the literature (Walter 1992, Dubath et al. 1996, Covino et al. 1997, Neuhäuser & Comerón 1999), from Guenther et al. (in prep., see Joergens & Guenther 2001) and from our own measurements based on UVES spectra for M-type T Tauri stars. The compiled 25 T Tauri stars (spectral types M5–G2) have a mean RV of 14.7 km s$^{-1}$ and a RV dispersion in terms of standard deviation of 1.3 km s$^{-1}$. The dispersion given in terms of fwhm is 3.1 km s$^{-1}$. Compared to the previous studies (Joergens & Guenther 2001, Joergens 2003), the T Tauri sample and RVs have been also revised by taking into account new UVES-based RVs obtained by us in 2002 and 2004 as well as overlooked RV measurements by Walter (1992). Furthermore, based on a recent census of Cha I (Luhman 2004) an up-to-date check of membership status was possible. The new standard deviation of 1.3 km s$^{-1}$ differs only slightly from the previous one of 1.5 km s$^{-1}$.

The resulting RV distributions of the T Tauri stars and the brown dwarfs are displayed in Fig. 2 in form of a histogram. The mean RV of the M6–M8 type brown dwarfs and very low-mass stars (15.7 km s$^{-1}$) is consistent within the errors with that of the T Tauri stars in the same field (14.7 km s$^{-1}$) and with that of the molecular gas of the surrounding (15.3 km s$^{-1}$, Mizuno et al. 1999). This confirms the membership of the substellar population to the Cha I star forming cloud.

Furthermore, we find that the RV dispersion of the brown dwarfs (0.9 km s$^{-1}$) is slightly smaller than that of the T Tauri stars in the same field (1.3 km s$^{-1}$) and slightly larger than that of the surrounding molecular gas (the fwhm of 1.2 km s$^{-1}$ given by Mizuno et al. (1999) translates to a standard deviation of 0.5 km s$^{-1}$ using the above given formula). For the comparison of the kinematic properties of brown dwarfs and stellar objects we have set the dividing line between the two populations arbitrarily at 0.1 $M_\odot$ knowing that this is not exactly the substellar border. However, the process forming brown dwarfs is expected to work continuously into the regime of very low-mass stars and it is a priori unknown if and for which mass range a different formation mechanism is operating. In a more detailed study (Joergens et al 2005b), we calculate the dispersion also for different choices of this dividing line but see no significant differences in the results.

RVs are tracing only space motions in the direction of the line-of-sight and the studied brown dwarfs could have a larger three-dimensional velocity dispersion. However, the studied brown dwarfs in Cha I have an age of 1–5 Myr and occupy a field of less than $12' \times 12'$ at a distance of 160 pc. Brown dwarfs born within this field and ejected during the early accretion phase in directions with a significant fraction perpendicular to the line-of-sight, would have flown out of the field a long time ago for velocities of 0.5 km s$^{-1}$ or larger (Joergens et al. 2003a). Therefore, the measurement of RVs is a suitable method to test if objects born and still residing in this field have significantly high velocities due to dynamical interactions during their formation process.

Numerical simulations of brown dwarf formation by the ejection scenario differ in their predictions of the resulting velocities over a wide range. In general, the ejection velocity scales with the inverse of the square root of the distance of the closest approach in the encounter that led to the ejection (Armitage & Clarke 1997). The resulting velocity distribution depends also on the gravitational potential of the cluster since it defines if an ejected object with a certain velocity can escape or is ejected only in a wide eccentric orbit (e.g. Kroupa & Bouvier 2003). Furthermore, the kinematic signature of the ejection can be washed out if only part of the brown dwarfs in a cluster are formed according to the ejection scenario while a significant fraction is formed by other mechanisms (disk instabilities alone or direct collapse above the fragmentation limit), or when, on the other hand, all (sub)stellar objects in a cluster undergo significant dynamical interactions.

Hydrodynamic calculations by Bate et al. (2003) and Bate & Bonnell (2005) predict velocity dispersions of 2.1 km s$^{-1}$ and 4.3 km s$^{-1}$, resp., and no kinematic difference between brown dwarfs and T Tauri stars. These calculations are performed for much denser star forming regions than Cha I and an extrapolation of their results to the low-density Cha I cloud might lead to a consistency with our observed RV dispersion of 0.9 $km\,s^{-1}$ for brown dwarfs and only slightly larger for T Tauri stars. However, comparison with Delgado-Donate et al. (2004) indicates that the dependence of the velocity dispersion on the stellar density is not yet well established. N-body simulations by Sterzik & Durisen (2003) predict that 25% of brown dwarf singles have a velocity smaller than 1 $km\,s^{-1}$, that is much lower than our observations of 67% of brown dwarfs in Cha I having RVs smaller than 1 $km\,s^{-1}$. Also the high velocity tail found by the authors of 40% single brown dwarfs having higher velocities than 1.4 $km\,s^{-1}$ and 10% $>$5 $km\,s^{-1}$ is not seen in our data, were none has a RV deviating by the mean by more than 1.4 $km\,s^{-1}$. Recent N-body calculations by Umbreit et al. (2005) showed that the ejection velocities depend strongly on accretion and by assuming different accretion models and rates, the authors predict an even more pronounced high-velocity tail with 60% to 80% single brown dwarfs having velocities larger than 1 $km\,s^{-1}$. This is also much larger than found by our observations, where only about 30% have velocities $>$ 1 $km\,s^{-1}$. One might argue that the non-detection of a high velocity tail in our data can be attributed to the relatively small size of our sample, on the other hand, the current N-body simulations do not take into account the gravitational potential of the cluster, which might cause a suppression of the highest velocities.

To conclude, the observed values of our kinematic study provide the first observational constraints for the velocity distribution of a group of very young brown dwarfs and show that they have a RV dispersion of 0.9 $km\,s^{-1}$, no high-velocity tail and their RVs are not more dispersed than that of T Tauri stars in the same field. These observed velocities are smaller than any of the theoretical predictions for brown dwarfs formed by the ejection scenario, which might be attributed to the lower densities in Cha I compared to some model assumptions, to shortcomings in the models, like neglection of feedback processes (Bate et al. 2003, Bate & Bonnell 2005, Delgado-Donate et al. 2004) or of the cluster potential (Sterzik & Durisen 2003, Umbreit et al. 2005), or to the fact that our sample is statistically relatively small. The current conclusion is that either the brown dwarfs in Cha I have been formed by ejection but with smaller velocities as theoretically predicted or they have not been formed in that way.

# 5 RV Survey for planets and brown dwarf companions with UVES

Based on the precise RV measurements in time resolved UVES spectra (Sect. 3.2), we are carrying out a RV survey for (planetary and brown dwarf) companions to the young brown dwarfs and very low-mass stars in the Cha I cloud.

The detection of planets around brown dwarfs as well as the detection of young spectroscopic brown dwarf binaries would be an important clue towards the formation of brown dwarfs. So far, no planet is known orbiting a brown dwarf (the recent publication of a direct image of a candidate for a 5 $M_{Jup}$ planet around a 25 $M_{Jup}$

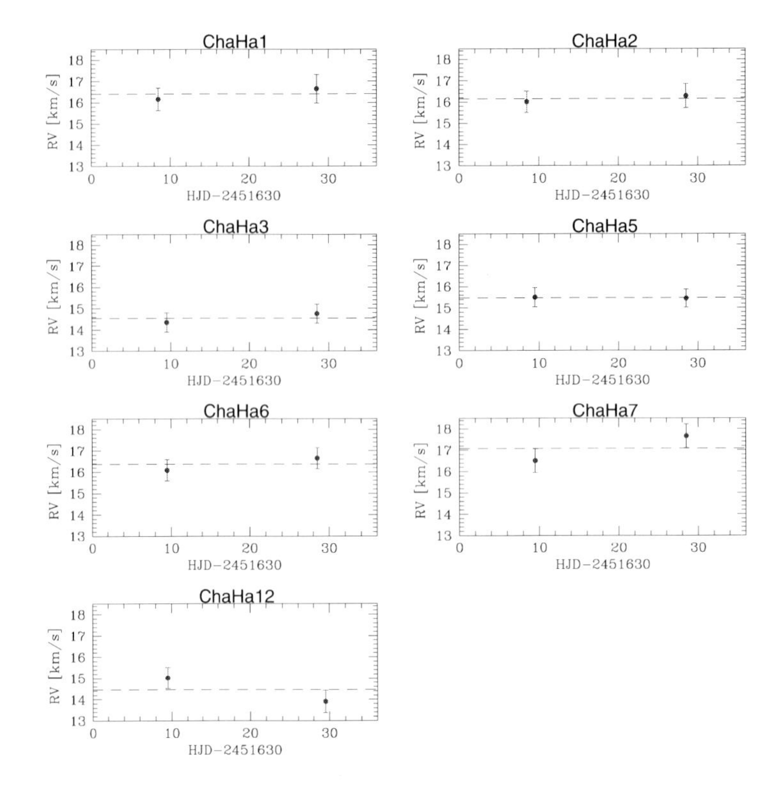


Figure 3: RV constant objects: RV vs. time in Julian days for brown dwarfs and very low-mass stars (M6–M8) in Cha I based on high-resolution UVES/VLT spectra. Error bars indicate 1 $\sigma$ errors.

brown dwarf by Chauvin et al. (2004) is very exciting but still very tentative since it might very well be a background object). There have been detected several brown dwarf binaries, among them there are three spectroscopic, and hence close systems (Basri & Martín 1999, Guenther & Wuchterl 2003). However, all known brown dwarf binaries are fairly old and it is not yet established if the typical outcome of the brown dwarf formation process is a binary or multiple brown dwarf system or a single brown dwarf.

Furthermore, the search for planets around very young as well as around very low-mass stars and brown dwarfs is interesting since the detection of *young* planets as well as a census of planets around stars of all spectral types, and maybe even around brown dwarfs, is an important step towards the understanding of planet formation. It would provide empirical constraints for planet formation time scales. Furthermore, it would show if planets can exist around objects which are of considerably lower mass and surface temperature than our sun.

## 5.1 RV constant objects

The RVs for the brown dwarfs and very low-mass stars Cha H$\alpha$ 1, 2, 3, 4, 5, 6, 7, 12 and B 34 are constant within the measurements uncertainties of 2 $\sigma$ for Cha H$\alpha$ 4 and of 1 $\sigma$ for all others, as displayed in Figs. 3, 4. From the non-detections of variability, we have estimated upper limits for the projected masses $M_2 \sin i$ of hypothetical companions for each object[2]. They range between 0.1 $M_{\mathrm{Jup}}$ and 1.5 $M_{\mathrm{Jup}}$ assuming a circular orbit, a separation of 0.1 AU between companion and primary object and adopting primary masses from Comerón et al. (1999, 2000). The used orbital separation of 0.1 AU corresponds to orbital periods ranging between 30 and 70 days for the masses of these brown dwarfs and very low-mass stars.

That means, that these nine brown dwarfs and very low-mass stars with spectral types M5–M8 and masses $\leq 0.12\,\mathrm{M}_{\odot}$ show no RV variability down to Jupitermass planets. There is, of course, the possibility that present companions have not been detected due to non-observations at the critical orbital phases. Furthermore, long-period companions may have been missed since for all of them but Cha H$\alpha$ 4, the time base of the observations does not exceed two months.

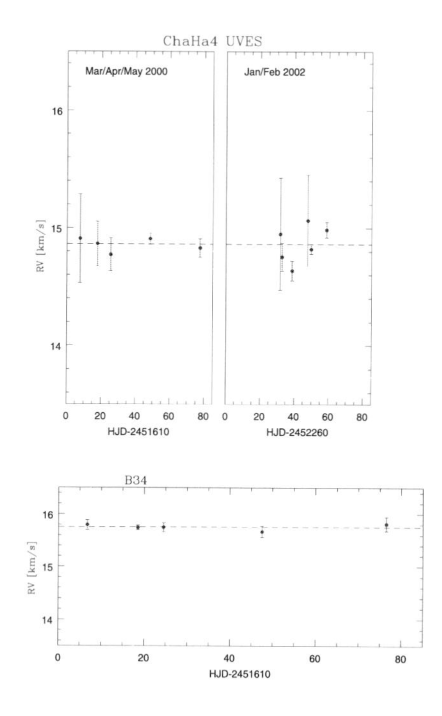


Figure 4: RV constant objects (continued): RV vs. time in Julian days for very low-mass stars ($\sim$0.1 $\mathrm{M}_{\odot}$) in Cha I based on high-resolution UVES/VLT spectra. Error bars indicate 1 $\sigma$ errors.

[2]Spectroscopic detections of companions allow in general no absolute mass determination but only the derivation of a lower limit of the companion mass $M_2 \sin i$ due to the unknown inclination $i$.

## 5.2 RV variable objects

For three objects, we have found significant RV variations, namely for the brown dwarf Cha H$\alpha$ 8 and the low-mass stars CHXR 74 ($\sim$0.17 $M_{\odot}$) and Sz 23 ($\sim$0.3 $M_{\odot}$) as shown in Fig. 5. The variability characteristic differs among the three objects. Sz 23 shows variability on time scales of days with no difference in the mean values of RVs recorded in 2000 and in 2004. On the other hand, Cha H$\alpha$ 8 and CHXR 74 show only very small amplitude variations or no variations at all on time scales of days to weeks, whereas the mean RV measured in 2000 differs significantly from the one measured years later, namely in 2002 for Cha H$\alpha$ 8 and in 2004 for CHXR 74, respectively, hinting at variability periods of the order of months or longer.

One possibility for the nature of these RV variations is that they are the Doppler-shift caused by the gravitational force of orbiting companions. The poor sampling does not allow us to determine periods of the variations but based on the data we suggest that the period for Cha H$\alpha$ 8 is 150 days or longer. A 150 d period would corresponds to a 6 $M_{Jup}$ planet orbiting at a separation of 0.2 AU around Cha8. For longer periods the orbital separations as well as the mass of the companion would be larger. For CHXR74, a period of about 200 days would be able to explain the RV data of 2000 and 2004 corresponds to a 15 $M_{Jup}$ brown dwarf orbiting CHXR74 at a separation of 0.4 AU.

The other possibility is that they are caused by surface activity since prominent surface spots can cause a shifting of the photo center at the rotation period (see also Sect. 6.2). The upper limits for the rotational periods of Cha H$\alpha$ 8, CHXR 74 and Sz 23 are 1.9 d, 4.9 d and 2.1 d, respectively, based on projected rotational velocities $v \sin i$ (Sect. 6, Joergens & Guenther 2001). Thus, the time-scale of the RV variability of Sz 23 is of the order of the rotation period and could be a rotation-induced phenomenon. In contrast to the other RV variable objects, Sz 23 is also displaying significant emission in the CaII IR triplet lines, which is an indicator for chromospheric activity. A further study of time variations of these lines is underway.

The RV variability of Cha H$\alpha$ 8 and CHXR 74 on time scales of months to years cannot be explained by being rotational modulation. If caused by orbiting companions, the detected RV variations of CHXR 74 and Cha H$\alpha$ 8 correspond to giant planets of a few Jupiter masses with periods of several months.

## 5.3 Discussion: Multiplicity, separations, RV noise

Fig. 6 displays for all targets of the RV survey the measured RV semiamplitude (for RV constant objects the upper limit for it) versus their mass, as adopted from Comerón et al. (1999, 2000). The three RV variable objects have RV amplitudes above 1 km s$^{-1}$ (top three data points in Fig. 6) and are clearly separated from the RV constant objects. Interestingly, the RV constant objects follow a clear trend of decreasing RV amplitude with increasing mass. On one hand, this reflects simply the depending of the RV precision on the signal-to-noise of the spectra. On the other hand, it is an interesting finding by itself, since it shows that this group of brown dwarfs and very low-mass stars with masses of 0.12 $M_{\odot}$ and below display no significant RV noise due to surface spots, which would cause systematic RV errors with a RV amplitude increasing with mass.

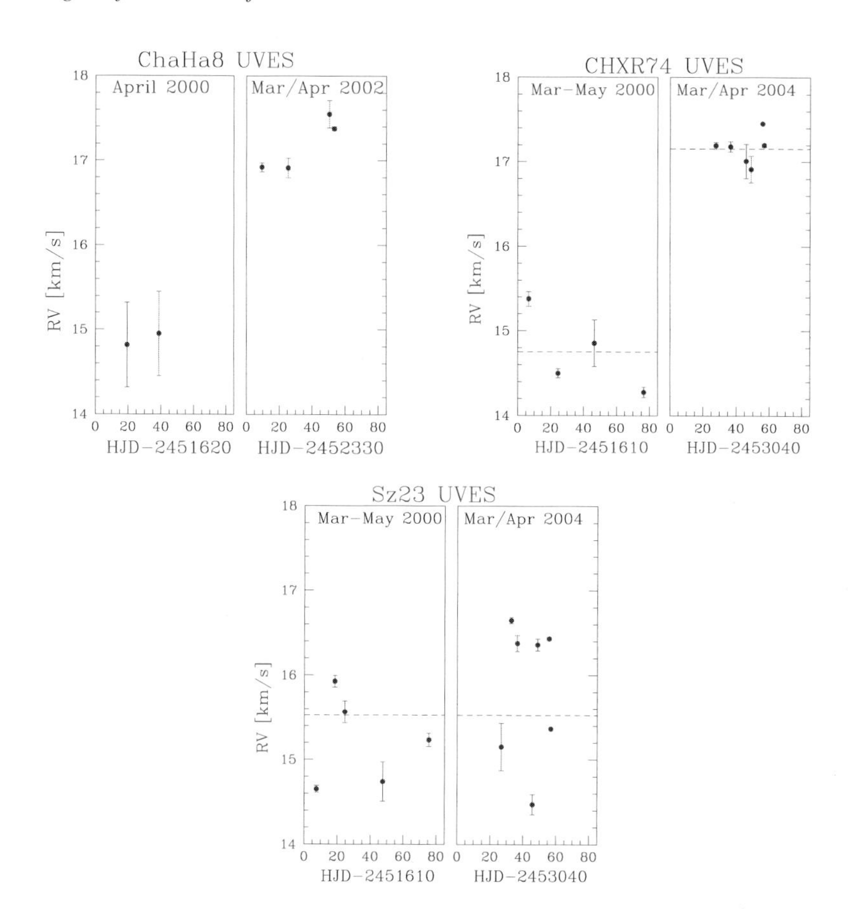


Figure 5: RV variable objects: RV vs. time in Julian days for the brown dwarf candidate Cha Ha 8 (M6.5) and the low-mass T Tauri stars CHXR 74 and Sz 23 ($\sim$0.2–0.3 $M_\odot$) based on high-resolution UVES/VLT spectra. Error bars indicate 1 $\sigma$ errors.

Among the subsample of ten brown dwarfs and very low-mass stars with masses $\leq 0.12\,M_\odot$ in this survey, only one (Cha H$\alpha$ 8) shows signs of RV variability, while the others are RV quiet with respect to both companions and spots in our observations. That hints at a very small multiplicity fraction of 10% or less. When considering also CHXR 74, i.e. eleven objects ($M \leq 0.17\,M_\odot$ and spectral types M4.5–M8), nine have constant RVs in the presented RV survey. Interestingly, Cha H$\alpha$ 8 is RV constant and CHXR 74 shows only small amplitude variations on time scales of days/weeks but both reveal larger amplitude RV variability only on longer time scales of at least several months. The fact that all other objects in this mass range do not show RV noise due to activity suggests that the sources for RV variability due to activity are also weak for Cha H$\alpha$ 8. Furthermore, the timescales of the variability are much too long for being caused by rotational modulation since the rotational period is of the order of 2 days. The only other explanation would be a companion

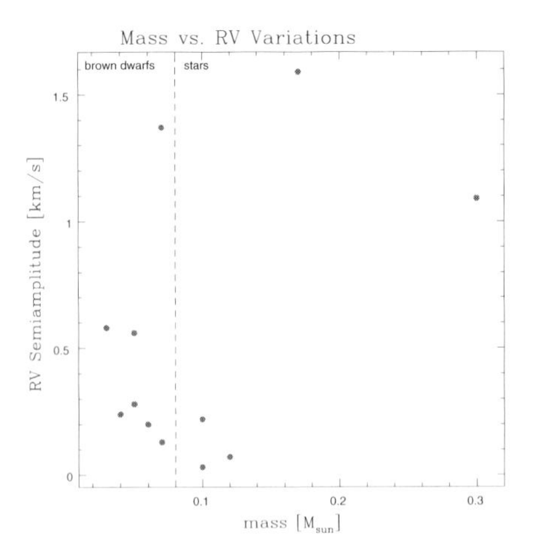


Figure 6: RV semiamplitude vs. object mass. The upper three data points represent the RV variable objects with amplitudes above 1 km s$^{-1}$. The remaining data points represent RV constant objects, which clearly follow a trend of decreasing RV amplitude with increasing mass. indicating that they are displaying no significant RV noise due to surface spots down to the precision required to detect Jupitermass planets, which would cause and increasing RV amplitude with mass.

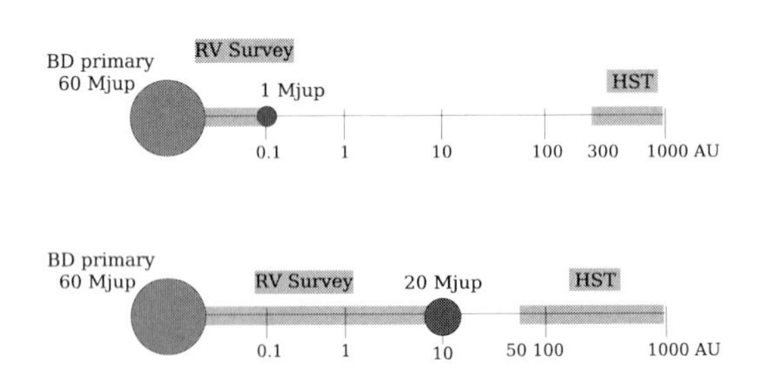


Figure 7: Separation ranges, which can be covered based on the achieved RV precision by our RV survey and by a direct imaging survey with the HST (Neuhäuser et al. 2002). An example is given for a one Jupitermass planet and a 20 Jupitermass brown dwarf orbiting a 60 Jupitermass brown dwarf. Due to limited time base not the whole possible range for the RV survey has been covered yet.

with a mass of several Jupitermasses or more, i.e. a supergiant planet or a brown dwarf. These observations give hints that companions to young brown dwarfs and very low-mass stars might have periods of several months.

The RV survey probes the regions close to the central objects in respect of the occurrence of companions. Fig. 7 shows the separation ranges, which can be covered based on the achieved RV precision. For example, a 20 Jupitermass brown dwarf in orbit around a 60 Jupitermass brown dwarf would be detectable out to 10 AU if the time base is long enough. For smaller companion masses the covered separation ranges are correspondingly smaller. At the current stage, the limits in the covered separation range is set by the time base rather than the RV precision. Therefore, further 3rd epoch RV measurements are planned. The found small multiplicity fraction of the brown dwarfs and very low-mass stars in Cha I at small separations in this RV survey, is also supported by the results of a direct imaging search for wide (planetary or brown dwarf) companions to mostly the same targets, Cha H$\alpha$ 1–12, by Neuhäuser et al. (2002, 2003), who find a multiplicity fraction of $\leq$10%. The separation ranges covered by this HST survey are also indicated in Fig. 7.

# 6 Rotation

Measurements of rotation rates of young brown dwarfs are important to determine the evolution of angular momentum in the substellar regime in the first several million years of their lifetime, during which rapid changes are expected due to the contraction on the Hayashi track, the onset of Deuterium burning and possible magnetic interaction with a circumstellar disk. Rotation speeds can be determined in terms of projected rotational velocities $v \sin i$ based on the line broadening of spectral features (Sect. 6.1) or, if an object exhibits prominent surface features, which modulate the brightness as the object rotates, the absolute rotation period can be determined by a light curve analysis (Sect. 6.2). Both techniques have been applied to the brown dwarfs and (very) low-mass stars in Cha I.

## 6.1 Projected rotational velocities $v \sin i$

Projected rotational velocities $v \sin i$ have been measured based on the line broadening of spectral lines in UVES spectra. We found that the $v \sin i$ values of the bona fide and candidate brown dwarfs in Cha I with spectral types M6–M8 range between 8 km s$^{-1}$ and 26 km s$^{-1}$ . The spectroscopic rotational velocities of the (very) low-mass stars Cha H$\alpha$ 4, Cha H$\alpha$ 5, B 34, CHXR 74 and Sz 23 are 14–18 km s$^{-1}$ and, therefore, lie also within the range of that of the studied substellar objects. These measurements provided the first determination of projected rotational velocities for very young brown dwarfs (Joergens & Guenther 2001). To compare them with $v \sin i$ values of older brown dwarfs, we consider that the late-M type brown dwarfs in Cha I at and age of 1–5 Myrs will further cool down and develop into L dwarfs at some point between and age of 100 Myr and 1 Gyr (Burrows et al. 2001) and later into T dwarfs. $v \sin i$ values for old L dwarfs range between 10 and 60 km s$^{-1}$ with the vast majority rotating faster than 20 km s$^{-1}$ (Mohanty & Basri

2003). Thus, the brown dwarfs in Cha I rotate on average slower than old L dwarfs in terms of rotational velocities $v \sin i$. These results for brown dwarfs in Cha I are in agreement with $v \sin i$ values determined for five brown dwarf candidates in Taurus (7–14 km s$^{-1}$, White & Basri 2003) and for one brown dwarf in $\sigma$ Ori, which has a $v \sin i$ of 9.4 km s$^{-1}$ (Muzerolle et al. 2003) and an absolute rotational velocity of 14$\pm$4 km s$^{-1}$ (Caballero et al. 2004).

Projected rotational velocities $v \sin i$ are lower limits of the rotational velocity $v$ since the inclination $i$ of the rotation axis remains unknown. Based on $v \sin i$ and the radius of the object an upper limit of the rotational period P/sin $i$ can be derived. We estimated the radii of the brown dwarfs and very low-mass stars in Cha I by means of the Stefan-Boltzmann law from bolometric luminosities and effective temperatures given by Comerón et al. (1999, 2000). The approximate upper limits for their rotational periods range between one and three days for all studied brown dwarfs and (very) low-mass stars with the exception of the low-mass star CHXR 74 (M4.5, $\sim$0.17 $M_\odot$), for which P/sin $i$ is five days.

## 6.2 Absolute rotational periods from lightcurve modulations

An object exhibiting prominent surface features distributed inhomogeneously over its photosphere provides a way to measure its absolute rotation period since the surface spot(s) cause a periodic modulation of the brightness of the object as it rotates. See Fig. 8 for an illustration.

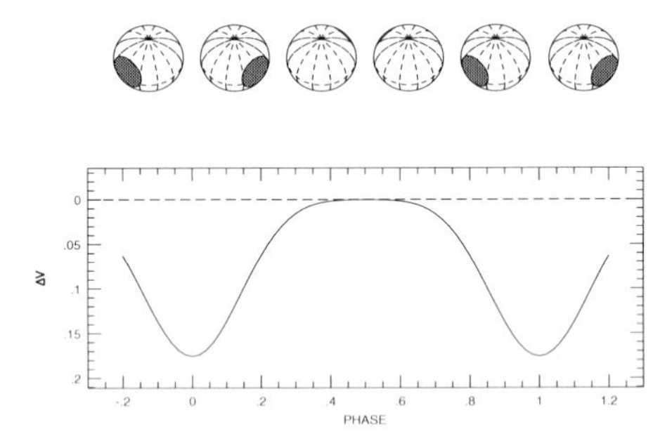


Figure 8: **Rotational brightness modulation caused by surface spot.** Courtesy of G. Torres. A cool spot on the surface of a star / brown dwarf causes a periodic dimming of the total brightness at the rotation period. Plotted are the relative V magnitudes $\Delta$V over the orbital phase.

Based on photometric observations in the R and i band filter (see Sect. 3.3), we searched for periodic variations in the light curves of the targets with the string-length method (Dworetsky 1983). Periodic brightness variations were found for the three brown dwarf candidates Cha H$\alpha$ 2, Cha H$\alpha$ 3, Cha H$\alpha$ 6 and two very low-mass stars B 34 and CHXR 78C. The original light curves are shown in Fig. 9. In addition to i and R band data, we have analysed J-band monitoring data of the targets (Carpenter

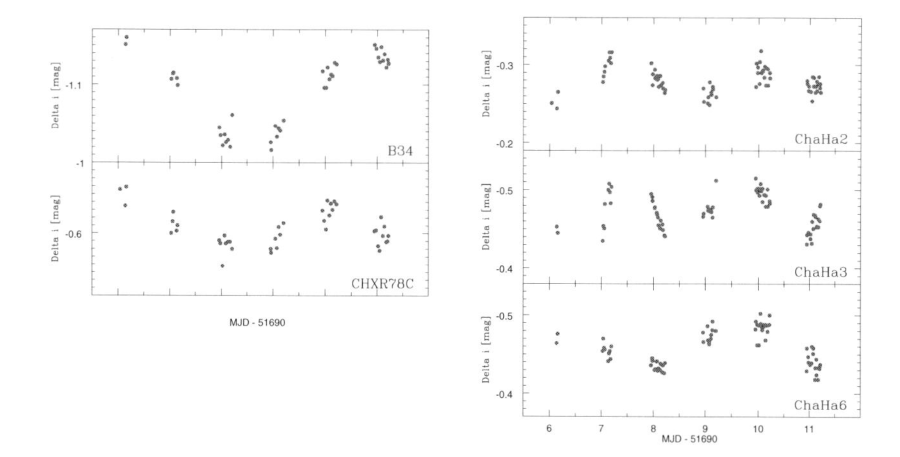


Figure 9: **Light curves for two very low-mass stars (left panel) and three brown dwarf candidates (right panel).** From top to bottom and left to right are displayed i-band light curves for the very low-mass stars B34 (M5, 0.12 $M_\odot$) with a period of 4.5 days and CHXR78C (M5.5, 0.09 $M_\odot$) with a period of 3.9 days, and for the three brown dwarf candidates Cha H$\alpha$ 2 (M6.5, 0.07 $M_\odot$) with a period of 2.8 days, Cha H$\alpha$ 3 (M7, 0.06 $M_\odot$) with a period of 2.2 days and Cha H$\alpha$ 6 (M7, 0.05 $M_\odot$) with a period of 3.4 days.

et al. 2002), which have been taken a few weeks earlier and confirm the periods found in the optical data. The determined periods are interpreted as rotation periods based on a consistency with the $v \sin i$ values from UVES spectra (previous section), recorded color variations in agreement with the expectations for spots on the surface and the fact that with effective temperatures of more than 2800 K, the objects are young and still hot and, therefore, their atmospheric gas is very likely sufficiently ionized for the formation of spots. Additionally, variability due to clouds, which could occur on time scales of the formation and evolution of clouds, can be excluded because their temperatures are too high for significant dust condensation (e.g. Tsuji et al. 1996a,b, Allard et al. 1997, Burrows & Sharp 1999).

The found rotational periods of the three brown dwarf candidates Cha H$\alpha$ 2, Cha H$\alpha$ 3 and Cha H$\alpha$ 6 are 3.2 d, 2.2 d and 3.4 d, respectively (Joergens et al. 2003b). The results show that brown dwarfs at an age of 1–5 Myr display an inhomogeneous surface structure and rotate slower than old brown dwarfs (rotational periods below one day, e.g. Bailer-Jones & Mundt 2001, Martín et al. 2001). It is known that Cha H$\alpha$ 2 and 6 have optically thick disks (Comerón et al. 2000), therefore magnetic braking due to interactions with the disk may play a role for them. This is suggested by the fact, that among the three brown dwarf candidates with determined periods, the one without a detected disk, Cha H$\alpha$ 3, has the shortest period.

### 6.3 Rotation of young brown dwarfs

The brown dwarfs in Cha I have rotation periods of 2–3 days and projected rotation velocities $v \sin i$ of 8–26 km s$^{-1}$ as shown by our observations. Their rotation periods are significantly larger than those for old brown dwarfs (below one day, e.g. Bailer-Jones & Mundt 2001) and their rotational velocities are on average smaller than for old brown dwarfs (10–60 km s$^{-1}$, Mohanty & Basri 2003). This is in agreement with the idea that they are in an early contracting stage and will further spin up and contract.

Periodic photometric variabilities have been also detected for two dozen substellar $\sigma$ Ori and $\epsilon$ Ori members and member candidates (Bailer-Jones & Mundt 2001, Zapatero Osorio et al. 2003, Scholz & Eislöffel 2004a,b, Caballero et al. 2004). Including our periods in Cha I, the to-date known photometric periods for young brown dwarfs cover a wide range from 46 min to 3.4 days. Apart from rotational modulation due to chromospheric spots, several other possible processes have been suggested to account for the periodic variabilities, like accretion phenomena, formation and evolution of dust coverage for the cooler ones and eclipsing companions, which could all occur on various time scales. Among all known periods for young brown dwarfs, only four have been confirmed as rotational periods by spectroscopic measurements of their rotational velocities $v \sin i$ (Cha H$\alpha$ 2, Cha H$\alpha$ 3, Cha H$\alpha$ 6, S Ori 25, Joergens & Guenther 2001, Joergens et al. 2003b, Muzerolle et al. 2003, Caballero et al. 2004), they all lie between 1.7 and 3.5 days. Several periods of the order of a few hours have been also interpreted as rotational periods. However, as noticed by Joergens et al. (2003b), rotation periods of a few hours for very young brown dwarfs imply extreme rotation with rotational velocities of 100 km s$^{-1}$ or more given the still large radii at this young age. These speeds come close to break-up velocities and for the 46 min period, Zapatero Osorio et al. (2003) found that it is indeed above break-up velocity and cannot be a rotation period. The finding of a rotation of 100 km s$^{-1}$ or more for brown dwarfs at a few million years would also be surprising in respect of the fact that they are in an early contracting stage and the expectation of a further spin-up in their future evolution. A check of the proposed extreme rapid rotation by $v \sin i$ measurements would be desirable. So far, all $v \sin i$ determinations for young brown dwarfs in Cha I, Taurus and $\sigma$ Ori indicate rotational velocities below 26 km s$^{-1}$ with the vast majority below 20 km s$^{-1}$ (Joergens & Guenther 2001, White & Basri 2003, Muzzerolle et al. 2003).

## 7 Summary and conclusions

The presented work reports about observations of a population of very young brown dwarfs and low-mass stars close to the substellar borderline in the Cha I star forming region. At an age of only a few million years, their exploration allows insights into the formation and early evolution of brown dwarfs. The targets were studied in terms of their kinematic properties, the occurrence of multiple systems among them as well as their rotational characteristics.

We are carrying out a RV survey for planets and brown dwarf companions to the targets with UVES at the VLT. The achieved RV precision allows us to search for companions down to Jupiter mass planets. The analysis of the high-resolution spectra reveals very constant RVs on time scales of weeks to months for the majority of the targets as well as RV variability for three sources (Joergens et al. 2005a). The RV constant objects are six brown dwarfs and three very low-mass stars ($M \leq 0.12\,M_\odot$, spectral types M5–M8), for which we estimate upper limits for masses of hypothetical companions to lie between $0.1\,M_{Jup}$ and $1.5\,M_{Jup}$. This group shows a relation of decreasing RV amplitude with increasing mass. This reflects simply a higher RV precision for more massive targets due to a better signal-to-noise, whereas the effect of RV errors caused by surface features would be the opposite. This demonstrates that brown dwarfs and very low-mass stars ($M \leq 0.12\,M_\odot$) in Cha I display no significant RV noise due to surface spots down to the precision necessary to detect Jupitermass planets. Thus, they are suitable targets to search for planets with the RV technique.

Three objects exhibit significant RV variations with peak-to-peak amplitudes of 2–3 $km\,s^{-1}$: the brown dwarf Cha H$\alpha$ 8 and the low-mass stars CHXR 74 ($\sim 0.17\,M_\odot$) and Sz 23 ($\sim 0.3\,M_\odot$). For Sz 23, which is the highest mass object in our sample, we have indications that the RV variations are caused by surface spots from the variability time scale and from significant CaII IR triplet emission. Cha H$\alpha$ 8 and CHXR 74 show a different variability behaviour with displaying only very small amplitude or no variations on time scales of days to weeks but significant RV variations on times scales of months or longer, which cannot be explained by being rotational modulation. If caused by orbiting companions, the detected RV variations of CHXR 74 and Cha H$\alpha$ 8 correspond to giant planets of a few Jupiter masses with periods of several months. In order to explore the nature of the detected RV variations follow-up observations of CHXR 74 and Cha H$\alpha$ 8 are planned. If confirmed as planetary systems, they would be unique because they would contain not only the lowest mass primaries and the first brown dwarf with a planet but with an age of a few million years also the by far youngest extrasolar planets found to date. That would provide empirical constraints for planet formation time scales as well as for the formation of brown dwarfs.

The found multiplicity fraction in this survey is obviously very small. Considering the subsample of the ten brown dwarfs and very low-mass stars with masses $\leq 0.12\,M_\odot$, only one of them (Cha H$\alpha$ 8) shows signs of RV variability, while the others are RV quiet with respect to both companions and spots in our observations. That hints at a very small multiplicity fraction of 10% or less. However, the RV variable brown dwarf Cha H$\alpha$ 8 and also the higher mass CHXR 74 (not included in the above considered subsample) are hinting at the possibility that companions to young brown dwarfs and very low-mass stars have periods of several months and such a time scale was not covered for all targets. Thus, our results of small multiplicity reflects so far mainly separations of around 0.1 AU. Further 3rd epoch RV determinations are planned. At much larger separations, a direct imaging search for wide (planetary or brown dwarf) companions to mostly the same targets also found a very small multiplicity fraction of $\leq$10% (Neuhäuser et al. 2002, 2003). There still remains a significant gap in the studied separation ranges, which will partly be probed by the planned follow-up 3rd epoch RV measurements and partly is only accessible with high-resolving AO imaging (NACO / VLT) or even requires interferometric techniques (e.g. AMBER at the VLTI). For the already studied separations, the overall picture is a multiplicity fraction significantly smaller than for T Tauri stars even when taking into account the smaller available mass range for the companions. This hints at differences in the formation processes of brown dwarfs and T Tauri stars.

In order to test the proposed ejection scenario for the formation of brown dwarfs, we explored the kinematic properties of our substellar targets based on absolute mean RVs derived within the presented RV survey. We find that the brown dwarfs in Cha I form also kinematically a very homogeneous group. They have very similar absolute RVs with a RV dispersion in terms of standard deviation of only 0.9 km $s^{-1}$ (Joergens & Guenther 2001, Joergens et al. 2005b). A study of T Tauri stars in the same field showed that there are no indications for a more violent dynamical evolution, like more frequent ejections, for the brown dwarfs compared to the T Tauri stars since the RV dispersion of the T Tauri stars (1.3 km $s^{-1}$) was determined to be even slightly larger than that for the brown dwarfs. This is the first observational constraint for the velocity distribution of a homogeneous group of closely confined very young brown dwarfs and therefore a first *empirical* upper limit for ejection velocities.

Theoretical models of the ejection scenario have been performed by several groups in recent years. Some of them are hinting at the possibility of a only small or of no mass dependence of the velocities. Thus, the fact that we do not find a larger velocity dispersion for the brown dwarfs than for the T Tauri stars does not exclude the ejection scenario. However, we observe smaller velocities than any of the theoretical predictions for brown dwarfs formed by the ejection scenario. This might be partly attributed to the fact that our sample is statistically relatively small, or it might be explained by the lower densities in Cha I compared to some model assumptions, or by shortcomings in the models, like neglection of feedback processes (Bate et al. 2003, Bate & Bonnell 2005, Delgado-Donate et al. 2004) or of the cluster potential (Sterzik & Durisen 2003, Umbreit et al. 2005). The current conclusion is that either the brown dwarfs in Cha I have been formed by ejection but with smaller velocities as theoretically predicted or they have not been formed in that way. We are planning to enlarge the sample in the future to put the results on an improved statistical basis.

Finally, we studied the rotational properties of the targets in terms of projected rotational velocities $v \sin i$ measured in UVES spectra as well as in terms of absolute rotational periods derived from light curve analysis. We found that the $v \sin i$ values of the bona fide and candidate brown dwarfs in Cha I range between 8 km $s^{-1}$ and 26 km $s^{-1}$. These were the first determinations of rotational velocities for very young brown dwarfs (Joergens & Guenther 2001). Furthermore, we have determined rotational periods, consistent with $v \sin i$ values, by tracing modulations of the light curves due to surface spots at the rotation period. We found periods for the three brown dwarf candidates Cha H$\alpha$ 2, Cha H$\alpha$ 3 and Cha H$\alpha$ 6 of 2–3 days and for two very low-mass stars B 34 and CHXR78C ($M \leq 0.12\,M_\odot$) of 4–5 days (Joergens et al. 2003b). Magnetic braking due to interactions with a circum-stellar disk may play a role for some of them since the ones with detected disks are the slower rotators.

The emerging picture of the rotation of young brown dwarfs at an age of a few million years based on rotational velocities for brown dwarfs in Cha I, Taurus and $\sigma$ Ori (Joergens & Guenther 2001, White & Basri 2003, Muzzerolle et al. 2003) and on spectroscopically confirmed rotation periods (our three brown dwarfs in Cha I plus one $\sigma$ Ori brown dwarf, Caballero et al. 2004) indicates that young brown dwarfs rotate with periods of the order of a few days and speeds of 7 to 26 km s$^{-1}$. Their rotation periods are significantly larger than those for old brown dwarfs (below one day, e.g. Bailer-Jones & Mundt 2001) and their rotational velocities are on average smaller than for old brown dwarfs (10–60 km s$^{-1}$, e.g. Mohanty & Basri 2003). This is in agreement with the idea that they are in an early contracting stage and will further spin up and contract before they reach a final radius when their interior electrons are completely degenerate.

Despite the fact that the origins of brown dwarfs are still shrouded in mist, we think that the presented comprehensive observations of very young brown dwarfs in Cha I and the determination of their fundamental parameters brought us an important step forward in revealing the details of one of the main open issues in stellar astronomy and in the origins of solar systems.

## Acknowledgements

It is a pleasure to acknowledge fruitful collaborations in the last years on the subject of this article with Ralph Neuhäuser, Eike Guenther, Matilde Fernández, Fernando Comerón and Günther Wuchterl. Furthermore, I am grateful for a grant from the 'Deutsche Forschungsgemeinschaft' (Schwerpunktprogramm 'Physics of star formation') during my PhD time as well as current support by the European Union through a Marie Curie Fellowship under contract number FP6-501875.

## References

Allard F., Hauschildt P.H., Alexander D.R., Starrfield S. 1997 ARA&A 35, 137

Armitage, P. J.; Clarke, C. J. 1997, MNRAS, 285, 540

Bailer-Jones C.A.L., Mundt R. 2001, A&A 367, 218

Basri G., Martín E.L. 1999, ApJ 118, 2460

Basri G. 2000, ARA&A 38, 485

Bate M.R., Bonnell I.A., Bromm V. 2003, MNRAS 339, 577

Bate M.R., Bonnell I.A. 2005, MNRAS 356, 1201

Béjar V.J.S., Zapatero Osorio M.R., Rebolo R. 1999, ApJ 521, 671

Burrows A., Sharp C.M. 1999, ApJ 512, 843

Burrows A., Hubbard W.B., Lunine J.I., Liebert J. 2001, Rev. Mod. Phys. 73, 3

Caballero J.A., Béjar V.J.S., Rebolo R., Zapatero Osorio M.R. 2004, A&A 424, 857

Carpenter J.M., Hillenbrand L.A., Skutskie M.F., Meyer M.R. 2002, AJ 124, 1001

Chauvin G., Lagrange A.-M., Dumas C. et al. 2004, A&A 425, L29

Comerón F., Rieke G.H., Neuhäuser R. 1999, A&A 343, 477

Comerón F., Neuhäuser R., Kaas A.A. 2000, A&A 359, 269

Covino E., Alcalá J.M, Allain S. et al. 1997, A&A 328, 187

Dekker H., D'Odorico S., Kaufer A., Delabre B., Kotzlowski H. 2000, In: SPIE Vol. 4008, p. 534, ed. by M.Iye, A.Moorwood

Delfosse X., Tinney C.G., Forveille T. et al. 1997, A&A 327, L25

Delgado-Donate E.J., Clarke C.J., Bate M.R. 2003, MNRAS 342, 926

Delgado-Donate E.J., Clarke C.J., Bate M.R. 2004, MNRAS 347, 759

Dubath P., Reipurth B., Mayor M. 1996, A&A 308, 107

Durisen R.H., Sterzik M.F., Pickett B.K. 2001, A&A 371, 952

Dworetsky M.M. 1983, MNRAS 203, 917

Ghez A.M., Neugebauer G., Matthews K. 1993, AJ 106, 2005

Guenther E.W., Wuchterl G. 2003, A&A 401, 677

Hawley S.L., Covey K.R., Knapp G.R. et al. 2002, AJ 123, 3409

Joergens V., Guenther E. 2001, A&A 379, L9

Joergens V. 2003, PhD thesis, Ludwigs-Maximilians Universität München

Joergens V., Fernández M., Carpenter J.M., Neuhäuser R. 2003b, ApJ 594, 971

Joergens V., Neuhäuser R., Guenther E.W., Fernández M., Comerón F., In: IAU Symposium No. 211, Brown Dwarfs, ed. by E. L. Martín, Astronomical Society of the Pacific, San Francisco, 2003a, p.233

Joergens V. et al. 2005a, in prep.

Joergens V. et al. 2005b, in prep.

Kirkpatrick J.D., Reid I.N., Liebert J. et al. 2000, AJ 120, 447

Köhler R., Kunkel M., Leinert Ch., Zinnecker H. 2000, A&A 356, 541

Kroupa, P., Bouvier, J. 2003, MNRAS 346, 369

Kumar S. 1962, AJ 67, 579

Kumar S. 1963, ApJ 137, 1121

Leinert Ch., Zinnecker H., Weitzel N. et al. 1993, A&A 278, 129

Low C., Lynden-Bell D. 1976, MNRAS 176, 367

Luhman K. 2004, ApJ 602, 816

Martín E.L., Brandner W., Bouvier J. et al. 2000 ApJ 543, 299

Martín, E. L., Zapatero Osorio M. R. & Lehto H.J. 2001, ApJ, 557, 822

Mayor M., Queloz D. 1995, Nature 378, 355

Mayor, M., Udry, S., Naef, D. et al. 2003, A&A 415, 391

Mizuno A., Hayakawa T., Tachihara K. et al. 1999, PASJ 51, 859

Mohanty S., Basri G. 2003, ApJ 583, 451

Muzerolle J., Hillenbrand L., Calvet N., Briceño C., Hartmann L. 2003, ApJ, 592, 266

Nakajima T., Oppenheimer B.R., Kulkarni S.R. et al. 1995, Nature 378, 463

Neuhäuser R., Comerón F. 1998, Science 282, 83

Neuhäuser R., Comerón F. 1999, A&A 350, 612

Neuhäuser R., Brandner W., Alves J., Joergens V., Comerón F. 2002, A&A 384, 999

Neuhäuser R., Brandner W., Guenther E. 2003, IAUS 211, 309, ed. by E. Martín

Oppenheimer B.R., Kulkarni S.R., Matthews K., Nakajima T. 1995 Science 270, 1478

Rebolo R., Martín E.L., Magazzú A. 1992, ApJ 389, L83

Rebolo R., Zapatero Osorio M.R., Martín E.L. 1995, Nature 377, 129

Reipurth B., Clarke C. 2001, ApJ 122, 432

Santos N.C., Bouchy F., Mayor M., Pepe F., Queloz D. et al. 2004, A&A 426, L19

Scholz A., Eislöffel J. 2004a, A&A 419, 249

Scholz A., Eislöffel J. 2004b, A&A in press, astro-ph/0410101

Shu F.H., Adams F.C., Lizano S. 1987, ARA&A 25, 23

Stauffer J.R., Hamilton D., Probst R. 1994, AJ 108, 155

Sterzik M.F., Durisen R.H. 1995, A&A 304, L9

Sterzik M.F., Durisen R.H. 1998, A&A 339, 95

Sterzik M.F., Durisen R.H. 2003, A&A, 400, 1031

Tsuji T., Ohnaka K., Aoki W. 1996a, A&A 305, L1

Tsuji T., Ohnaka K., Aoki W., Nakajima T. 1996b, A&A 308, L29

Umbreit S., Burkert A., Henning T., Mikkola S., Spurzem R. 2005, A&A, in press

Valtonen M., Mikkola S. 1991, ARA&A 29, 9

Walter F.M. 1992, AJ 104, 758

White, R. J. & Basri, G. 2003, ApJ 582, 1109

Whitworth, A.P., Zinnecker, H. 2004, A&A 427, 299

Wuchterl G., Guillot T., Lissauer J.J. 2000. In: *Protostars and Planets IV*, ed. by Mannings V., Boss A.P., Russell S.S., Univ. of Arizona Press, Tucson, p.1081

Zapatero Osorio M.R., Caballero J.A., Béjar V.J.S., Rebolo R. 2003, A&A 408, 663

# Index of Contributors

# General Table of Contents

## Volume 1 (1988): Cosmic Chemistry



## Volume 2 (1989)

## Volume 3 (1990): Accretion and Winds

Volume 4 (1991)

## Volume 5 (1992): Variabilities in Stars and Galaxies



## Volume 6 (1993): Stellar Evolution and Interstellar Matter

## Volume 7 (1994)



## Volume 8 (1995): Cosmic Magnetic Fields

## Volume 9 (1996): Positions, Motions, and Cosmic Evolution

## Volume 10 (1997): Gravitation

## Volume 11 (1998): Stars and Galaxies



## Volume 12 (1999): Astronomical Instruments and Methods at the Turn of the 21st Century

## Volume 13 (2000): New Astrophysical Horizons

## Volume 14 (2001): Dynamic Stability and Instabilities in the Universe



## Volume 15 (2002): JENAM 2001 – Five Days of Creation: Astronomy with Large Telescopes from Ground and Space

## Volume 16 (2003): The Cosmic Circuit of Matter



## Volume 17 (2004): The Sun and Planetary Systems – Paradigms for the Universe

## Volume 18 (2005): From Cosmological Structures to the Milky Way

# General Index of Contributors